Beyond the Quantum

Beyond the Quantum

Editors

Theo M. Nieuwenhuizen
Institute for Theoretical Physics, University of Amsterdam, the Netherlands

Václav Špička
Institute of Physics, Academy of Sciences, the Czech Republic

Bahar Mehmani
Institute for Theoretical Physics, University of Amsterdam, the Netherlands

Maryam J. Aghdami
Department of Physics, Sharif University, Iran

Andrei Yu. Khrennikov
International Center for Mathematical Modeling in Physics, Engineering and Cognitive Science, Växjö University, Sweden

World Scientific

NEW JERSEY • LONDON • SINGAPORE • BEIJING • SHANGHAI • HONG KONG • TAIPEI • CHENNAI

Published by

World Scientific Publishing Co. Pte. Ltd.
5 Toh Tuck Link, Singapore 596224
USA office: 27 Warren Street, Suite 401-402, Hackensack, NJ 07601
UK office: 57 Shelton Street, Covent Garden, London WC2H 9HE

British Library Cataloguing-in-Publication Data
A catalogue record for this book is available from the British Library.

BEYOND THE QUANTUM

ISBN-13 978-981-277-117-9
ISBN-10 981-277-117-4

Printed in Singapore by B & JO Enterprise

PREFACE

A centenary after Einstein's *annus mirabilis* it is timely to reconsider the foundations of physical theories. Quantum mechanics, our best theory, works wherever it has been applied, in fields ranging from the solid state and quantum chemistry to high energy physics and the early Universe. Its most modern application is Quantum String Theory.

Despite all the success, there remains the old question: what is this theory actually stands for? On the foundational level, it has come hardly much further than the Feynman quote "nobody understands quantum mechanics". Up to this date, quantum effects such as uncertainty, interference and entanglement have become experimental facts, but their explanation remains puzzling. It was Einstein's dream that one day quantum theory would appear to arise from physics at a deeper level, more precisely, as the statistics of such a world.

Indeed, none of the present theories is capable to describe, even in principle, individual measurements or individual events. Though it was long agreed that "such questions should not be posed" their relevance is getting more and more acknowledged. In its July 2005 issue, the journal *Science* selects among the Top 25 questions that face scientific inquiry over the next quarter-century: *Do Deeper Principles Underlie Quantum Uncertainty and Nonlocality?*, together with the probably related questions *What is the Universe Made of?* and *Can the Laws of Physics be Unified?*

During the last couple of years various new results have been reported supporting the possibility of an underlying more deterministic structure: Various arguments in favor of the statistical interpretation of quantum mechanics, deriving rather than postulating the von Neumann collapse and the Born rule, loopholes in nonlocality arguments, quantum gravity approaches, demonstration of the compatibility of quantum theory and contextual Kolmogorov probability theory, Bohmian mechanics, stochastic electrodynamics, collapse models, the Empiricist interpretation of quantum mechanics, occurrence of entanglement in classical physics. Stochastic optics has proposed a local and realistic explanation of entanglement for experiments with photons. Intriguing ideas have been published comparing over-extremal Kerr-Newman solutions in electrogravity with charged, spinning elementary particles, which invites to consider topological features of these solutions.

Some of these questions may lead to tests in quantum optics, where entangled Bell-pairs are routinely made.

A workshop that would put together and compare these many different approaches was deemed very timely. Progress can be hoped for by combining the insights from the communities of the quantum gravity and statistical/empirical interpretations of quantum mechanics with the communities from Stochastic Electrodynamics, Stochastic Optics, Stochastic Collapse and Bohm, to focus on questions such as: Problems and paradoxes in ordinary quantum mechanics and quantum field theory, including quantum state reduction in relativistic quantum theory; Is there experimental evidence to go beyond the ordinary quantum theory? What have we learned from quantum gravity and string theory for these problems? Such questions were confronted to experimental, mathematical and philosophical insights, and these proceedings are a written testimony of this confrontation.

In the opening address of the workshop it was stressed that individual measurements give individual outcomes. The statistics of the outcomes of measurements is described by quantum theory. But since it is the task of theoretical physicists to describe Nature, they have to find a theoretical description for the underlying individual events.

To open the proceedings, we may recall the last opening words of the workshop:

Bohr
closed the door
but we now open the floor
for the quantum and much more

Theo M. Nieuwenhuizen, Roger Balian, Ana María Cetto,
Gerard 't Hooft and Andrei Yu. Khrennikov

Walter Philipp presenting his "Beyond the Quantum" lecture, May 2006.

In Memoriam: Walter Philipp (1936–2006)

Walter Philipp was born on Dec. 14, 1936 in Vienna, Austria. He spent most of his scientific carreer in Urbana, Illinois but kept close contacts always with his friends in the mathematics department of Vienna. He worked hard and made friends easily. During the last decade of his life, Walter took up one of the central problems of quantum mechanics, the Bell inequality. He demanded a clear-headed and critical analysis of the statistics surrounding this important result, because he found the standard treatments deficient in fundamental aspects. Throughout the years, his insights became deeper and his arguments simpler, drawing others closer to his perspective.

As a person, Walter was easy going and a wise father. He understood well that it would take time before others joined his insights and was always willing to review in detail any aspect of his thinking with others. Walter loved to walk. In view of his age, this was remarkable: with big steps he would easily outwalk others decades younger than himself. Walter always talked about his hiking trips in his beloved motherland, Austria. It was here, during his last trip, where he passed away on July 19, 2006, shortly after having participated in the Lorentz workshop "Beyond the Quantum" in Leiden.

Walter will be missed for his positive, pleasant and cheerful character and willingness to transmit mathematical knowledge to others.

Karl Hess, Gregg Jaeger, Andrei Khrennikov and Theo Nieuwenhuizen.

CONTENTS

PART A

Introductions

THE MATHEMATICAL BASIS FOR DETERMINISTIC QUANTUM MECHANICS

G. 't HOOFT

Institute for Theoretical Physics, Utrecht University,
Leuvenlaan 4, 3584 CC Utrecht, the Netherlands, and
Spinoza Institute
Postbox 80.195, 3508 TD Utrecht, the Netherlands
E-mail: g.thooft@phys.uu.nl

If there exists a classical, *i.e. deterministic theory underlying quantum mechanics, an explanation must be found of the fact that the Hamiltonian, which is defined to be the operator that generates evolution in time, is bounded from below. The mechanism that can produce exactly such a constraint is identified in this paper. It is the fact that not all classical data are registered in the quantum description. Large sets of values of these data are assumed to be indistinguishable, forming equivalence classes. It is argued that this should be attributed to information loss, such as what one might suspect to happen during the formation and annihilation of virtual black holes.*

The nature of the equivalence classes is further elucidated, as it follows from the positivity of the Hamiltonian. Our world is assumed to consist of a very large number of subsystems that may be regarded as approximately independent, or weakly interacting with one another. As long as two (or more) sectors of our world are treated as being independent, they all must be demanded to be restricted to positive energy states only. What follows from these considerations is a unique definition of energy in the quantum system in terms of the periodicity of the limit cycles of the deterministic model.

Keywords: Quantum mechanics; Information loss; Hamiltonian; Deterministic model.

1. Introduction

There may exist different versions of deterministic theories underlying Quantum Mechanics, usually referred to as "hidden variable theories". For instance, one may or may not assume the occurrence of information loss at tiny distance scales. One may suspect some sort of cellular automaton or a classical system of continuous fields, or even classical loops, D-branes, or whatever. Instead of quantizing such systems in the usual manner, we here consider what we will refer to as *pre-quantization*. With this term we mean that the physical system is not modified, as in usual quantization schemes, which depend on a new constant $\hbar$, but only rephrased in a language suitable for quantum mechanical manipulations at a later stage.

Pre-quantization may be useful when complex systems are handled probabilistically. The probability distribution W is then re-written as the absolute square of a wave function. The wave function obeys all the axioms of conventional quantum mechanics, allowing us to perform all the mathematical tricks known from Quantum Mechanics and Quantum Field Theory, such as group representation theory and renormalization group transformations.

We suspect that our world can be understood by starting from a pre-quantized classical, or 'ontological', system. However, a serious difficulty is then encountered: one indeed gets Quantum Mechanics, but the Hamiltonian is not naturally bounded from below. If time would be assumed to be discrete, the Hamiltonian eigenvalues would turn out to be periodic, so one might limit oneself to eigenvalues E with $0 \leq E < 2\pi/\delta t$, where δt is the duration of a fundamental time step, but then the choice of a vacuum state is completely ambiguous, unlike the situation in the real world that one might want to mimic. If time is continuous, the Hamiltonian eigenvalues tend to spread over the real line, from $-\infty$ to ∞.

In realistic theories, one therefore must impose a "superselection rule", projecting out a subspace of Hilbert space where all energies are non-negative. How exactly to do this will be described here. At first sight, the freedom to choose phase factors in wave functions allows one to make such a selection without loss of generality. This observation, however, is not the solution to the positivity problem of the Hamiltonian, since positivity must also be demanded to hold for subsystems, and when such systems interact, the suppression of negative energy states might cause the violation of unitarity, or locality, or both.

In this paper, we derive the plausibility of our assumptions from first principles. First, the formalism is displayed in Section 2. Deterministic systems are shown to be accessible by quantum mechanical procedures, although this does not turn them into acceptable quantum mechanical models just yet, because the Hamiltonian is not bounded from below. Then, we demonstrate that the most basic building blocks of any deterministic theory consists of units that would evolve with periodicity if there were no interaction (Section 3). We use the empirically known fact that the Hamiltonians are all bounded from below both before introducing the interaction and after having included the interaction. This necessitates our introduction of equivalence classes (Section 4), such that neither the quantum mechanical nor the macroscopic observer can distinguish the elements within one equivalence class, but they can distinguish the equivalence classes.

This procedure is necessary in particular when two systems are considered together *prior* to considering any interaction. We are led to the discovery that, besides the Hamiltonian, there must be a classical quantity E that also corresponds to energy, and is absolutely conserved as well as positive (Section 5). It allows us to define the equivalence classes. We end up discovering a precise definition of the quantum wave function for a classical system (both amplitude and phase), and continue our procedure from there.

Physical and intuitive arguments were displayed in Ref.[1] In that paper, it was argued that any system with information loss tends to show periodicity at small scales, and quantization of orbits. It was also argued that some lock-in mechanism was needed to relate the Hamiltonian with an ontologically observable quantity E

that is bounded from below. The lock-in mechanism was still not understood; here however we present the exact mathematical treatment and its relation to information loss. Interaction can be introduced in a rather direct manner (Section 6), by assuming energy not to be directly additive, but then it is difficult to understand how different energy sectors of the theory can be related to one another.

A more satisfactory picture emerges if one realizes that energy is not directly locally observable, but determined by the periods of the limit cycles. This is explained in Section 7. We think that this interpretation is imperative, and it sheds an interesting new light on the phenomenon we call quantum mechanics. After a discussion of our results (Section 8) an appendix follows in which we discuss the 'random automaton'. It allows us to estimate the distribution of its limit cycles, though we immediately observe that the quantum models it generates are not realistic because the energy will not be an extensive quantity. The deterministic models that might reproduce observed quantum field systems must be very special.

This paper was written while these facts were being discovered, so that it represents an original train of thought, which may actually be useful for the reader.

2. Variables, Beables and Changeables

Any classical, deterministic system will contain some set of degrees of freedom $\vec{q}$ that follow some orbit $\vec{q}(t)$ in time. Time might be defined as a discrete variable or a continuous one, but this distinction is not as fundamental as one might think. If time is discrete, then the set $\vec{q}$ will have to include a clock that gives a tick at every time step $t_n = n\,\delta t$, or,

$$\frac{\mathrm{d}}{\mathrm{d}t}\, q_{\text{clock}} = 1 \; ; \tag{1}$$

$$q_i \to q_i'(\vec{q}) \;\; \text{at} \;\; q_{\text{clock}} = 0 \bmod \delta t \; , \quad \forall i \neq \text{clock} \; . \tag{2}$$

It is not difficult to ascertain that this is just a special case of a more general equation of motion,

$$\frac{\mathrm{d}}{\mathrm{d}t}\, \vec{q} = \vec{f}(\vec{q}) \; . \tag{3}$$

For simplicity we therefore omit specific references to any clock[a]. In general, the orbit $\vec{q}(t)$ will be dictated by an equation of motion of the form (3).

In the absence of information loss, this will correspond to a Hamiltonian

$$H = \sum_i p_i f_i(\vec{q}) + g(\vec{q}) \; , \tag{4}$$

where $p_i = -i\partial/\partial q_i$ is the *quantum* momentum operator. It will be clear that the quantum equations of motion generated by this Hamiltonian will exactly correspond

---

[a]Thus, we do, as yet, use an absolute notion of time. Special and general relativistic transformations are left for future studies.

to the *classical* equation (3). The function $g(\vec{q})$ is arbitrary, its imaginary part being adjusted so as to ensure hermiticity:

$$H - H^\dagger \;=\; -i\vec{\nabla}\cdot\vec{f}(\vec{q}) + 2\,i\,\mathrm{Im}(g(\vec{q})) \;=\; 0\ , \tag{5}$$

Any observable quantity $A(\vec{q})$, not depending on operators such as p_i, and therefore commuting with all q_i, will be called a *beable.* Through the time dependence of $\vec{q}$, the beables will depend on time as well. Any pair of beables, A and B, will commute with one another at all times:

$$[A(t_1),\, B(t_2)] = 0\ , \quad \forall\, t_1,\, t_2\ . \tag{6}$$

A *changeable* is an operator not commuting with at least one of the q_i's. Thus, the operators p_i and the Hamiltonian H are changeables. Using beables and changeables as operators,[1] we can employ all standard rules of quantum mechanics to describe the classical system (3). At this point, one is tempted to conclude that the classical systems form just a very special subset of all quantum mechanical systems.

This, however, is not quite true. Quantum mechanical systems normally have a Hamiltonian that is bounded from below; the Hamiltonian (4) is not. At first sight, one might argue that all we have to do is *project out* all negative energy states[1].[2] We might obtain a physically more interesting Hilbert space this way, but, in general, the commutator property (6) between two beables is lost, if only positive energy states are used as intermediate states. As we will see, most of the beables (6) will not be observable in the quantum mechanical sense, a feature that they share with non-gauge-invariant operators in more conventional quantum systems with Yang-Mills fields. The projection mechanism that we need will be more delicate. As we will see, only the beables describing equivalence classes will survive as quantum observables.

We will start with the Hamiltonian (4), and only later project out states. Before projecting out states, we may observe that many of the standard manipulations of quantum mechanics are possible. For instance, one can introduce an integrable approximation $f_i^{(0)}(\vec{q})$ for the functions $f_i(\vec{q})$, and write

$$f_i(\vec{q}) = f_i^{(0)}(\vec{q}) + \delta f_i(\vec{q})\ , \tag{7}$$

after which we do perturbation expansion with respect to the small correction terms δf_i. However, the *variation principle* in general does not work at this level, because it requires a lowest energy state, which we do not have.

3. The Harmonic Oscillator

We assume that a theory describing our world starts with postulating the existence of sub-systems that in some first approximation evolve independently, and then are assumed to interact. For instance, one can think of independent local degrees of freedom that are affected only by their immediate neighbors, not by what happens at a distance, baring in mind that one may have to expand the notion of immediate neighbors to include variables that are spatially separated by distances of the order of the Planck length. Alternatively, one may think of elementary particles that, in a first approximation, behave as free particles, and are then assumed to interact.

Temporarily, we switch off the interactions, even if these do not have to be small. Every sub-system then evolves independently. Imagine furthermore that some form of *information loss* takes place. Then, as was further motivated in Ref.,[1] we suspect that the evolution in each domain will become *periodic*.

Thus, we are led to consider the case where we have one or more independent, periodic variables $q_i(t)$. Only at a later stage, coupling between these variables will have to be introduced in order to make them observable to the outside world. Thus, the introduction of periodic variables is an essential ingredient of our theory, in addition to being just a useful exercise.

Consider a single periodic variable:

$$\frac{\partial q}{\partial t} = \omega \ , \tag{8}$$

while the state $\{q = 2\pi\}$ is identified with the state $\{q = 0\}$. Because of this boundary condition, the associated operator $p = -i\partial/\partial q$ is quantized:

$$p \ = \ 0, \pm 1, \pm 2, \cdots \ . \tag{9}$$

The inessential additive coefficient $g(q)$ of Eq. (4) here has to be real, because of Eq. (5), and as such can only contribute to the unobservable phase of the wave function, which is why we permit ourselves to omit it:

$$H = \omega p = \omega n \ ; \qquad n \ = \ 0, \pm 1, \pm 2, \cdots \ . \tag{10}$$

If we would find a way to dispose of the negative energy states, this would just be the Hamiltonian of the quantum harmonic oscillator with internal frequency ω (apart from an inessential constant $\frac{1}{2}\omega$).

Theorem 3.1.
Consider any probability distribution $W(q)$ that is not strictly vanishing for any value of q, that is, a strictly positive, real function of q. Then a complex wave function $\psi(q)$ can be found such that

$$W(q) = \psi^*(q)\psi(q) \ , \tag{11}$$

and $\psi(q)$ is a convergent linear composition of eigenstates of H with non-negative eigenvalues only.

The proof is simple mathematics. Write

$$\psi(q) = \exp(\alpha(q) + i\beta(q)) \ , \qquad z = e^{iq} \ . \tag{12}$$

Choose $\alpha+i\beta$ to be en entire function within the unit circle of z. Then an elementary exercise in contour integration yields,

$$\alpha(q) = \frac{1}{2}\log(W(q)) \ ; \quad \beta(q) = \beta_0 - \mathcal{P}\oint \frac{\mathrm{d}q'}{2\pi} 1 + \frac{\cos(q'-q)}{\sin(q'-q)}\alpha(q') \ , \tag{13}$$

where $\mathcal{P}$ stands for the principal value, and β_0 is a free common phase factor. In fact, Eq. (13) is not the only function obeying our theorem, because we can choose any number of zeros for $\psi(z)$ inside the unit circle and then again match (11). One concludes from this theorem that no generality in the function W is lost by limiting ourselves to positive energy eigenfunctions only.

In fact, we may match the function W with a wave function ψ that has a zero of an arbitrary degree at the origin of z space. This way, one can show that the lowest energy state can be postulated to be at any value of E.

In this paper, however, we shall take a different approach. We keep the negative energy states, but interpret them as representing the bra states $\langle\psi|$. These evolve with the opposite sign of the energy, since $\langle\psi(t)| = e^{+iHt}\langle\psi(0)|$. As long as we keep only one single periodic variable, it does not matter much what we do here, since energy is absolutely conserved. The case of two or more oscillators is more subtle, however, and this we consider in the next section.

In this bra-ket formalism, it will be more convenient to tune the energy of the lowest ket state at $\frac{1}{2}\omega$. The kets $|n\rangle$ and bras $\langle n|$ have $E_n = (n + \frac{1}{2})\omega$. The time evolution of the bras goes as if $E_n = -(n + \frac{1}{2})\omega$, so that we have a sequence of energy values ranging from $-\infty$ to ∞.

4. Two (or more) Harmonic Oscillators

As was explained at the beginning of Section 3, we expect that, when two periodic variables interact, again periodic motion will result. This may seem to be odd. If the two periods, ω_1 and ω_2 are incommensurate, an initial state will never exactly be reproduced. Well, this was before we introduced information loss. In reality, periodicity will again result. We will show how this happens, first by considering the quantum harmonic oscillators to which the system should be equivalent, according to Section 3, and then by carefully interpreting the result.

In Fig. 1, the states are listed for the two harmonic oscillators combined. Let their frequencies be ω_1 and ω_2. The kets $|n_1, n_2\rangle = |n_1\rangle|n_2\rangle$ have $n_1 \geq 0$ and $n_2 \geq 0$, so they occupy the quadrant labelled I in Fig. 1. The bra states, in view of their time dependence, occupy the quadrant labelled III. The other two quadrants contain states with mixed positive and negative energies. Those must be projected away. If we would keep those states, then any interaction between the two oscillators would result in inadmissible mixed states, in disagreement with what we know of ordinary quantum mechanics. So, although keeping the bra states is harmless because total energy is conserved anyway, the mixed states must be removed. This is very important, because now we see that the joint system cannot be regarded as a direct product. Some of the states that would be allowed classically, must be postulated to disappear. We now ask what this means in terms of the two periodic systems that we thought were underlying the two quantum harmonic oscillators.

First, we wonder whether the spectrum of combined states will still be discrete. The classical, non interacting system would only be periodic if the two frequencies

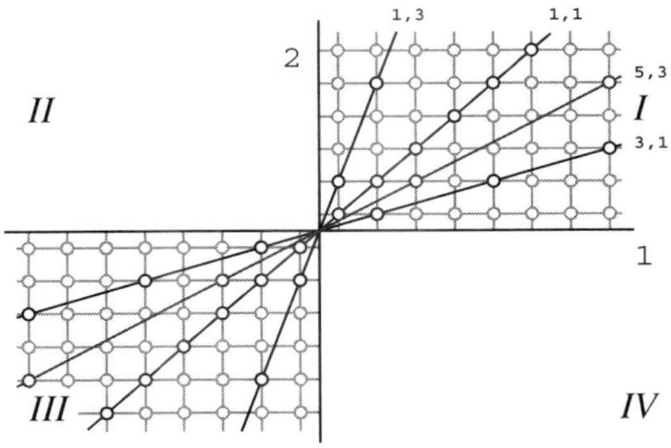

Fig. 1. Combining two harmonic oscillators. Tilted lines show sequences of spectral states again associated to harmonic oscillators. For further explanation, see text.

have a rational ratio: $p\,\omega_1 - q\,\omega_2 = 0$, where p and q are relative primes. The smallest period would be $T = 2\pi q/\omega_1 = 2\pi p/w_2$, so that we would expect equally spaced energy levels with spacings $\omega_1/q = \omega_2/p$. Indeed, at high energies, we do get such spacings also in the quantum system, with increasing degeneracies, but at lower energies many of these levels are missing. If the frequencies have an irrational ratio, the period of the classical system is infinite, and so a continuous spectrum would have to be expected.

When two quantum harmonic oscillators are considered together, this does not happen. The spectrum is always discrete. In Fig. 1, it is indicated how to avoid having missing states and variable degeneracies. We see that actually full series of equally spaced energy levels still exist:

> *At any given choice of a pair of odd relative primes p and q, we have a unique series of bra- and ket states with energies $\omega_{pq}(n + \frac{1}{2})$,* with $\omega_{pq} = p\,\omega_1 + q\,\omega_2$.

It is easy to see that these sequences are not degenerate, that all odd relative prime pairs of integers $(p,\, q)$ occur exactly once, and that all states are represented this way:

$$E_{n_1,\,n_2} = (n_1 + \frac{1}{2})\,\omega_1 + (n_2 + \frac{1}{2})\,\omega_2 = (n + \frac{1}{2})(p\,\omega_1 + q\,\omega_2)\ ; \tag{14}$$

$$\frac{2n_1 + 1}{2n_2 + 1} = \frac{p}{q}\ . \tag{15}$$

Some of these series are shown in the Figure.

We see that, in order to reproduce the quantum mechanical features, that is, to avoid the unphysical states where one energy is positive and the other negative, we have to combine two periodic systems in such a way that a new set of periodic systems arises, with frequencies ω_{pq}. Only then can one safely introduce interactions of some form. Conservation of total energy ensures that the bra and ket states cannot mix. States where one quantum oscillator would have positive energy and one has negative energy, have been projected out.

But how can such a rearrangement of the frequencies come about in a pair of classical periodic systems? Indeed, why are these frequencies so large, and why are they labelled by odd relative primes? In Fig. 2 the periodicities are displayed in configuration space, $\{q_1, q_2\}$. The combined system evolves as indicated by the arrows. The evolution might not be periodic at all. Consider now the (5, 3) mode. We can explain its short period $T_{53} = 2\pi/\omega_{53}$ only by assuming that the points form equivalence classes, such that different points within one equivalence class are regarded as forming the same 'quantum' state. If all points on the lines shown in Fig. 2 (the ones slanting downwards) form one equivalence class, then this class evolves with exactly the period of the oscillator whose frequency is ω_{53}.

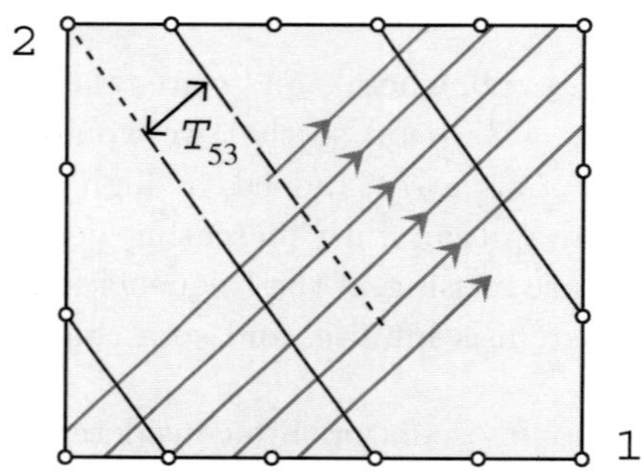

Fig. 2. The equivalence class for oscillators in the case $(p, q) = (5, 3)$. Lines with arrows pointing right and up: time trajectories of individual points. Solid and broken lines going downwards: (part of) the $(5, 3)$ equivalence class at $t = 0$. For further explanation, see text.

In principle, this equivalence class can be formed in one of two ways: the information concerning the location of a point on this line is lost, either because there is an inherent information loss mechanism, implying that two different states may actually evolve to become the same state, or it could simply be that this information cannot be transmitted to macroscopically observable quantities. One could imagine a renormalization group technique that relates microscopic states to states at much larger distance scales. Not all data are being faithfully transmitted in the procedure. This latter option will later be dismissed as being impractical; it is highly revealing to assume explicit information loss.

For the time being, imagine that information loss takes place by means of processes that are random, uncontrollable or impossible to follow in detail, that cause our data point to fluctuate along the line of its equivalence class. The line itself moves with the deterministic speed of the original oscillators.

Observe in Figure 2 that, in case $(p, q) = (5, 3)$, due to these fluctuations, five points of system 1 alone now form a single equivalence class, and three points in system 2. This is because we have assigned 5 quanta of energy to system 1 for every three quanta of energy of system 2. More generally, we could represent this situation with the wave function

$$\psi_{pq} = e^{i\,(n+\frac{1}{2})\,(p\,q_1+q\,q_2-\omega_{pq}t)}\,e^{-\frac{1}{2}\,i(q_1+q_2)}\ , \tag{16}$$

where both variables $q_{1,2}$ were taken to be periodic with periods 2π. The $(p,\,q)$ equivalence classes appear to be defined by the condition

$$p\,q_1 + q\,q_2 \;=\; \text{Constant}\,, \tag{17}$$

and this means that the n-dependent part of the wave function (16) has a the same phase all over the entire equivalence class, if we may assume that the second term in Eq. (16), arising from the vacuum fluctuations $\frac{1}{2}\,\omega\,t$, may be ignored.

To describe the equivalence classes it is helpful to introduce time variables t_a for the subsystems $a = 1, 2, \cdots$ in terms of their *unperturbed* evolution law, $q_a = \omega_a t_a$. Then, writing $E_1 = p\,\omega_1\,,\ E_2 = q\,\omega_2$, one can characterize the equivalence classes as

$$E_1\,\delta\,t_1 + E_2\,\delta\,t_2 \;=\; 0\ , \tag{18}$$

which means that the reactions that induce information loss cause q_a to speed up or slow down by an amount $\pm\delta\,t_a$ obeying this equation. One can easily generalize this result for many coexisting oscillators. They must form equivalence classes such that fluctuating time differences occur that are only constrained by

$$\sum_a E_a\,\delta\,t_a = 0\ , \tag{19}$$

which also are the collections of points that have the same phase in their quantum wave functions. We conclude that, in the ontological basis $\{|\vec{q}\rangle\}$, *all states* $|\vec{q}\rangle$ *which have the same phase in the wave function* $\langle\vec{q}|\psi\rangle$ (apart from a fixed, time independent term), *form one complete equivalence class.*

5. Energy and Hamiltonian

In the previous section, it was derived that the energies of the various oscillators determine the shape of the equivalence classes that are being formed. However, this would require energy to be a beable, as defined in Section 2. Of course, the Hamiltonian, being the generator of time evolution, cannot be a beable. It is important to notice here, that the parameters p and q defining the equivalence classes as in Section 4, are not exactly the energies of q_1 and q_2; the Hamiltonian eigenvalues are

$$H_1 = (n+\frac{1}{2})p\,\omega_1\ ; \qquad H_2 = (n+\frac{1}{2})q\,\omega_2\ , \tag{20}$$

with a common integral multiplication factor $n+\frac{1}{2}$. This n indeed defines the Hamiltonian of the orbit of the equivalence class. Generalizing this, the relation between the energies E_i in Eqs. (18) and (19) and the Hamiltonian H, is

$$H = (n+\frac{1}{2})E\ , \tag{21}$$

where n defines the evolution of a single clock that monitors the evolution of the entire universe.

Now that the relative primes p and q have become beables, we may allow for the fact that the periods of q_1 and q_2 depend on p and q as a consequence of some non-trivial interaction. But there is more. We read off from Fig. 2, that p points on the orbit of q_1 in fact belong to the same equivalence class. Assuming that the systems 1 and 2 that we started off with, had been obtained again by composing other systems, we must identify these points. But this forces us to redefine the original periods by dividing these by p and q, respectively, and then we end up with two redefined periodic systems that are combined in the one and only allowed way:

$$p = q = 1 \ . \tag{22}$$

Only a single line in Fig. 1 survives: the diagonal.

The picture that emerges is the following. We are considering a collection of variables q_a, each being periodic with different periods $T_a = 2\pi/\omega_a$. They each are associated with a positive beable E_a, such that $E_a = \omega_a$. The interactions will be such that the total energy $E = \sum_a E_a$ is conserved. Now as soon as these variables are observed together (even if they do not interact), an uncontrollable mixing mechanism takes place in such a way that the variables are sped up or slowed down by time steps δt_a obeying Eq. (19), so that, at any time t, all states obeying

$$\sum_a E_a t_a = \Big(\sum_a E_a\Big) t \ , \tag{23}$$

form one single equivalence class.

The evolution and the mixing mechanism described here are entirely classical, yet we claim that such a system turns into an acceptable quantum mechanical theory when handled probabilistically. However, we have not yet introduced interactions.

6. Interactions

We are now in a position to formulate the problem of interacting systems. Consider two systems, labelled by an index $a = 1,\, 2$. System a is characterized by a variable $q_a \in [0,\, 2\pi)$ and a discrete index $i = 1,\, \cdots,\, N_a$, which is a label for the spectrum of states the system can be in. Without the interaction, i stays constant. Whether the interaction will change this, remains to be seen.

The frequencies are characterized by the values $E^i_a = \omega^i_a$, so that the periods are $T^i_a = 2\pi/\omega^i_a$. Originally, as in Section 4, we had $\omega^i_a = p_a\omega_a$, where $p_1 = p$ and $p_2 = q$ were relative primes (and both odd), but the periods ω_a are allowed to depend on p_a, so it makes more sense to choose a general spectrum to start with.

The non-interacting parts of the Hamiltonians of the two systems, responsible for the evolution of each, are described by

$$H^0_a|n_a,\, i\rangle = (n_a + \frac{1}{2})E^i_a|n_a,\, i\rangle \ , \tag{24}$$

where the integer $n_a = -\infty, \cdots, +\infty$ is the changeable generating the motion along the circle with angular velocity ω_a. We have

$$n_a = -i\partial/\partial q_a \ . \tag{25}$$

The total Hamiltonian describing the evolution of the combined, unperturbed, system is not $H_1^0 + H_2^0$, but

$$H_{\rm tot}^0 = (n_{\rm tot} + \frac{1}{2})(E_1^i + E_2^j) \ , \tag{26}$$

where i, j characterize the states 1 and 2, but we have a single periodic variable $q_{\rm tot} \in [0, 2\pi)$, and

$$n_{\rm tot} = -i\partial/\partial q_{\rm tot} \ . \tag{27}$$

In view of Eq. (17), which here holds for $p = q = 1$, we can define

$$q_{\rm tot} = q_1 + q_2 \ , \tag{28}$$

while $q_1 - q_2$ has become invisible. We can also say,

$$n_1 = n_2 = n_{\rm tot} \ . \tag{29}$$

An interacting system is expected to have perturbed energy levels, so that its Hamiltonian should become

$$H^0 + H^{\rm int} = (n_{\rm tot} + \frac{1}{2})(E_1^i + E_2^j + \delta E^{ij}) \ , \tag{30}$$

where δE^{ij} are correction terms depending on both i and j. This is realized simply by demanding that the beables E_1^i and/or E_2^j get their correction terms straight from the other system. This is an existence proof for interactions in this framework, but, at first sight, it appears not to be very elegant. It means that the velocity ω_1^{ij} of one variable q_1 depends on the state j that the other variable is in, but no matrix diagonalization is required. Indeed, we still have no transitions between the different energy states i. It may seem that we have to search for a more general interaction scheme. Instead, the scheme to be discussed next differs from the one described in this section by the fact that the energies E cannot be read off directly from the state a system is in, even though they are beables. The indices i, j are locally unobservable, and this is why we usually work with superimposed states.

7. Limit Cycles

Consider the evolution following from a given initial configuration at $t = t_0$, having an energy E. Let us denote the state at time t by $F(t)$. The equivalence classes $|\psi(t)\rangle$ defined in Section 4 are such that the state $F(t)$ is equivalent to the state $F(t + \Delta t)$, where

$$\Delta t = h/E \ , \tag{31}$$

in which h is Planck's constant, or:

$$|\psi(t)\rangle \stackrel{\text{def}}{=} \{F(t + n\Delta t)\,,\ \forall n|_{n\geq n_1}\}\ , \tag{32}$$

for some n_1. Let us now assume that the equivalence indeed is determined by information loss. That the states in Eq. (32) are equivalent then means that there is a smallest time t_1 such that

$$F(t_1 + n\Delta t) = F(t_1)\ ,\ \forall n \geq 0\ . \tag{33}$$

Thus, the system ends in a *limit cycle* with period Δt.

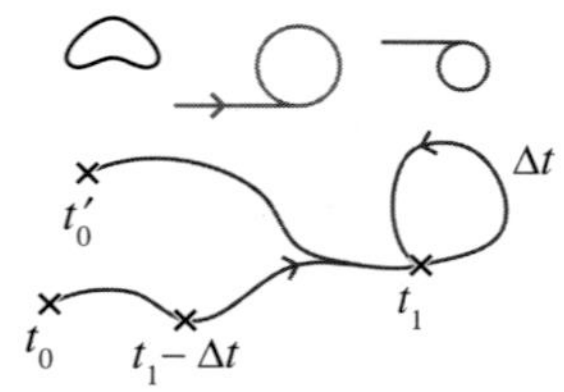

Fig. 3. Configuration space showing the limit cycles of an evolving system, indicating the times t_0, t_1 and the period Δt of a limit cycle. The points at t_0 and t_0' form one equivalence class.

One now may turn this observation around. A closed system that can only be in a finite number of different states, making transitions at discrete time intervals, would necessarily evolve back into itself after a certain amount of time, thus exhibiting what is called a Poincaré cycle. If there were no information loss, these Poincaré cycles would tend to become very long, with a periodicity that would increase exponentially with the size of the system. If there is information loss, for instance in the form of some dissipation effect, a system may eventually end up in Poincaré cycles with much shorter periodicities. Indeed, time does not have to be discrete in that case, and the physical variables may form a continuum; there could be a finite set of stable orbits such that, regardless the initial configuration, any orbit is attracted towards one of these stable orbits; they are the limit cycles. The energy of a state is then simply defined to be given by Eq. (31), or $E \stackrel{\text{def}}{=} h/P$, where $P = \Delta t$ is the period of the limit cycle, and h is Planck's constant.

Since this period coincides with the period of the wave function, we now deduce a physical interpretation of the phase of the wave function: *The phase of a wave function* (in the frame of energy eigenstates) *indicates where in the limit cycle the state will be.*

In general, we will have a superposition of many possibilities, and so we add to this the interpretation of the amplitude of the wave function: the absolute value of the amplitude in the frame of energy eigenstates indicates the probability that a particular limit cycle will be reached. Thus, we have reached the exact physical meaning of a quantum wave function.

We identified any deterministic system having information loss, with a quantum mechanical system evolving with a Hamiltonian defined by Eq. (21). However, in order to obtain a realistic model, one has to search for a system where the energy is extrinsic. With this, we mean that the universe consists of subsystems that are weakly coupled. Uncoupled systems are described as in Section 4; weakly coupled systems must be such that the limit cycles of the combination (12) of two systems (1) and (2) must have periods P_{12} obeying

$$\frac{1}{P_{12}} \approx \frac{1}{P_1} + \frac{1}{P_2} , \tag{34}$$

which approaches the exact identity in the limit of large systems being weakly coupled. This is the energy conservation law.

8. Discussion

When we attempt to regard quantum mechanics as a deterministic system, we have to face the problem of the positivity of the Hamiltonian, as was concluded earlier in Refs[12].[3] There, also, the suspicion was raised that information loss is essential for the resolution of this problem. In this paper, the mathematical procedures have been worked out further, and we note that the deterministic models that we seek must have short limit cycles, obeying Eq. (31). Short limit cycles can easily be obtained in cellular automaton models with information loss, but the problem is to establish the addition rule (34), which suggests the large equivalence classes defined by Eq. (17). We think that the observations made in this paper are an important step towards the demystification of quantum mechanics.

We found that the energy eigenstates of a quantum system correspond to the limit cycles of the deterministic model. If P is the period of the limit cycle, then the energy E of this state is $E = h/P$ (see Eq. 31).

In models with more or less random evolution laws, one can guess the distribution of the periods of the limit cycles. In the Appendix, we derive the distribution of limit cycles with periods $\Delta t = P$ for a "completely random" model, which we define to be a model where the mapping $F(t) \to F(t+1)$ is chosen completely randomly for every $F(t)$, independently of how many other states $F'(t)$ might map into the same state $F(t+1)$. It is found that the distribution of the cycles may then be expected to be logarithmic, which leads to a logarithmic energy spectrum: the energy eigenstates are a Poisson distribution on a logarithmic scale:

$$\varrho(E)\mathrm{d}E = \mathrm{d}E/E = \mathrm{d}\log E , \tag{35}$$

with cutoffs at

$$E_{\mathrm{min}} = \mathcal{O}(h/\sqrt{2\mathcal{N}}) , \qquad E_{\mathrm{max}} = \mathcal{O}(1/\delta t) , \tag{36}$$

where $\mathcal{N} = \mathcal{O}(e^V)$ is the total number of possible states and δt is some cutoff in time that our system may have. This is not a realistic energy spectrum for a quantum field theory, so we must conclude that realistic models will have to be far from random.

Cellular automaton models can be written down that show a rapid convergence towards small limit cycles, starting from any state $F(0)$. Conway's "game of life"[4] is an example, although that also features 'glider solutions', which are structures that are periodic, but they move forward when released in an empty region, so that they are not limit cycles in the strict sense. It must be emphasized, however, that Conway's game of life will not serve as a model generating quantum mechanics. In a model generating quantum mechanics, the vacuum state is the state with the *longest* limit cycle, since it has the lowest energy. Thus, the empty state in Conway's game of life would carry more energy than its glider solutions.

The limit cycles of a random model are too long, those of cellular automata such as Conway's Game of Life are too short. In the real universe we have a small number of massless particle species, and many more massive ones, generating a rich spectrum of energies all very low compared to the Planck energy. If we assume that the Planck time would be Nature's natural time scale, then we observe that there must be many limit cycles whose periods are very long compared to the Planck time. Our universe appears to be built in such a way that, as soon as several of these cyclic limiting solutions are allowed to interact, new limit cycles will be reached with shorter periods, due to information loss.

In cellular automaton models, one might be able to mimic such a situation best by introducing a nearly conserved, positive quantity resembling energy, which can be seen statistically to decrease slowly on the average, so that the most chaotic initial states relax into more organized states that can easily end up in a limit cycle. The more chaotic the initial state, the smaller the period of its eventual limit cycle is expected to be, but there are many special initial states with very long limit cycles: the low energy states.

States of interest, with which we might attempt to describe the universe as we know it, must be very far away from any limit cycle. They are also far away from the strictly stationary eigenstates of the Hamiltonian. This means that we *do not yet know* which of the numerous possible limit cycles our universe will land into. This is why we normally use wave functions that have a distribution of amplitudes in the basis of the Hamiltonian eigenmodes. The squares of these amplitudes indicate the probability that any particular limit cycle will be reached. Also note that, according to General Relativity, taking into account the negative energies in the gravitational potentials, the total energy of the universe should vanish, which means that the entire universe might never settle for any limit cycle, as is indeed suggested by what we know of cosmology today: the universe continues to expand. The limit cycles mentioned in this paper refer to idealized situations where small sections of the universe are isolated from the rest, so as to be able to define their energies exactly. Only when a small part of the universe is sealed off from the rest, it is destined to end up in a limit cycle.

It may be of importance to note that our definition of energy, as being the inverse of the period of the limit cycle, supplies us with an absolute scale of energy: it is not allowed to add a constant to it. Moreover, the energy E in Eq. (21), as opposed to the Hamiltonian H, is a beable. Thus it is allowed to couple it to gravity by imposing Einstein's equations. If indeed the vacuum has a limit cycle with a large period, it carries a very low energy, and this is why we suspect that the true resolution of the cosmological constant problem[5] will come from deterministic quantum mechanics rather than some symmetry principle.[6] Earlier, the cosmological constant has been considered in connection with deterministic quantum mechanics by Elze.[7] The fact that the observed cosmological constant appears to be non-vanishing implies that a finite volume V of space will have a largest limit cycle with period

$$P = 8\pi \frac{hG}{\Lambda V} \,, \tag{37}$$

which is of the order of a microsecond for a volume of a cubic micron. If Λ were negative we would have had to assume that gravity does not exactly couple to energy.

Lorentz transformations and general coordinate transformations have not been considered in this paper. Before doing that, we must find models in which the Hamiltonian is indeed extensive, that is, it can be described as the integral of an Hamiltonian density $T_{00}(\vec{x}, t)$ over 3-space, as soon as the integration volume element $\mathrm{d}^3\vec{x}$ is taken to be large compared to the 'Planck volume'. When that is achieved, we will be only one step away from generating locally deterministic quantum field theories.

What can be said from what we know presently, is that a particle with 4-momentum p_μ must represent an equivalence class that contains all translations $x^\mu \to x^\mu + \Delta x^\mu$ with $p_\mu \, \Delta x^\mu = n\, h$, where n is an integer. Note that a limit cycle having this large transformation group as an invariance group is hard to imagine, which probably implies that this particular limit cycle will take an infinite time to be established. Indeed, a particle in a fixed momentum state occupies the entire universe, and we already observed that the entire universe will never reach a limit cycle.

In the real world, we have only identified the observable quantum states, which we now identify with the equivalence classes of ontological states. We note that the physical states of the Standard Model in fact are also known to be gauge equivalence classes, Local gauge transformations modify our description of the dynamical variables, but not the physical observations. It is tempting to speculate that these gauge equivalence classes (partly) coincide with the equivalence discussed in this paper, although our equivalence classes are probably a lot larger, which may mean that many more local gauge symmetries are still to be expected.

It is even more tempting to include here the gauge equivalence classes of General Relativity: perhaps local coordinate transformations are among the dissipative transitions. In this case, the underlying deterministic theory might not be invariant under local coordinate transformations, and here also one may find novel approaches towards the cosmological constant problem and the apparent flatness of our universe.

Our reason for mentioning virtual black holes being sources of information loss might require further explanation. Indeed, the *quantum mechanical* description of a black hole is not expected to require information loss (in the form of quantum decoherence); it is the corresponding classical black hole that we might expect to play a role in the ontological theory, and that is where information loss is to be expected, since classical black holes do not emit Hawking radiation. As soon as we turn to the quantum mechanical description in accordance to the theory explained in this paper, a conventional, fully coherent quantum description of the black hole is expected.

Although we do feel that this paper is bringing forward an important new approach towards the interpretation of Quantum Mechanics, there are many questions that have not yet been answered. One urgent question is how to construct explicit models in which energy can be seen as extrinsic, that is, an integral of an energy density over space. A related problem is how to introduce weak interactions between two nearly independent systems. Next, one would like to gain more understanding of the phenomenon of (destructive) interference, a feature typical for Quantum Mechanics while absent in other statistical theories.

Appendix A.

9. The random deterministic model

For simplicity, take the duration δt of one time step, in Eq. (2), to be 1. Consider a completely random mapping $F(t) \to F(t+1)$. This means that, using a random number generator, some rule has been established to define $F(t+1)$ for any given $F(t)$. The rule does not depend on t. In general, the mapping will not at all be one-to-one. Let the total number of states $F(t)$ be $\mathcal{N}$. In general, $\mathcal{N}$ wil grow exponentially with the volume V of the system. We claim that the distribution of limit cycles may be expected to be as described in Section 8. The argument goes as follows.

At $t = 0$, take an arbitrary element of the space of states, $F(0)$. The series $F(1),\ F(2),\ \cdots$ will end up in a limit cycle, which means that there is a time T that is the smallest time with the property that $F(T) = F(t_1)$ for some $0 \le t_1 < T$. The length of the limit cycle is $P = T - t_1$, see Fig. 3. We consider the case that $\mathcal{N} \gg 1$.

Let $Q(x)$ be the probability that $T > x-1$, given the state $F(0)$. The probability that $T = x$ is then equal to $(x/\mathcal{N})Q(x)$, so that

$$\frac{\mathrm{d}}{\mathrm{d}x}Q(x) = -\frac{x}{\mathcal{N}}\,Q(x)\,, \qquad Q(0) = 1 \qquad \to \qquad Q(x) = e^{-x^2/(2\mathcal{N})}\,. \quad \text{(A.1)}$$

In general, $t_1 > 0$, so that the state $F(0)$ itself is not a member of the limit cycle. Apparently the limit cycles contain only a small subset of all states. The probability that the element $F(0)$ actually happens to sit in the limit cycle with length P (to be called $C(P)$, is the probability that $t_1 = 0$, or

$$\mathcal{P}\Big(F(0) \in C(P)\Big) = \frac{1}{\mathcal{N}}\, Q(P)\ . \tag{A.2}$$

This is understood as follows: $Q(P)$ is the chance that the cycle, starting from $F(0)$ did not close earlier, and $1/\mathcal{N}$ is the probability that the Pth state happens to coincide with $F(0)$.

Since $C(P)$ contains P states, and the total number of states is $\mathcal{N}$, one derives that the expectation value of the number of limit cycles with length P is

$$\mathcal{E}(P) = \frac{1}{P}\, e^{-P^2/(2\mathcal{N})}\ . \tag{A.3}$$

From this, we derive the distribution of periods P to be

$$\varrho(P)\mathrm{d}P = \frac{\mathrm{d}P}{P}\, e^{-P^2/(2\mathcal{N})}\ . \tag{A.4}$$

For all periods P that are small compared to $\sqrt{\mathcal{N}}$, the exponent can be ignored, and since $E = h/P$, Eq. (35) follows. The largest period is of order $\sqrt{\mathcal{N}}$, and the smallest one is the fundamental time unit, which here was taken to be one.

References

1. G. 't Hooft, *Class. Quant. Grav.* **16** (1999) 3263, *Quantum Mechanics and determinism*, in Proceedings of the Eighth Int. Conf. on "Particles, Strings and Cosmology, Univ. of North Carolina, Chapel Hill, Apr. 10–15, 2001, P. Frampton and J. Ng, Eds., Rinton Press, Princeton, pp. 275–285.
2. H. Th. Elze, *Deterministic models of quantum fields.*
3. M. Blasone, P. Jizba and H. Kleinert, *Annals of Physics* **320** (2005) 468; id., *Braz. J. Phys.* **35** (2005).
4. M. Gardner, *Scientific American,* **223** (4), 120; (5), 118; (6), 114 (1970).
5. S. Nobbenhuis, *Foundations of Physics,* **36**, 613 (2006).
6. G. 't Hooft and S. Nobbenhuis, *Class. Quant. Grav.*, 3819 (2006).
7. H. Th. Elze, *Quantum fields, cosmological constant and symmetry doubling,* hep-th/0510267.

WHAT DID WE LEARN FROM QUANTUM GRAVITY?

A. ASHTEKAR

Institute for Gravitational Physics and Geometry, Physics Department, Penn State, University Park, PA 16802-6300
and Institute for Theoretical Physics, University of Utrecht, Princetonplein5, 3584 CC Utrecht, The Netherlands
E-mail: ashtekar@gravity.psu.edu

The goal of the article is to provide some glimpses into the challenges and successes of quantum gravity. After a general introduction, for concreteness I focus on a specific approach which goes under the name *loop quantum gravity*. The underlying ideas are first summarized and recent advances are then illustrated by applying these ideas to cosmology. Quantum effects of geometry resolve the big-bang singularity of classical general relativity. Quantum physics does not break down at the big-bang. In simple models where details have been fully worked out, there is a pre-big-bang branch joined to the current post-big-bang branch by well-defined quantum evolution.

Keywords: Quantum gravity; Loop quantum gravity; Cosmology; General relativity; Big-bang.

1. Introduction

My task at this conference was simplified by the organizers: they not only advised me on the scope of the talk but also provided the title. However, the subject of quantum gravity is rather old and consequently developments in the field are simply too numerous to be covered in a single talk. (For a brief historical account see, e.g., Ref. [1].) Therefore, I will be able to provide only a flavor of the conceptual challenges and recent advances through illustrative examples.

Let us begin with a brief introduction to the subject. General relativity and quantum theory are among the greatest intellectual achievements of the 20th century. Each of them has profoundly altered the conceptual fabric that underlies our understanding of our physical world. Furthermore, each has been successful in describing the physical phenomena in its own domain to an astonishing degree of accuracy. And yet, they offer us *strikingly* different pictures of physical reality. Our past experience in physics tells us that these two pictures must be approximations, special cases that arise as appropriate limits of a single, universal theory. That theory must therefore represent a synthesis of general relativity and quantum mechanics. This would be the quantum theory of gravity that we invoke when faced with phenomena, such as the big bang and the final state of black holes, where the worlds of general relativity and quantum mechanics must unavoidably meet.

Remarkably, the necessity of a quantum theory of gravity was pointed out by Einstein already in 1916. In a paper in the Preussische Akademie Sitzungsberichte he wrote:

> *Nevertheless, due to the inneratomic movement of electrons, atoms would have to radiate not only electromagnetic but also gravitational energy, if only in tiny amounts. As this is hardly true in Nature, it appears that quantum theory would have to modify not only Maxwellian electrodynamics but also the new theory of gravitation.*

Ninety years later, our understanding of the physical world is vastly richer but a fully satisfactory unification of general relativity with quantum physics still eludes us. Indeed, the problem has now moved to the center-stage of fundamental physics.

A key reason why the issue is still open is the lack of experimental data with direct bearing on quantum gravity. As a result, research is necessarily driven by theoretical insights on what the key issues are and what will 'take care of itself' once this core is understood. As a consequence, there are distinct starting points which seem natural. Such a diversity of theoretical approaches existed in the early stages of our understanding of other fundamental forces as well. However, clear cut experiments soon became available to weed out ideas which, in spite of their theoretical appeal, fail to be realized in Nature. We do not yet have this luxury in quantum gravity. But then, in absence of strong experimental constraints, one would expect a rich variety of internally consistent theories. Why is it then that we do not have a single one? The reason, I believe, lies the deep conceptual difference between the description of gravity in general relativity and that of non-gravitational forces in other fundamental theories. In those theories, space-time is given a priori, serving as an inert background, a stage on which the drama of evolution unfolds. General relativity, on the other hand, is not only a theory of gravity, it is also a theory of space-time structure. Indeed, in general relativity, gravity is encoded in the very geometry of space-time. Therefore, a quantum theory of gravity has to simultaneously bring together *gravity, geometry and the quantum.* This is a band new adventure and our past experience with other forces can not serve as a reliable guide.

For concreteness and brevity, I will focus on *loop quantum gravity* (LQG), an approach that attempts to face this challenge squarely (for details, see, e.g., Refs. [2–4]). Recall that Riemannian geometry provides the appropriate mathematical language to formulate the physical, kinematical notions as well as the final dynamical equations of any classical theory of relativistic gravity. This role is now assumed by *quantum* Riemannian geometry. Thus, in LQG both matter and geometry are quantum mechanical 'from birth'.

In the classical domain, general relativity stands out as the best available theory of gravity. Therefore, it is natural to ask: *Does quantum general relativity, coupled to suitable matter* (or supergravity, its supersymmetric generalization) *exist as consistent theories non-perturbatively?* In particle physics circles the answer is often assumed to be in the negative, not because there is concrete evidence which rules out this possibility, but because of the analogy to the theory of weak interactions. There, one first had a 4-point interaction model due to Fermi which works quite well at low energies but which fails to be renormalizable. Progress occurred not by looking for non-perturbative formulations of the Fermi model but by going to the Glashow-Salam-Weinberg renormalizable theory of electro-weak interactions, in which the 4-point interaction is replaced by $W^{\pm}$ and Z propagators. Therefore, it is often assumed that perturbative non-renormalizability of quantum general relativity points in a similar direction. However this argument overlooks a crucial and qualitatively new element of general relativity. Perturbative treatments pre-suppose that space-time is a smooth continuum *at all scales* of interest to physics under consideration. This assumption is safe for weak interactions. In the gravitational case, on the other hand, the scale of interest is *the Planck length* and there is no physical basis to pre-suppose that the continuum approximation should be valid down to that scale. The failure of the standard perturbative treatments may largely be due to this grossly incorrect assumption and a non-perturbative treatment which correctly incorporates the physical micro-structure of geometry may well be free of these inconsistencies.

Are there any situations, outside LQG, where such physical expectations are borne out by detailed mathematics? The answer is in the affirmative. There exist quantum field theories (such as the Gross-Neveu model in three dimensions) in which the standard perturbation expansion is not renormalizable although the theory is *exactly soluble*! Failure of the standard perturbation expansion can occur because one insists on perturbing around the trivial, Gaussian point rather than the more physical, non-trivial fixed point of the renormalization group flow. Interestingly, thanks to the recent work by Lauscher, Reuter, Percacci, Perini and others, there is now growing evidence that situation may be similar with general relativity (see [[5]] and references therein). Impressive calculations have shown that pure Einstein theory may also admit a non-trivial fixed point. Furthermore, the requirement that the fixed point should continue to exist in presence of matter constrains the couplings in physically interesting ways.[6]

Let me conclude this introduction with an observation. There is no a priori reason for a non-perturbative quantum gravity theory to be the 'final' theory of all known physics. In particular, as is the case with classical general relativity, while requirements of background independence and general covariance do restrict the form of interactions between gravity and matter fields and among matter fields themselves, such a theory would not have a built-in principle which *determines* these interactions. However, just as general relativity has had powerful implications in spite of this limitation in the classical domain, quantum general relativity should have qualitatively new predictions, pushing further the existing frontiers of physics. In section 3 we will see an illustration of this possibility.[a]

[a]In my talk, I also provided another illustrative application: statistical mechanical definition of

2. Quantum Riemannian Geometry

A central lesson of general relativity is that gravity is coded in the very geometry of space-time. The resulting physical nature of geometry constitutes a driving force for LQG. In the twentieth century we learned that matter and radiation has atomic structure and their microscopic structure is best described in terms of appropriate quanta. Since geometry is a physical entity like matter and radiation, we are led to ask: What are the fundamental quanta of geometry? What is the Hilbert space of its states? What are the self-adjoint operators that represent geometrical observables? Do they have discrete eigenvalues? If so, the space-time continuum underlying today's physics can only be an approximation, obtained by suitable coarse graining. These issues were analyzed in detail in the mid-nineties.[2–4,10–16] Here I will provide a brief summary, emphasizing the aspects that are important for the application discussed in section 3.

The starting point of LQG is a Hamiltonian formulation of general relativity based on spin connections.[17] Since all other basic forces of nature are also described by theories of connections, this formulation naturally leads to an unification of all four fundamental forces at a *kinematical* level. Specifically, the phase space of general relativity is the same as that of a Yang-Mills theory. The difference lies in dynamics: whereas in the standard Yang-Mills theory the Minkowski metric features prominently in the definition of the Hamiltonian, there are no background fields whatsoever once gravity is switched on.

Let us focus on the gravitational sector of the theory. Then, the phase space $\mathbf{\Gamma}_{\rm grav}$ consists of canonically conjugate pairs (A_a^i, P_a^i), where A_a^i is a connection on a 3-manifold M and P_a^i a vector density of weight one, both of which take values in the Lie-algebra su(2). The connection A enables one to parallel transport chiral spinors (such as the left handed fermions of the standard electro-weak model) along curves in M. Its curvature is directly related to the electric and magnetic parts of the space-time *Riemann tensor*. P_i^a plays a double role. Being the momentum canonically conjugate to A, it is analogous to the Yang-Mills electric field. In addition, $E_i^a := 8\pi G\gamma P_i^a$, has the interpretation of a frame or an orthonormal triad (with density weight 1) on M, where γ is the so-called 'Barbero-Immirzi parameter' representing a quantization ambiguity. Each triad E_i^a determines a positive definite 'spatial' 3-metric q_{ab}, and hence the Riemannian geometry of M. This dual role of P is a reflection of the fact that now SU(2) is the (double cover of the) group of rotations of the orthonormal spatial triads on M itself rather than of rotations in an 'internal' space associated with M.

To pass to quantum theory, one first constructs an algebra of 'elementary' functions on $\mathbf{\Gamma}_{\rm grav}$ (analogous to the phase space functions x and p in the case of a particle) which are to have unambiguous operator analogs. In LQG, the configu-

black hole entropy. This is a fascinating subject because it brings together general relativity, quantum physics and statistical mechanics. However it turned out to be too difficult to cover it within the page limitation. For summaries see e.g. Refs. [2,7].

ration variables are the holonomies h_e built from A_a^i which enable us to parallel transport chiral spinors along edges e. The momentum functions are the fluxes $E_{S,f}$ of 'electric fields' or 'triads' (smeared with test fields f) across 2-surfaces S. This choice is motivated by the absence of background fields: every connection A_a^i can be naturally integrated along 1-dimensional edges e to define holonomies h_e and the smeared vector densities $E_i^a f_i$ can be naturally integrated along two dimensional surfaces, both without reference to a background metric.

These functions generate a certain algebra $\mathfrak{a}$ (analogous to the algebra generated by operators $\widehat{\exp i\lambda x}$ and $\hat{p}$ in quantum mechanics). The first principal task is to find representations of this algebra. In that representation, *quantum* Riemannian geometry can be probed through the triad operators $\hat{E}_{S,f}$, which stem from classical orthonormal triads. Quite surprisingly, the requirement of diffeomorphism covariance on M suffices to single out a *unique* representation of $\mathfrak{a}$[8,9]! This recent result is the quantum geometry analog to the seminal results by Segal and others that characterized the Fock vacuum in Minkowskian field theories. However, while that result assumes not only Poincaré invariance of the vacuum but also specific (namely free) dynamics, it is striking that the present uniqueness theorems make no such restriction on dynamics. The requirement that there be a diffeomorphism invariant state is surprisingly strong and makes the 'background independent' quantum geometry framework surprisingly tight.

This unique representation was in fact introduced already in the mid-nineties[10–14] and has been extensively used in LQG since then. As in familiar quantum mechanics, the underlying Hilbert space is the space $\mathcal{H} = L^2(\bar{\mathcal{A}}, \mathrm{d}\mu_o)$ of square-integrable functions on the configuration space. However, as in quantum field theories, because of the presence of an infinite degrees of freedom, the quantum configuration $\bar{\mathcal{A}}$ is a certain *completion* of the classical configuration space $\mathcal{A}$ consisting of smooth connections on M. ($\bar{\mathcal{A}}$ is the Gel'fand spectrum of the holonomy algebra generated by the functions h_e on $\mathcal{A}$.) μ_o is a diffeomorphism invariant, faithful, regular Borel measure on $\bar{\mathcal{A}}$. The holonomy (or configuration) operators $\hat{h}_e$ act just by multiplication. In the classical theory, by taking suitable derivatives in M of holonomies h_e along arbitrary edges e, one can recover the connection from which the holonomy is built. However, in the quantum theory, the operators $\hat{h}_e$ are discontinuous in their dependence on e and there is no operator $\hat{A}$ corresponding to the connection itself. The momentum operators $\hat{P}_{S,f}$ act as Lie-derivatives. Since the fields E_i^a represent orthonormal triads in the classical theory, quantum geometry is built out of the momentum operators $\hat{P}_{S,f}$. Given a surface S and a region R one can express the area A_S and volume V_R using the triads. Although they are non-polynomial functions of triads, the operators $\hat{A}_S$ and $\hat{V}_R$ are well-defined and also have *discrete* eigenvalues. In this precise sense, Riemannian geometry is quantized.

Finally, the Hilbert space $\mathcal{H}$ and the associated holonomy and (smeared) triad operators constitute the *kinematical* framework —the quantum analog of the full phase space. Like the classical phase space, this kinematic setup provides a home for *formulating* quantum dynamics. In the Hamiltonian framework, the dynamical content of any background independent theory is contained in its constraints —in our case the quantum Einstein equations. Thus, to provide quantum dynamics, one has to first obtain expressions of the constraint operators on $\mathcal{H}$. Physical states belong to the kernel of these operators. The final task is to endow the space of solutions with the structure of a Hilbert space, identify physically interesting operators and interpret the resulting theory. In the next section, I will illustrate this procedure in the cosmological setting.

3. Application: Homogeneous Isotropic Cosmology

As emphasized in Sec. 1, a central feature of general relativity is its encoding of the gravitational field in the Riemannian geometry of space-time. This encoding is directly responsible for the most dramatic ramifications of the theory: the big-bang, black holes and gravitational waves. However, it also leads one to the conclusion that space-time itself must end and classical physics must come to a halt at the big-bang and black hole singularities. A central question is whether the situation improves when gravity is treated quantum mechanically. Analysis of models within LQG strongly suggests that the answer is in the affirmative. Because of space limitation, I will restrict myself to the big bang singularity and that too only in the simplest setting of homogeneous, isotropic cosmology.[b]

Let us begin with a list of some of the long-standing questions that any satisfactory quantum gravity theory is expected to answer:

- How close to the big bang does a smooth space-time of general relativity make sense? In particular, can one show from first principles that this approximation is valid at the onset of inflation?
- Is the big-bang singularity naturally resolved by quantum gravity? Or, is some external input such as a new principle or a boundary condition at the big bang essential?
- Is the quantum evolution across the 'singularity' deterministic? Since one needs a fully non-perturbative framework to answer this question in the affirmative, in the pre-big-bang[18] and Ekpyrotic/Cyclic[19,20] scenarios, for example, so far the answer is in the negative.
- If the singularity is resolved, what is on the 'other side'? Is there just a 'quantum foam', far removed from any classical space-time, or, is there another large, classical universe?

For many years, these and related issues had been generally relegated to the 'wish list' of what one would like the future, satisfactory quantum gravity theory to even-

[b]These considerations have been extended to allow for anisotropies and a non-zero cosmological constant, as well as to the $k = 1$, closed cosmology. More general models, including those with inhomogeneities are being investigated.

tually address. However, Since LQG is a background independent, non-perturbative approach, it is well-suited to address them. Indeed, starting with the seminal work of Bojowald some five years ago,[21] notable progress has been made in the context of symmetry reduced, minisuperspaces. In this section I will summarize the state of the art, emphasizing recent developments. For a comprehensive review of the older work see, e.g., Ref. [[22]].

Consider the spatially homogeneous, isotropic, $k{=}0$ cosmologies with a massless scalar field. It is instructive to focus on this model because *every* of its classical solutions has a singularity. There are two possibilities: In one the universe starts out at the big bang and expands, and in the other it contracts into a big crunch. The question is if this unavoidable classical singularity is naturally tamed by quantum effects. This issue can be analyzed in the geometrodynamical framework used in older quantum cosmology. Unfortunately, the answer turns out to be in the negative. For example, if one begins with a semi-classical state representing an expanding classical universe at late times and evolves it back via the Wheeler-DeWitt equation, one finds that it just follows the classical trajectory into the big bang singularity.[23,24]

In LQC, the situation is very different.[23–25] This may seem surprising at first. For, the system has only a finite number of degrees of freedom and von Neumann's theorem assures us that, under appropriate assumptions, the resulting quantum mechanics is unique. The only remaining freedom is factor-ordering and this is generally insufficient to lead to qualitatively different predictions. However, for reasons we will now explain, LQC does turn out to be qualitatively different from the Wheeler-DeWitt theory.[26]

Because of spatial homogeneity and isotropy, one can fix a fiducial (flat) triad ε_i^a and its dual co-triad ω_a^i. The SU(2) gravitational spin connection A_a^i used in LQG has only one component c which furthermore depends only on time; $A_a^i = c\, \omega_a^i$. Similarly, the triad E_i^a (of density weight 1) has a single component p; $E_i^a = p\, (\det \omega)\, \varepsilon_i^a$. p is related to the scale factor a via $a^2 = |p|$. However, p is not restricted to be positive; under $p \to -p$ the metric remains unchanged but the spatial triad flips the orientation. The pair (c, p) is 'canonically conjugate' in the sense that the only non-zero Poisson bracket is given by:

$$\{c,\, p\} = \frac{8\pi G\gamma}{3}\,, \tag{1}$$

where as before γ is the Barbero-Immirzi parameter.

Since a precise quantum mechanical framework was not available for full geometrodynamics, in the Wheeler-DeWitt quantum cosmology one focused just on the reduced model, without the benefit of guidance from the full theory. A major difference in Loop quantum cosmology (LQC) is that although the symmetry reduced theory has only a finite number of degrees of freedom, quantization is carried out by closely mimicking the procedure used in *full* LQG, outlined in section 2. Key differences between LQC and the older Wheeler-DeWitt theory can be traced back to this fact.

Recall that in full LQG diffeomorphism invariance leads one to a specific kinematical framework in which there are operators $\hat{h}_e$ representing holonomies and $\hat{P}_{S,f}$ representing (smeared) momenta but there is no operator(-valued distribution) representing the connection A itself.[8,9] In the cosmological model now under consideration, it is sufficient to evaluate holonomies along segments $\mu\, \varepsilon_i^a$ of straight lines determined by the fiducial triad ε_i^a. These holonomies turn out almost periodic functions of c —i.e. are of the form $N_{(\mu)}(c) := \exp i\mu(c/2)$— where the word 'almost' refers to the fact that μ can be any real number. These functions were studied exhaustively by the mathematician Harold Bohr, Niels' brother. In quantum geometry, the $N_{(\mu)}$ are the LQC analogs of the spin-network functions of full LQG. In quantum theory, then, we are led to a representation in which operators $\hat{N}_{(\mu)}$ and $\hat{p}$ are well-defined, but there is *no* operator corresponding to the connection component c. This seems surprising because our experience with quantum mechanics suggests that one should be able to obtain the operator analog of c by differentiating $N(\mu)$ with respect to the parameter μ. However, in the representation of the basic quantum algebra that descends to LQC from full LQG, although the $\hat{N}_{(\mu)}$ provide a 1-parameter group of unitary transformations, it fails to be weakly continuous in μ. Therefore one can not differentiate and obtain the operator analog of c.

In quantum mechanics, this would be analogous to having well-defined (Weyl) operators corresponding to the classical functions $\exp i\mu x$ but no operator $\hat{x}$ corresponding to x itself. This violates one of the assumptions of the von-Neumann uniqueness theorem. New representations (of the Weyl algebra) then become available which are *inequivalent* to the standard Schrödinger one. In quantum mechanics, these representations are not of direct physical interest because we need the operator $\hat{x}$. In LQC, on the other hand, full LQG naturally leads us to a new representation, i.e., to *new quantum mechanics.* This theory is inequivalent to the Wheeler-DeWitt type theory already at a kinematical level. In particular, just as we are led to complete the space $\mathcal{A}$ of smooth connections to the space $\bar{\mathcal{A}}$ of generalized connections in LQG, in LQC we are led to consider the Bohr compactification $\bar{\mathbb{R}}_{\rm Bohr}$ of the real line representing the 'c-axis'. The gravitational part of the Hilbert space is now $L^2(\bar{\mathbb{R}}_{\rm Bohr}, {\rm d}\mu_{\rm Bohr})$, rather than the standard $L^2(\mathbb{R}, {\rm d}\mu)$ used in the Wheeler-DeWitt theory[26] where ${\rm d}\mu_{\rm Bohr}$ is the LQC analog of the measure ${\rm d}\mu_o$ selected by the uniqueness results[8,9] in full LQG. While in the semi-classical regime LQC is well approximated by the Wheeler-DeWitt theory, important differences manifest themselves at the Planck scale. These are the hallmarks of quantum geometry.[2,22]

The new representation also leads to a qualitative difference in the structure of the Hamiltonian constraint operator: the gravitational part of the constraint is a *difference* operator, rather than a differential operator as in the Wheeler-DeWitt theory. The derivation[23,24,26] can be summarized briefly as follows. In the classical theory, the gravitational part of the constraint is given by $\int d^3x\, \epsilon^{ijk} e^{-1} E_i^a E_j^b F_{ab\,k}$ where $e = |\det E|^{1/2}$ and F_{ab}^k the curvature of the connection A_a^i. The part of this operator involving triads can be quantized[21,26] using a standard procedure introduced by Thiemann in the full theory.[4] However, since there is no operator corresponding to the connection itself, one has to express F_{ab}^k as a limit of the holonomy around a loop divided by the area enclosed by the loop, as the area shrinks to zero. Now, quantum geometry tells us that the area operator has a minimum non-zero eigenvalue, Δ, and in the quantum theory it is natural to shrink the loop only till it attains this minimum. It is the existence of this 'area gap' Δ that leads one to a difference equation.

Let us represent states as functions $\Psi(v, \phi)$, where ϕ is the scalar field and the dimensionless real number v represents geometry. Specifically, $|v|$ is the eigenvalue of the operator $\hat{V}$ representing volume (essentially the cube of the scale factor):

$$\hat{V}\rangle v = K\,(\frac{8\pi\gamma}{6})^{\frac{3}{2}}\ |v|\,\ell_{\rm Pl}^3\,\rangle v \quad \text{where} \quad K = \frac{3\sqrt{3\sqrt{3}}}{2\sqrt{2}} \tag{2}$$

Then, the LQC Hamiltonian constraint assumes the form:

$$\begin{aligned}\partial_\phi^2 \Psi(v,\phi) &= [B(v)]^{-1}\ \big(C^+(v)\,\Psi(v+4,\phi) + C^o(v)\,\Psi(v,\phi) + C^-(v)\,\Psi(v-4,\phi)\big) \\ &=: -\Theta\,\Psi(v,\phi)\end{aligned} \tag{3}$$

where the coefficients $C^{\pm}(v)$, $C^o(V)$ and $B(v)$ are given by:

$$\begin{aligned} C^+(v) &= \frac{3\pi KG}{8}\,|v+2|\ \big||v+1| - |v+3|\big| \\ C^-(v) &= C^+(v-4) \quad \text{and} \quad C^o(v) = -C^+(v) - C^-(v) \\ B(v) &= \left(\frac{3}{2}\right)^3 K\ |v| \left| |v+1|^{1/3} - |v-1|^{1/3} \right|^3 . \end{aligned} \tag{4}$$

Eq. (3) is the quantum Einstein equation —the LQC analog of the Wheeler-DeWitt differential equation of older quantum cosmology. Our task is to unravel quantum physics from its solutions.

Now, in each classical solution to the standard Einstein's equation, ϕ can be shown to be a globally monotonic function of time; it can therefore be taken as the dynamical variable representing an *internal* clock. In quantum theory there is no space-time metric, even when the equations of motion are satisfied. But since the quantum constraint (3) dictates how $\Psi(v,\phi)$ 'evolves' as ϕ changes, it is convenient to regard the argument ϕ in $\Psi(v,\phi)$ as *emergent time* and v as the physical degree of freedom. A convenient complete set of observables is then provided by the constant of motion $\hat{p}_\phi$ and operators $\hat{v}|_{\phi_o}$ determining the value of v at the 'instant' $\phi = \phi_o$.

Physical states are the (suitably regular) solutions to Eq. (3). The map $\hat{\Pi}$ defined by $\hat{\Pi}\,\Psi(v,\phi) = \Psi(-v,\phi)$ corresponds just to the flip of orientation of the spatial triad (under which geometry remains unchanged); $\hat{\Pi}$ is thus a large gauge transformation on the space of solutions to Eq. (3). One is therefore led to divide physical states into sectors, each providing an irreducible, unitary representation of this symmetry. Physical considerations[23,24] imply that we should consider the symmetric sector, with eigenvalue +1 of $\hat{\Pi}$.

To endow this space with the structure of a Hilbert space, one can proceed along one of two paths. In the first, one defines the action of the observables on the space of suitably regular solutions to the constraints and selects the inner product by demanding that these operators be self-adjoint.[27,28] A more systematic procedure is the 'group averaging method'.[29] The technical implementation[23,24] of both these procedures is greatly simplified by the fact that the difference operator Θ on the right side of (3) is independent of ϕ and can be shown to be self-adjoint and positive definite on the Hilbert space $L^2(\bar{\mathbb{R}}_{\rm Bohr}, B(v)\mathrm{d}\mu_{\rm Bohr})$.

The final result can be summarized as follows. Since Θ is a *difference* operator, the physical Hilbert space $\mathcal{H}_{\rm phy}$ has sectors $\mathcal{H}_\epsilon$ which are superselected; $\mathcal{H}_{\rm phy} = \oplus_\epsilon \mathcal{H}_\epsilon$ with $\epsilon \in (0,2)$. The overall predictions are insensitive to the choice of a specific sector (for details, see [[23,24]]). States $\Psi(v,\phi)$ in $\mathcal{H}_\epsilon$ are symmetric, $\Psi(v,\phi) = \Psi(-v,\phi)$, and have support on points $v = |\epsilon| + 4n$ where n is an integer. Wave functions $\Psi(v,\phi)$ in a generic sector solve (3) and are of positive frequency with respect to the 'internal time' ϕ: they satisfy the 'positive frequency' square root

$$-i\partial_\phi \Psi = \sqrt{\Theta}\, \Psi\,. \tag{5}$$

of Eq. (3). The physical inner product is given by:

$$\langle \Psi_1 \,|\, \Psi_2 \rangle = \sum_{v\in\{|\epsilon|+4n\}} B(v)\, \bar{\Psi}_1(v,\phi_o)\Psi_2(v,\phi_o) \tag{6}$$

and is 'conserved', i.e., is independent of the 'instant' ϕ_o chosen in its evaluation. On these states, the Dirac observables act in the expected fashion:

$$\begin{aligned} \hat{p}_\phi \Psi &= -i\hbar \partial_\phi \Psi \\ \hat{v}|_{\phi_o}\, \Psi(v,\phi) &= e^{i\sqrt{\Theta}(\phi-\phi_o)}\, v\, \Psi(v,\phi_o) \end{aligned} \tag{7}$$

To construct semi-classical states and for numerical simulations, it is convenient to express physical states as linear combinations of the eigenstates of $\hat{p}_\phi$ and Θ. To carry out this step, it is convenient to consider the older Wheeler-DeWitt theory first sine it provides a familiar setting with differential operators. Let us begin with the observation that, for $v \gg 1$, there is a precise sense[24] in which the difference operator Θ approaches the Wheeler DeWitt differential operator $\underline{\Theta}$, given by

$$\underline{\Theta}\underline{\Psi}(v,\phi) = 12\pi G\; v\partial_v \big(v\partial_v \underline{\Psi}(v,\phi)\big) \tag{8}$$

Thus, if one ignores the quantum geometry effects, Eq. (3) reduces to the Wheeler-DeWitt equation

$$\partial_\phi^2 \underline{\Psi} = -\underline{\Theta}\, \underline{\Psi}. \tag{9}$$

Note that the operator $\underline{\Theta}$ is positive definite and self-adjoint on the Hilbert space $L^2_s(\mathbb{R}, \underline{B}(v)\mathrm{d}v)$ where the subscript s denotes the restriction to the symmetric eigenspace of Π and $\underline{B}(v) := Kv^{-1}$ is the limiting form of $B(v)$ for large v. Its eigenfunctions $\underline{e}_k$ with eigenvalue $\omega^2 (\geq 0)$ are 2-fold degenerate on this Hilbert space. Therefore, they can be labelled by a real number k:

$$\underline{e}_k(v) := \frac{1}{\sqrt{2\pi}}\, e^{ik\ln|v|} \tag{10}$$

where k is related to ω via $\omega = \sqrt{12\pi G}|k|$. They form an orthonormal basis on $L^2_s(\mathbb{R}, \underline{B}(v)dv)$. A 'general' positive frequency solution to Eq. (9) can be written as

$$\underline{\Psi}(v, \phi) = \int_{-\infty}^{\infty} dk\, \tilde{\underline{\Psi}}(k)\, \underline{e}_k(v) e^{i\omega\phi} \tag{11}$$

for suitably regular $\tilde{\underline{\Psi}}(k)$.

Let us now return to LQC. The complete set of eigenfunctions $e_k(v)$ of the discrete operator Θ is also labelled by a real number k and detailed numerical simulations show that $e_k(v)$ are well-approximated by $\underline{e}_k(v)$ for $v \gg 1$. The eigenvalues $\omega^2(k)$ of Θ are again given by $\omega = \sqrt{12\pi G}|k|$. Finally, the $e_k(v)$ satisfy the standard orthonormality relations $< e_k|e'_k >= \delta(k, k')$. A physical state $\Psi(v, \phi)$ can therefore be expanded as:

$$\Psi(v, \phi) = \int_{-\infty}^{\infty} dk\, \tilde{\Psi}(k)\; e_k^{(s)}(v)\; e^{i\omega(k)\phi} \tag{12}$$

where $\tilde{\Psi}(k)$ is any suitably regular function of k, and $e_k^{(s)}(v) = (1/\sqrt{2})(e_k(v) + e_k(-v))$. Thus, as in the Wheeler-DeWitt theory, each physical state is characterized by a free function $\tilde{\Psi}(k)$. The difference between the two theories lies in the functional forms of the eigenfunctions $e_k(v)$ of Θ and $\underline{e}_k(v)$ of $\underline{\Theta}$. While $e_k(v)$ is well approximated by $\underline{e}_k(v)$ for large v, the differences are very significant for small v and they lead to very different dynamics near the big-bang.

We can now construct states which are semi-classical at late times —e.g., now— and evolve them numerically 'backward in time'. There are three natural constructions to implement this idea in detail, reflecting the freedom in the notion of semi-classical states. In all cases, the main results are the same.[23,24] Here I will report on the results obtained using the strategy that brings out the contrast with the Wheeler DeWitt theory most sharply.

As noted before, p_ϕ is a constant of motion. For the semi-classical analysis, we are led to choose a large value $p_\phi^\star$ ($\gg \hbar$ in the classical units $G = c = 1$). In the closed model, for example, this condition is necessary to ensure that the universe can expand out to a macroscopic size. Fix a point $(v^\star, \phi_o)$ on the corresponding classical trajectory which starts out at the big bang and then expands, choosing $v^\star \gg 1$. We want to construct a state which is peaked at $(v^\star, p_\phi^\star)$ at a 'late initial time' $\phi = \phi_o$ and follow its 'evolution' backward. At 'time' $\phi = \phi_o$, consider then the function

$$\Psi(v, \phi_o) = \int_{-\infty}^{\infty} dk\, \tilde{\Psi}(k)\, \underline{e}_k(v)\, e^{i\omega(\phi_o - \phi^\star)}, \quad \text{where } \tilde{\Psi}(k) = e^{-\frac{(k-k^\star)^2}{2\sigma^2}} \tag{13}$$

where $k^\star = -p_\phi^\star/\sqrt{12\pi G\hbar^2}$ and $\phi^\star = -\sqrt{1/12\pi G}\, \ln(v^\star) + \phi_o$. In the Wheeler-DeWitt theory one can easily evaluate the integral in the approximation $|k^*| \gg 1$ and calculate mean values of the Dirac observables and their fluctuations. One finds that, as required, the state is sharply peaked at values $v^\star, p_\phi^\star$. The above construction is closely related to that of coherent states in non-relativistic quantum

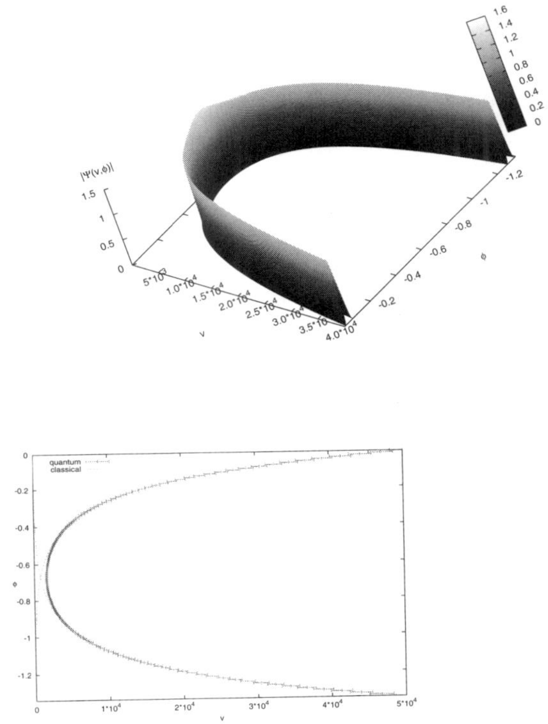

Fig. 1. The figure on left shows the absolute value of the wave function Ψ as a function of ϕ and v. Being a physical state, Ψ is symmetric under $v \rightarrow -v$. The figure on the right shows the expectation values of Dirac observables $\hat{v}|_\phi$ and their dispersions. They exhibit a quantum bounce which joins the contracting and expanding classical trajectories marked by fainter lines. In this simulation, the parameters in the initial data are: $v^\star = 5 \times 10^4$, $p_\phi^\star = 5 \times 10^3 \sqrt{G}\hbar$ and $\Delta p_\phi / p_\phi = 0.0025$.

mechanics. The main difference is that the observables of interest are not v and its conjugate momentum but rather v and p_ϕ —the momentum conjugate to 'time', i.e., the analog of the Hamiltonian in non-relativistic quantum mechanics. Now, one can evolve this state backwards using the Wheeler-DeWitt equation Eq. (9). It follows immediately from the form (11) of the general solution to Eq. (9) and the fact that p_ϕ is large that this state would remain sharply peaked at the chosen classical trajectory and simply follow it into the big-bang singularity.

In LQC, we can use the restriction of (13) to points $v = |\epsilon| + 4n$ as the initial data and evolve it backwards numerically. Now the evolution is qualitatively different (see Fig. 1). The state does remains sharply peaked at the classical trajectory till the matter density reaches a critical value:

$$\rho_{\rm crit} = \frac{\sqrt{3}}{16\pi^2\gamma^3 G^2\hbar}\,, \tag{14}$$

which is about 0.8 times the Planck density. However, *then it bounces.* Rather than following the classical trajectory into the singularity as in the Wheeler-DeWitt theory, the state 'turns around'. What is perhaps most surprising is that it again becomes semi-classical and follows the 'past' portion of a classical trajectory, again with $p_\phi = p_\phi^\star$, which was headed towards the big crunch. Let us summarize the forward evolution of the full quantum state. In the distant past, the state is peaked on a classical, contracting pre-big-bang branch which closely follows the evolution dictated by Friedmann equations. But when the matter density reaches the Planck regime, quantum geometry effects become significant. Interestingly, they make gravity *repulsive*, not only halting the collapse but turning it around; the quantum state is again peaked on the classical solution now representing the post-big-bang, expanding universe. Since this behavior is so surprising, a very large number of numerical simulations were performed to ensure that the results are robust and not an artifact of the special choices of initial data or of the numerical methods used to obtain the solution.[23,24]

For states which are semi-classical at late times, the numerical evolution in exact LQC can be well-modelled by an effective, modified Friedman equation :

$$\frac{\dot{a}^2}{a^2} = \frac{8\pi G}{3}\,\rho\left[1 - \frac{\rho}{\rho_{\rm crit}}\right] \tag{15}$$

where, as usual, a is the scale factor. In the limit $\hbar \to 0$, $\rho_{\rm crit}$ diverges and we recover the standard Friedmann equation. Thus the second term is a genuine quantum correction. Eq. (15) can also be obtained analytically from (3) by a systematic procedure.[30] But the approximations involved are valid only well outside the Planck domain. It is therefore surprising that the bounce predicted by the exact quantum equation (3) is well approximated by a naive extrapolation of (15) across the Planck domain. While the essential reason behind this seemingly 'unreasonable success' of the effective equation Eq. (15) is now understood, further work is still needed to have a full control on this issue.

Finally let us return to the questions posed in the beginning of this section. In the model, LQC has been able to answer all of them. One can deduce from first principles that classical general relativity is an excellent approximation till very early times, including the onset of inflation in standard scenarios. Yet quantum geometry effects have a profound, global effect on evolution. In particular, the singularity is naturally resolved without any external input and there is a classical space-time also in the pre-big-bang branch. LQC provides a deterministic evolution which joins the two branches.

4. Discussion

The problem of unifying general relativity with quantum physics is quite old. However, in the last two decades there has been considerable progress in the field especially in string theory and loop quantum gravity. In my view, the two approaches are complementary. String theory has provided a fresh strategy for unification of all interactions while loop quantum gravity has brought new insights into the nature of quantum geometry and its physical ramifications. But both theories are still very incomplete. A more satisfactory theory will have to simultaneously address both sets of problems in a natural fashion. Finally, I would like to emphasize that there are a number of other approaches. Although progress in addressing physical issues has been somewhat slower, they contain a variety of fascinating and highly original ideas that I could not discuss. These include causal dynamical triangulations,[31] Euclidean quantum gravity,[5,6] discrete approaches,[32] causal sets,[33] twistor theory[34] and the theory of H-spaces,[35] asymptotic quantization[36] and non-commutative geometry.[37]

In this article, I focussed on quantum cosmology for a number of reasons. First, this application provides a technically simple setting in which one can face many of the deep conceptual problems of quantum gravity, most particularly: *how does one do physics in absence of a space-time continuum in the background?* We saw a concrete example in which the 'emergent time paradigm' could be realized and used very effectively. The second reason is that the Big-Bang singularity offers an outstanding opportunity to see quantum gravity effects in action. Do quantum effects tame the singularity? The classical space-time ends and classical physics stops there. What about the quantum space-time and quantum physics? We found that quantum geometry effects do resolve the singularity and enable us to 'evolve' the quantum state to the past of the putative classical singularity. Finally, in quantum cosmology there can be no external observers. Therefore, it provides an ideal setting to further develop quantum mechanics of closed systems, ideas associated with decoherence and mechanisms for spontaneous reduction of the wave function. Because of this, the discussion of section III may be practically useful for experts on foundations of quantum mechanics that this workshop brought together.

Acknowledgments

This work was supported in part by the NSF grants PHY-0354932 and PHY-0456913, the Alexander von Humboldt Foundation, the Krammers Chair program of the University of Utrecht and the Eberly research funds of Penn State.

References

1. A. Ashtekar,*New J. Phys.*7, 198-231 (2005); *arXiv:gr-qc/0410054.*
2. A. Ashtekar and J. Lewandowski *Class. Quant. Grav.* **21**, R53-R152 (2004).
3. C. Rovelli *Quantum Gravity*, (CUP, Cambridge, 2004).
4. T. Thiemann, *Introduction to Modern Canonical Quantum General Relativity*, (CUP, Cambridge, at press).

5. O. Luscher and M. Reuter *arXiv:hep-th/0511260.*
6. R. Percacci and D. Perini *Phys. Rev.* **D68**, 044018 (2003).
7. A. Ashtelar and K. Krasnov,*Black Holes, Gravitational Radiation and the Universe*(Kluwer, Dodrecht, 1999); *gr-qc/9804039.*
8. J. Lewandowski, A. Okolow, H. Sahlmann and T. Thiemann, *Commun. Math. Phys.* (at Press); *arXiv:gr-qc/0504147.*
9. C. Fleishchack, *arXiv:math-ph/0407006.*
10. A. Ashtekar and J. Lewandowski, Representation theory of analytic holonomy algebras, in *Knots and Quantum Gravity*, ed J. Baez, (Oxford U. Press, Oxford, 1994); *Jour. Math. Phys.* **36**, 2170 (1995); A. Ashtekar and J. Lewandowski, *Jour. Geo. & Phys.* **17**, 191 (1995)
11. J. C. Baez, *Lett. Math. Phys.* **31**, 213 (1994).
12. D. Marolf and J. Mourão, *Commun. Math. Phys.* **170**, 583 (1995).
13. C. Rovelli and L. Smolin, *Phys. Rev.* **D52**, 5743 (1995).
14. J. C. Baez, *The Interface of Knots and Physics* (American Mathematical Society, Providence) pp. 167–203 (1996).
15. C. Rovelli and L. Smolin, *Nucl. Phys.* **B442**, 593 (1995) ; Erratum: *Nucl. Phys.* **B456**, 753 (1995).
16. A. Ashtekar and J. Lewandowski, *Class. Quant. Grav.* **14**, A55 (1997); *Adv. Theo. Math. Phys.* **1**, 388 (1997).
17. A. Ashtekar, *Phys. Rev. Lett.***57**, 2244 (1986); *Phys. Rev.* **D36**, 1587 (1987).
18. M. Gasperini and G. Veneziano, *Phys. Rept.* **373**, 1 (2003).
19. J. Khoury, B. A. Ovrut, P. J. Steinhardt, N. Turok, *Phys. Rev.* **D64** 123522 (2001).
20. J. Khoury, B. Ovrut, N. Seiberg, P. J. Steinhardt, N. Turok, *Phys. Rev.* **D65**, 086007 (2002).
21. M. Bojowlad, *Phys. Rev. Lett.* **86**, 522730 (2001); *Class. Quantum. Grav.* **19**, 2717 (2002).
22. M. Bojowald, *Liv. Rev. Rel.* **8**, 11 (2005), *arXiv:gr-qc/0601085.*
23. A. Ashtekar, T. Pawlowski and P. Singh, *Phys. Rev.* **D73**, 124038 (2006).
24. A. Ashtekar, T. Pawlowski and P. Singh, *Phys. Rev.* **D** (at press); *gr-qc/0607039.*
25. A. Ashtekar, T. Pawlowski and P. Singh, *Phys. Rev. Lett.* **96**, 141301 (2006).
26. A. Ashtekar, M. Bojowald, J. Lewandowski, *Adv. Theo. Math. Phys.* **7**, 233 (2003).
27. A. Ashtekar, *Lectures on non-perturbative canonical gravity*, (World Scientific, Singapore, 1991).
28. A. Rendall, *Class. Quantum Grav.* **10**, 2261 (1993).
29. D. Marolf, *arXives:gr-qc/9508015;Class. Quant. Grav.* **12**, 1199 (1995).
30. J. Willis, *On the low energy ramifications and a mathematical extension of loop quantum gravity*, (Ph.D. Dissertation, The Pennsylvaina State University 2004); A. Ashtekar, M. Bojowald and J. Willis, *Corrections to Friedmann equations induced by quantum geometry*, (IGPG preprint 2004).
31. R. Loll, *Living Rev. Rel.* **1**, 13 (1998).
32. R. Gambini and J. Pullin, *100 Years of Relativity; Space-time Structure: Einstein and Beyond*, (World Scientific, Singapore, 2005).
33. F. Dowker, *100 Years of Relativity; Space-time Structure: Einstein and Beyond, op.cit..*
34. R. Penrose, *100 Years of Relativity; Space-time Structure: Einstein and Beyond, op.cit..*
35. M. Ko, M. Ludvigsen, E. T. Newman and P. Tod, *Phys. Rep.* **71**, 51 (1981).
36. Ashtekar A 1984 *Asymptotic quantization* (Bibliopolis, Naples); also available at *http://cgpg.gravity.psu.edu/research/asymquant-book.pdf*
37. A. Connes *Non-commutative geometry* (Academic Press, New York, 1994).

BOSE–EINSTEIN CONDENSATES AND EPR QUANTUM NON-LOCALITY

F. LALOË

Laboratoire Kastler Brossel, ENS, CNRS, UPMC;
24 rue Lhomond, 75005, Paris, France

The EPR argument points to the existence of additional variables that are necessary to complete standard quantum theory. It was dismissed by Bohr because it attributes physical reality to isolated microscopic systems, independently of the macroscopic measurement apparatus. Here, we transpose the EPR argument to macroscopic systems, assuming that they are in spatially extended Fock spin states and subject to spin measurements in remote regions of space. Bohr's refutation of the EPR argument does not seem to apply in this case, since the difference of scale between the microscopic measured system and the macroscopic measuring apparatus can no longer be invoked.

In dilute atomic gases at very low temperatures, Bose–Einstein condensates are well described by a large population occupying a single-particle state; this corresponds, in the many particle Hilbert space, to a Fock state (or number state) with large number N. The situations we consider involve two such Fock states associated to two different internal states of the particles. The two internal states can conveniently be seen as the two eigenstates of the Oz component of a fictitious spin. We assume that the two condensates overlap in space and that successive measurements are made of the spins of single particles along arbitrary transverse directions (perpendicular to Oz).

In standard quantum mechanics, Fock states have no well defined relative phase: initially, no transverse spin polarization exists in the system. The theory predicts that it is only under the effect of quantum measurement that the states acquire a well defined relative phase, giving rise to a transverse polarization. This is similar to an EPR situation with pairs of individual spins (EPRB), where spins acquire a well defined spin direction under the effect of measurement - except that here the transverse polarization involves an arbitrary number of spins and may be macroscopic. We discuss some surprising features of the standard theory of measurement in quantum mechanics: strong effect of a small system onto an arbitrarily large system (amplification), spontaneous appearance of a macroscopic angular momentum in a region of space without interaction (non-locality at a macroscopic scale), reaction onto the measurement apparatus and angular momentum conservation (angular momentum version of the EPR argument). Bohr's denial of physical reality for microscopic systems does not apply here, since the measured system can be arbitrarily large. Since here we limit our study to very large number of particles, no Bell type violation of locality is obtained.

Keywords: Fock States; Quantum measurement; EPR; Bell inequality.

The famous Einstein-Podolsky-Rosen (EPR) argument[1] considers two correlated particles, located in two remote regions in space A and B, and focuses onto the "elements of reality" contained in these two regions. It starts from three ingredients:

realism, locality, and the assumption that the predictions of quantum mechanics concerning measurements are correct[a]; from these inputs it proves that, to provide a full description of physical reality, standard quantum mechanics must be completed with additional variables (often called "hidden variables" for historical reasons). The EPR argument was refuted by Bohr,[2] who did not accept the notion of realism introduced by EPR; we give more details on his reply in § 1. The purpose of the present article is to transpose the discussion to the macroscopic scale: we investigate situations that are similar to those considered by EPR but, instead of single particles, we study Bose–Einstein condensates made of many particles, which can be macroscopic. For dilute gases, these condensates can be represented as single quantum states populated with a large, but well defined, number of particles, in other words by Fock states (number states) with large N (large, but well defined, population).

Several authors[3–14] have studied the interference between two such condensates; since the phase of two Fock states is completely undefined according to standard quantum mechanics, the question is whether or not a well defined relative phase will be observed in the interference. These authors show that a well defined value of the relative phase can in fact emerge under the effect of successive quantum measurements. The value taken by this phase is random: it can be completely different form one realization of the experiment to the next. But, in a given realization, it becomes better and better defined while the measurements of the position of particles are accumulated. In other words, a perfectly clear interference pattern emerges from the measurements with a visible, but completely random and unpredictable, phase.

An interesting variant of this situation occurs if the two highly populated states correspond to two different internal states of the atoms.[15–17] As usual, these two states can conveniently be seen as the two eigenstates of the Oz component of a fictitious spin. One can then study for instance the situation shown schematically in Fig. 1, where the two different internal states with high populations are initially trapped in two different sites, and then released to let them expand and overlap. Many other situations are also possible; we will discuss some of them in this article. In the overlap region, measurements of the spin component of particles along directions in the xOy plane are sensitive to the relative phase of the two condensates. A free adjustable parameter for every measurement is the angle φ which defines the direction of measurement in this plane; as discussed in,[17] this introduces more flexibility in choosing a strategy for optimum determination of the phase. Otherwise the situation is similar to that with spinless particles: initially the relative relative phase is completely undefined, and nothing can be said about its value. But, as long as the results of the measurements accumulate, the phase becomes better an better determined under the very effect of the quantum measurement process. This

[a]More precisely, the EPR reasoning only requires that some predictions of quantum mechanics are correct, those concerning the complete correlations observed between remote mesurements performed on entangled particles.

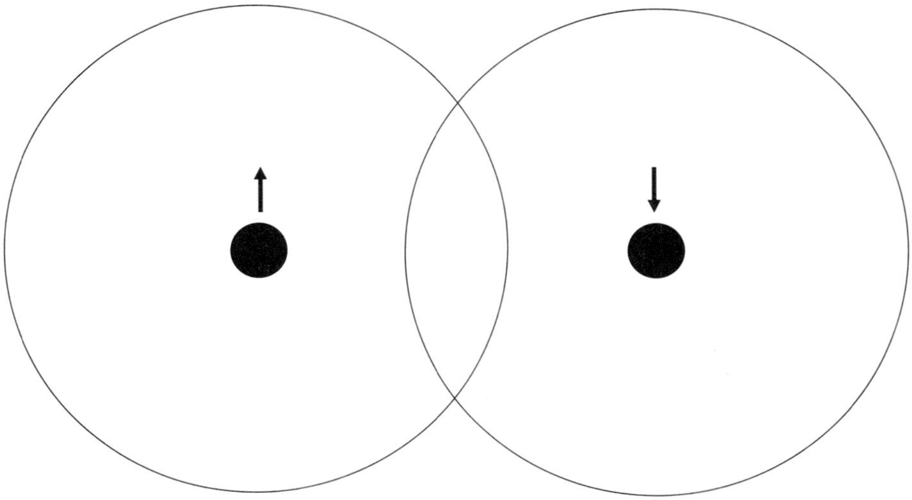

Fig. 1. Two condensates, corresponding to two different internal spin states, expand and overlap in a region of space. In this region, particle spin measurements are performed in transverse directions (perpendicular to the spin directions of the two initial condensates).

eventually creates a transverse spin polarization of the whole system, which can be macroscopic for large samples.

We consider situations where two single particle states associated with different internal state of atoms overlap in space, and assume that each of these states has a large population. Transverse spin measurements are then performed in the region of overlap. This happens if two Bose-Einstein condensates are trapped in different sites, and then released to let them expand.

Usually, in the quantum theory of measurement, one emphasizes the role of a classical macroscopic pointer, the part of the measurement apparatus that directly provides the information to the human observer. Here we have a curious case where it is the measured system itself that spontaneously creates a pointer made of a macroscopic number of parallel spins. Moreover, for condensates that are extended in space, we will see that this process can create instantaneously parallel pointers in remote regions of space, a situation is obviously reminiscent of the EPR argument in its spin version given by Bohm (often called EPRB).[18,19] We study in this article how the EPR argument can be transposed to this case, and show that the argument becomes stronger, mostly because the measured systems themselves are now macroscopic. Bohr's refutation, based on the denial of any physical reality for microscopic systems (cf. §1), then does not apply in the same way, if it still applies at all.

In § 1 we recall the main features of the EPR argument, which also gives us the opportunity to summarize Bohr's reply and emphasize his fundamental distinction between microscopic observed systems and macroscopic measurement apparatuses.

In § 2, we introduce the formalism and generalize the simple calculation of,[16] in particular to include the case where no particle is detected in the region of measurement. This provides us with the general expression of the joint probability for any sequence of spin measurements performed in the transverse direction, and any sequence of results. Then § 3 contains a discussion of the physics that is involved: amplification during quantum measurement, conservation of angular momentum and recoil effects of the measurements apparatus, quantum non-locality.

1. EPR Argument and Its Refutation by Bohr

The EPR argument[120] considers a physical system made of two correlated microscopic particles, assuming that they are located in two remote regions in space A and B where two physicists can perform arbitrary measurements on them. EPR specifically discuss situations where quantum mechanics predicts that the result of a first measurement performed in A is completely random, but nevertheless determines with complete certainty the result of another measurement performed in B. They introduce their "condition for the reality of a physical quantity" with the famous sentence: "if, without in any way disturbing a system, we can predict with certainty the value of a physical quantity, then there exists an element of physical reality corresponding to this physical quantity". As a consequence, just after the measurement in A (but before the measurement in B), since the result of the second experiment is already certain, an element of reality corresponding to this certainty must exist in region B. But, according to locality, an element of reality in B can not have been created by the first measurement performed in region A, at an arbitrarily large distance; the element of reality necessarily existed even before any measurement. Since standard quantum theory does not contain anything like such a pre-existing element of reality, it is necessarily an incomplete theory[b]; the state vector is not sufficient to describe a single realization of an experiment, but describes only a statistical ensemble of many realizations

Bohr, in his reply,[2] does not criticize the EPR reasoning, but the assumptions on which it is based, which he considers as unphysical. He states that the criterion of

[b]Here we give only the part of the EPR argument that is sometimes called EPR-1: we consider one type of measurement in each region of space, in other words only one experimental setup. This is sufficient to show that standard quantum mechanics is incomplete (if one accepts the EPR assumptions). This also justifies the introduction of statistical averages (or of a variable λ that is integrated over initial conditions) in order to prove the Bell theorem.
In their article, EPR go further and consider several incompatible types of measurements performed region A. They then prove that variables in region B can have simultaneous realities, even if they are considered as incompatible in standard quantum mechanics. This provides a second proof of incompleteness, sometimes called EPR-2. Bohr's refutation of the EPR argument also emphasizes the exclusive character of measurements of incompatible observables, and therefore concentrates onto EPR-2.
In addition, EPR show in their famous article that, in their views, quantum mechanics is not only incomplete but also redundant: it can represent the same physical reality in region B by several differents state vectors (EPR-3).

physical reality proposed by EPR "contains an essential ambiguity when applied to quantum phenomena" and that "their argumentation does not seem to me to adequately meet the actual situation with which we are faced in atomic physics" (here, "atomic" is presumably equivalent to "microscopic" in modern language). His text has been discussed by many authors (for an historical review, see for instance[21]), but still remains difficult to grasp in detail (see for instance Appendix I of[22]). Instead of concentrating his arguments on the precise situation considered by EPR, Bohr emphasizes in general the consistency of the mathematical formalism of quantum mechanics and the "impossibility of controlling the reaction of the object on the measuring instruments". But, precisely, the main point of the EPR argument is to select a situation where these unavoidable perturbations do not exist! EPR locality implies that a measurement performed in regions A can create no perturbation on the elements of reality in region B.

Only the second part of Bohr's article really deals with the EPR argument. After stating again that the words "without in any way disturbing the system" are ambiguous, he concedes that " there is of course in a case like that considered (by EPR) no question of a mechanical disturbance of the system under investigation during the last critical stage of the measuring procedure". Nevertheless, for him what EPR have overlooked is that " there is essentially the question of an influence on the very conditions which define the precise types of predictions regarding the future behavior of the system" — the sentence is central but difficult; he probably means "an influence *of the measurement performed in A* on the conditions which define the predictions on the future behavior of the system *in B*, or maybe *the whole system in both A and B"*. He then states that these conditions are an essential element of any phenomenon to which the terms "physical reality"can be attached, and concludes that the EPR proof of incompleteness is non valid.

J.S. Bell summarizes the reply by writing[22] that, in Bohr's view "there is no reality below some classical macroscopic level". For Bohr, it is incorrect to assign physical reality to one of the two particles, or even to the group of both particles; physical reality only has a meaning when macroscopic systems are involved, which here means the measurement apparatuses. He actually attaches physical reality only to the whole ensemble of the microscopic system and macroscopic measurement apparatuses, which extends over the two regions A and B of space, and not to subsystems. Then, the EPR reasoning, which focuses on B only, becomes incorrect. We remark in passing that Bohr's refutation hinges on the microscopic character of the measured system, the two quantum particles.

2. Detecting the Transverse Direction of Spins; calculation

We consider a system composed of particles having two internal states α and β, which can be seen as the eigenstates of the Oz component of their spin with eigenvalues $+\hbar/2$ and $-\hbar/2$. The particles populate two quantum states, $\mid u_a, \alpha >$ (orbital variables described by an orbital state $\mid u_a >$) and $\mid u_b, \beta >$ (orbital state

$| u_b >$). Initially, the quantum system is in a "double Fock state", with N_a particles populating the first single-particle-state and N_b populating the other:

$$| \Phi > \;=\; \left[(a_{u_a,\alpha})^\dagger\right]^{N_a} \left[(a_{u_b,\beta})^\dagger\right]^{N_b} | \text{vac.} > \tag{1}$$

where $a_{u_a,\alpha}$ and $a_{u_b,\beta}$ are the destruction operators associated with the two single particle states, and $|\text{vac.} >$ is the vacuum state. With the notation of occupation numbers, the same initial state can also be written:

$$| \Phi > \;=\; | N_a : u_a, \alpha \;;\; N_b : u_b, \beta \; > \tag{2}$$

As in[16c], we note $\Psi_\mu(\mathbf{r})$, with $\mu = \alpha, \beta$, the field operators associated with internal states α,β. The $\mathbf{r}$ dependent local density operator is then:

$$n(\mathbf{r}) = \;\; \Psi^\dagger_\alpha(\mathbf{r})\Psi_\alpha(\mathbf{r}) + \Psi^\dagger_\beta(\mathbf{r})\Psi_\beta(\mathbf{r}) \tag{3}$$

while the three components of the local spin density are:

$$\begin{aligned} \sigma_z(\mathbf{r}) &= \;\Psi^\dagger_\alpha(\mathbf{r})\Psi_\alpha(\mathbf{r}) - \Psi^\dagger_\beta(\mathbf{r})\Psi_\beta(\mathbf{r}) \\ \sigma_x(\mathbf{r}) &= \;\Psi^\dagger_\alpha(\mathbf{r})\Psi_\beta(\mathbf{r}) + \Psi^\dagger_\beta(\mathbf{r})\Psi_\alpha(\mathbf{r}) \\ \sigma_y(\mathbf{r}) &= \;i\left[\Psi^\dagger_\beta(\mathbf{r})\Psi_\alpha(\mathbf{r}) - \Psi^\dagger_\alpha(\mathbf{r})\Psi_\beta(\mathbf{r})\right] \end{aligned} \tag{4}$$

The spin component in the direction of plane xOy making an angle φ with Ox is:

$$\sigma_\varphi(\mathbf{r}) = e^{-i\varphi}\Psi^\dagger_\alpha(\mathbf{r})\Psi_\beta(\mathbf{r}) + \; e^{i\varphi}\Psi^\dagger_\beta(\mathbf{r})\Psi_\alpha(\mathbf{r}) \tag{5}$$

Suppose now that one measurement is made of the φ component of the spin of particles within a small region of space $\Delta_\mathbf{r}$ centered around point $\mathbf{r}$. The corresponding operator is:

$$A(\mathbf{r}, \varphi) = \int_{\Delta_\mathbf{r}} d^3r' \;\sigma_\varphi(\mathbf{r}') \tag{6}$$

If the volume Δ of domain $\Delta_\mathbf{r}$ is sufficiently small, the probability to find more than one particle in this volume is negligible, and $A(\mathbf{r}, \varphi)$ has only three eigenvalues, 0 and ± 1. The eigenstates corresponding to the eigenvalue $\eta = 0$ are all those where $\Delta_\mathbf{r}$ contains no particle; the eigenstates corresponding to the eigenvalues $\eta = \pm 1$ are those for which only one particle is within $\Delta_\mathbf{r}$, in a product state:

$$| \Delta_\mathbf{r}, \eta > \;=\; | \Delta_\mathbf{r} > \otimes \frac{1}{\sqrt{2}} \left[e^{-i\varphi/2} \mid \alpha > +\eta e^{i\varphi/2} \mid \beta >\right] \tag{7}$$

where $| \Delta_\mathbf{r} >$ denotes a single particle orbital state with wave function given by the characteristic function of domain $\Delta_\mathbf{r}$ (equal to 1 in this domain, 0 elsewhere). In the limit where the volume Δ tends to 0, one can ignore states with more than one particle in $\Delta_\mathbf{r}$, and the N particle states in question provide a quasi-complete basis. The projector onto eigenvalue 0 is:

$$P_{\eta=0}(\mathbf{r}) = \; 1 - \int_{\Delta_\mathbf{r}} d^3r' \; n(\mathbf{r}') = \int_{\Delta_\mathbf{r}} d^3r' \; \left[\frac{1}{\Delta} - n(\mathbf{r}')\right] \tag{8}$$

[c]To correct a sign error in this reference, here we interchange α and β.

On the other hand, the projectors for eigenvalues $\eta = \pm 1$ are:

$$P_{\eta=\pm 1}(\mathbf{r},\varphi) = \frac{1}{2}\int_{\Delta_{\mathbf{r}}} d^3r \; [n(\mathbf{r}) + \eta\sigma_\varphi(\mathbf{r})] \tag{9}$$

We now consider a series of K measurements, the first of a spin along direction φ_1 in volume $\Delta_{\mathbf{r}_1}$, the second of a spin along direction φ_2 in volume $\Delta_{\mathbf{r}_2}$, etc., corresponding to the sequence of operators:

$$A(\mathbf{r}_1,\varphi_1) \;\; ; \;\; A(\mathbf{r}_2,\varphi_2) \;\; ; \;\; A(\mathbf{r}_3,\varphi_3) \; ; ... \; ; \; A(\mathbf{r}_K,\varphi_K) \tag{10}$$

As in,[16] we assume assume that all $\mathbf{r}$'s are different and that the regions of measurement $\Delta_{\mathbf{r}_1}$, $\Delta_{\mathbf{r}_2}$, ..., $\Delta_{\mathbf{r}_K}$ do not overlap, so that all these operators commute; in addition, and as already mentioned, we assume that the sequence of measurements is sufficiently brief to ignore any intrinsic evolution of the system other than the effect of the measurements themselves. Under these conditions, the probability of any sequence of results:

$$\eta_1 = 0, \pm 1 \;\; ; \;\; \eta_2 = 0, \pm 1 \;\; ; \;\; ... \qquad \eta_K = 0, \pm 1 \tag{11}$$

is simply given by the average value in state $\mid \Phi >$ of the product of projectors:

$$< \Phi \mid P_{\eta_1}(\mathbf{r}_1,\varphi_1) \times P_{\eta_2}(\mathbf{r}_2,\varphi_2) \times ....P_{\eta_K}(\mathbf{r}_K,\varphi_K) \mid \Phi > \tag{12}$$

When the projectors are replaced by their expressions (8) and (9), with (5), we obtain the product of several terms, each containing various products of field operators. In each term, because of the commutation of the measurements, we can push all $\Psi^\dagger_{\alpha,\beta}(\mathbf{r})$'s to the left, all $\Psi_{\alpha,\beta}(\mathbf{r})$'s to the right. It is then useful to expand the field operators onto the annihilation operators for single-particle-states $\mid u_a, \alpha >$ and $\mid u_b, \beta >$:

$$\Psi_\alpha(\mathbf{r}) \;=\; u_a(\mathbf{r}) \times a_{u_a,\alpha} + ... \;\; ; \;\; \Psi_\beta(\mathbf{r}) \;=\; u_b(\mathbf{r}) \times a_{b,\beta} + ... \tag{13}$$

where the terms +.... symbolize sums over other orbital states that, together with $u_a(\mathbf{r})$, or $u_b(\mathbf{r})$, complete a basis in the orbital space state of a single particle. Since the destruction operators give zero when they act on states that have zero population, it is easy to see that all these additional terms simply disappear. Each term now contains between the bra $< \Phi \mid$ and the ket $\mid \Phi >$ a sequence of creation operators, $(a_{u_a,\alpha})^\dagger$ or $(a_{u_b,\beta})^\dagger$, followed by another sequence of destruction operators, $(a_{u_a,\alpha})$ or $(a_{u_b,\beta})$. If each state, u_a, α or u_b, β, does not appear exactly the same number of times in the sequence of creation operators and the sequence of destruction operators, one obtains the product of two orthogonal kets, which is zero. If they appear exactly the same number of times, every creation and destruction operator introduces a factor $\sqrt{(N_{a,b} - q)}$, where q depends on the term considered, but remains smaller than the number of measurements K. We assume that:

$$K \ll N_a \;,\; N_b \tag{14}$$

which allows us to approximate, as in,[16] all factors $\sqrt{(N_{a,b} - q)}$ by $\sqrt{N_{a,b}}$.

At this point, all operators of $\Psi^\dagger_{\alpha,\beta}(\mathbf{r})$ are simply replaced by $\sqrt{N_{a,b}}u^*_{a,b}(\mathbf{r})$, all $\Psi_{\alpha,\beta}(\mathbf{r})$ by the complex conjugate, but we still have to take into account the necessity for particle number conservation in each sequence. This can be done by using the mathematical identity:

$$\int_0^{2\pi} \frac{d\Lambda}{2\pi}\, e^{in\Lambda} \;=\; \delta_{n,0} \tag{15}$$

(where n is an integer): if we multiply each $\Psi_\alpha(\mathbf{r})$ (or $\sqrt{N_a}u_a(\mathbf{r})$) by $e^{i\Lambda}$, and each $\Psi^\dagger_\alpha(\mathbf{r})$ (or $\sqrt{N_a}u^*_a(\mathbf{r})$) by $e^{-i\Lambda}$, and then integrate Λ over 2π, we express the necessary condition and automatically ensure particle number conservation.

We remark that neither $n(\mathbf{r})$ nor $P_{\eta=0}(\mathbf{r})$ introduce exponentials of Λ, since they always contain matched pairs of creation and destruction operators; exponentials only appear in $\sigma_\varphi(\mathbf{r})$ that is in the projectors $P_\eta(\mathbf{r},\varphi)$ when $\eta = \pm 1$. We assume that volume Δ is sufficiently small to neglect the variations of the orbital wave functions over all $\Delta_\mathbf{r}$'s. The probability of the sequence of results (11) is then proportional to :

$$\begin{aligned}&\prod_i \left[1 - \Delta\left(N_a\left|u_a(\mathbf{r}_i)\right|^2 + N_b\left|u_b(\mathbf{r}_i)\right|^2\right)\right] \int_0^{2\pi} \frac{d\Lambda}{2\pi} \times \\ &\prod_j \left\{\Delta\left[N_a\left|u_a(\mathbf{r}_j)\right|^2 + N_b\left|u_b(\mathbf{r}_j)\right|^2 + \eta_j\sqrt{N_aN_b}\left(e^{i(\Lambda-\varphi_j)}u_a(\mathbf{r}_j)u^*_b(\mathbf{r}_j) + \text{c.c.}\right)\right]\right\},\end{aligned} \tag{16}$$

where c.c. means complex conjugate. In this expression, the first line corresponds to the contribution of all results $\eta = 0$ (no particle found in the volumes of detection) and has no Λ (or φ) dependence; the second line corresponds to all positive detections of spins of particles. It is convenient to introduce the relative phase of the two wave functions by:

$$\xi(\mathbf{r}) = \arg\left[u_a(\mathbf{r})/u_b(\mathbf{r})\right] \tag{17}$$

so that the brackets in the second line become:

$$\left[N_a\left|u_a(\mathbf{r}_j)\right|^2 + N_b\left|u_b(\mathbf{r}_j)\right|^2 + 2\eta_j\sqrt{N_aN_b}_a\left|u_a(\mathbf{r}_j)\right|\left|u_b(\mathbf{r}_j)\right|\cos\left(\Lambda + \xi(\mathbf{r}_j) - \varphi_j\right)\right] \tag{18}$$

The contrast of the interference pattern is maximal at points $\mathbf{r}_j$ where $\sqrt{N_a}\left|u_a(\mathbf{r}_j)\right| = \sqrt{N_b}\left|u_b(\mathbf{r}_j)\right|$, i.e. where the two boson fields have the same intensity.

These are is the result onto which our discussion below will be based; for a generalization to spin measurements that are not necessarily in the xOy plane, see the appendix of.[17]

3. Physical Discussion

3.1. *Role of the Λ Integral*

Suppose that we consider a sequence where only one spin is detected; the product over j in Eq. (16) then contains only one bracket, summed over Λ between 0 and

2π. The contribution of each value of Λ gives nothing but the probability of the two results, $\eta_j = \pm 1$, for a spin 1/2 that is described by a density matrix ρ given by:

$$\rho \propto \begin{pmatrix} N_a \left|u_a(\mathbf{r}_1)\right|^2 & \sqrt{N_a N_b} u_a^*(\mathbf{r}_j) u_b(\mathbf{r}_1) e^{-i\Lambda} \\ \sqrt{N_a N_b} u_a(\mathbf{r}_1) u_b^*(\mathbf{r}_1) e^{i\Lambda} & N_b \left|u_b(\mathbf{r}_1)\right|^2 \end{pmatrix} \tag{19}$$

This is easily checked by calculating the trace of the product of ρ by the projector:

$$\left[1 + \eta \left(e^{-i\varphi} \sigma_+ + e^{i\varphi} \sigma_-\right)\right] / 2 \tag{20}$$

The Oz component of the spin before measurement is then proportional to $N_a \left|u_a(\mathbf{r}_1)\right|^2 - N_b \left|u_b(\mathbf{r}_1)\right|^2$, its transverse component proportional to $2\sqrt{N_a N_b}_a \left|u_a(\mathbf{r}_1)\right| \left|u_b(\mathbf{r}_1)\right|$, with an azimuthal direction specified by angle $\Lambda - \xi(\mathbf{r}_1)$. Now, since Λ is summed between 0 and 2π, the off diagonal elements disappear from (19), meaning that this azimuthal direction is initially completely random; the spin loses its transverse orientation and keeps only its Oz component. This is natural since we are starting from Fock states with completely undetermined relative phase. Therefore, for the first transverse measurement the two results ± 1 always have the same probability, and the adjustable parameter φ_1 plays no role.

Now consider a sequence with two measurements and two results $\eta_{1,2} = \pm 1$. In Eq. (16), the Λ integral then introduces correlations. The result of the first measurement provides an information on the probabilities of the results of the second: this information is contained in a Λ distribution that is given by (18), with η_j replaced by the result ± 1 of the first measurement, and $\mathbf{r}_j$ replaced by the point $\mathbf{r}_1$ at which this measurement was made. The information is still not very precise, since the width of the Λ distribution is of the order of π; but, for instance, if the first result was $+1$ and if the two angles of measurements φ_1 and φ_2 are close or even equal, there is more chance to find again $+1$ than -1 for the result of the second measurement.

When more and more spin measurements are obtained, the Λ distribution becomes narrower and narrower, meaning that more and more information on the value of Λ is accumulated. Standard quantum mechanics considers that Λ has no physical existence at the beginning of the series of measurements, and that its determination is just the result of a series of random perturbations of the system introduced by the measurements. Nevertheless, Eq. (16) shows that all observations are totally compatible with the idea of a pre-existing value of Λ, which is perfectly well defined but unknown, remains constant, and is only revealed (instead of being created) by the measurements. For a more detailed discussion of the evolution of the Λ distribution, and of the optimum strategy concerning the choice of the angles of measurement φ_j to better determine Λ, see ref.[17]

It is interesting to find a situation where an additional (hidden) variable Λ emerges so naturally from a standard calculation in quantum mechanics. It appears mathematically as a way to express the conservation of number of particles. In other words, the role of the additional variable is, by integration, to ensure the conservation of the conjugate variable. This contrasts with usual theories with additional

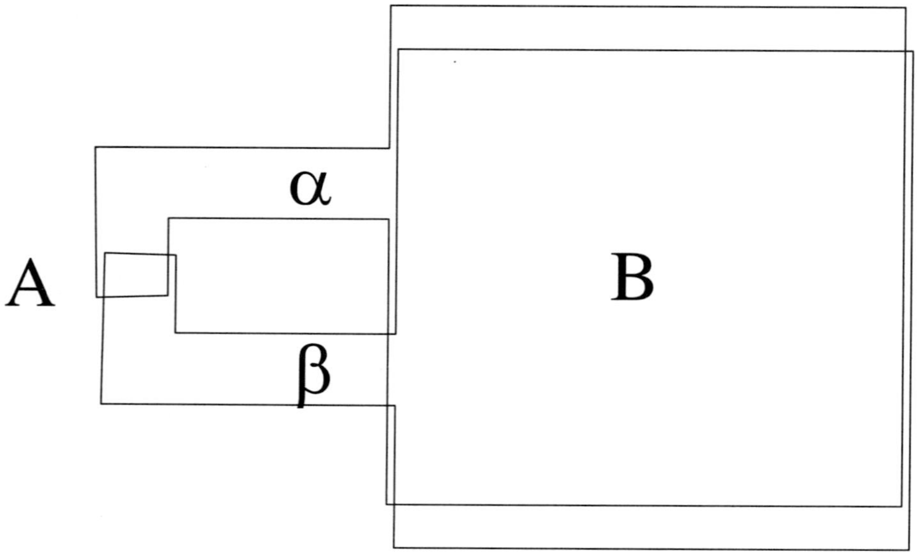

Fig. 2. Two different spin states ($\alpha = +$ and $\beta = -$) are associated with two orbital wave functions that overlap mostly in a large region B, but also have "fingers" that overlap in a much smaller region A. The two states are macroscopically populated. Under the influence of a few measurements of the spin of particles performed in region A, a macroscopic transverse spin polarization appears in region B.

variables, where they are introduced more or less arbitrarily, the only constraint being that the statistical average over the new variables reproduces the predictions of standard quantum mechanics.

3.2. *Small and Big Condensates; Amplification During Measurement*

In our calculation, we have made no special assumption concerning the orbital wave functions $u_a(\mathbf{r})$ and $u_b(\mathbf{r})$ associated with the two highly populated single particle quantum states; they can overlap much or little in space, and in many ways. We will consider situations where their configuration leads to the discussion of interesting physical effects. For simplicity, from now on we assume that the two wave functions have the same phase at every point of space $\mathbf{r}$, so that $\xi(\mathbf{r})$ vanishes. This simplification occurs if the two states correspond to stationary states trapped in a real potential, as often the case in experiments with Bose–Einstein condensates; it is convenient, but not essential[d].

---

[d]For instance, we exclude the case where two condensates are still expanding, as in figure Fig. 1 and as in the experiment described in ref.[24] Much of what we write can nevertheless be transposed to such cases, in terms of the phase of an helicoidal structure of the spin directions in space, instead of just parallel spins.

As a first example, we consider two states such as those represented schematically in figure Fig. 2. The states are mostly located in a region of space B where they strongly overlap, but also have "fingers" that overlap in another small region of space A, where the spin measurements are actually performed. We assume that A is not too small, and contains an average number of spins (100 for instance) that remains sufficient to perform several measurements and determine the relative phase of the two Fock states with reasonable accuracy. On the other hand, the number of spins in region B may be arbitrarily large, of the order of the Avogadro number for instance.

In this situation, our preceding calculation applies and predicts that the measurement of the spins in A will immediately create a spontaneous polarization in B that is parallel to the random polarization obtained in A. In other words, standard quantum mechanics predicts a giant amplification effect, where the measurement performed on a few microscopic particles induces a transverse polarization in a macroscopic assembly of spins. In itself, the idea is not too surprising, even in classical mechanics: one could see the assembly of spins in B as a metastable system, ready to be sensitive to the tiny perturbation of a microscopic system in A. In this perspective, the perturbations created by the measurement in A would propagate towards B and trigger its evolution towards a given spin direction. But this is not the context in which we have obtained the prediction: we have not assumed any evolution of the state vector of the system between one measurement and the next. In fact, what standard quantum mechanics describes here is not something that propagates along the state and has a physical mechanism (such as, for instance, the propagation of Bogolubov phonons in the condensates); it is just "something with no time duration" that is a mere consequence of the postulate of quantum measurement (wave packet reduction).

Leggett and Sols[25,26] discuss a similar situation in the context of two large superconductors, which acquire a spontaneous phase by the creation of a Josephson current between them, which in turn is measured by a tiny compass needle in order to obtain its phase. Here again we have a small system determining the state of a much larger system, without any physical mechanism. These authors comment the situation in the following terms: "can it really be that by placing, let us say, a minuscule compass needle next to the system, with a weak light beam to read off its position, we can force the system to realize a definite macroscopic value of the current? Common sense rebels against this conclusion, and we believe that in this case common sense is right". They then proceed to explain that the problem may arise because we are trying to apply to macroscopic objects quantum postulates that were designed 80 years ago for the measurements of microscopic objects, because other measurements were not conceivable then. In other words, we are trying to use present standard quantum mechanics beyond its range of validity. They conclude that what is needed in a new quantum measurement theory.

What is interesting to note, as we have already mentioned in the introduction, is that here we have a case where the measured system itself creates a macroscopic

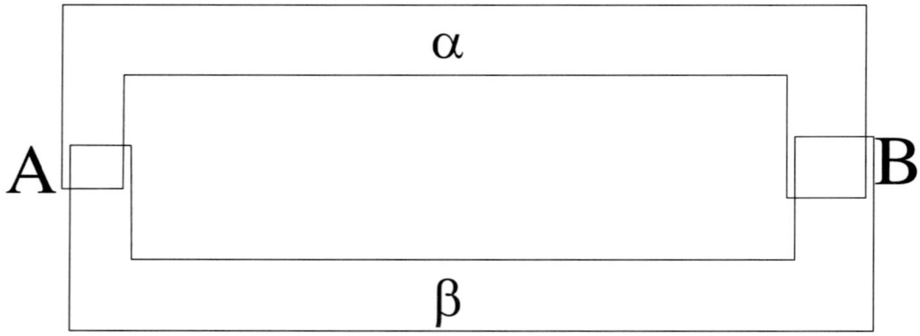

Fig. 3. Two highly populated Fock states associated with opposite spin direction ($\alpha = +$ and $\beta = -$) overlap in two remote regions A and B of space. A series of transverse spin measurements in A triggers the appearance of a well defined transverse orientation in A, and also that of a parallel macroscopic transverse orientation in B (quantum non locality). This corresponds to an angular momentum that seems to appear in B from nothing, with no interaction at all.

pointer, made of a large assembly of parallel spins, that directly "shows" the direction of the spins resulting from the measurements. Usually, in the theory of quantum measurement, this pointer is the last part of the measuring apparatus, not something that interacts directly with the measured system itself, or even less is part of it.

3.3. *EPR Non-locality with Fock States*

Now suppose that the two condensates have the shape sketched in figure Fig. 3, extending over a large distances, and overlapping only in two remote regions of space A and B. Again, the number of particles in both regions is arbitrary, and in particular can be macroscopic in B. We have a situation that is similar to the usual EPR situation: measurements performed in A can determine the direction of spins in both regions A and B. If we rephrase the EPR argument to adapt it to this case, we just have to replace the words "before the measurement in A" by "before the series of measurements in A", but all the rest of the reasoning remains exactly the same: since the elements of reality in B can not appear under the effect of what is done at an arbitrary distance in region A, these elements of reality must exist even before the measurements performed in A. Since the initial double Fock state of quantum mechanics does not contain any information on the direction of spins in B, this theory is incomplete.

What is new here is that the EPR elements of reality in B correspond to a system that is macroscopic. One can no longer invoke its microscopic character to deprive the system contained in B of any physical reality! The system can even be at our scale, correspond to a macroscopic magnetization that can be directly observable with a hand compass; is it then still possible to state that it has no intrinsic physical reality? When the EPR argument is transposed to the macroscopic world, it is clear that Bohr's refutation does no longer apply in the form written in his article; it has to be at least modified in some way.

Another curiosity, in standard quantum mechanics, is that it predicts the appearance of a macroscopic angular momentum in region B without any interaction. This seems to violate angular momentum conservation. Where does this momentum come from? Usually, when a spin is measured and found in some state, one considers that the angular momentum is taken as a recoil by the measurement apparatus. When the measured system is microscopic and the apparatus macroscopic, the transfer of angular momentum is totally negligible for the latter, so that there is no hope to check this idea; but, at least, one can use the idea as a theoretical possibility. Here, the situation is more delicate: what is the origin of the angular momentum that appears in B during measurement? Could it be that the apparatus in A, because the system in A is entangled with a macroscopic system in B, takes a macroscopic recoil, even if it measures a few spins only? A little analysis shows that this is impossible without introducing the possibility for superluminal communication: the recoil in A would allow to obtain information on B (if the states have been dephased locally for instance). So, it can not be the measurement apparatus in A that takes the angular momentum recoil corresponding to B. Then, if we believe that angular momentum can not appear in a region of space without interactions, even during operations that are considered as "measurements" in standard quantum mechanics, this leads us to an "angular momentum EPR proof": we are forced to conclude that the transverse polarization of the spins in B already existed before any measurement started[e]. Since this is not contained in the double Fock state, standard quantum mechanics is incomplete.

We can make the argument even more convincing by using the scheme sketched in figure Fig.4. We now assume that the two condensates in internal states α and β overlap in B but not in A, where they both overlap both with the same third condensate in a third internal state γ. We furthermore assume that angular momentum has matrix elements between α and β, but not between a and γ and between β and γ (for instance, the parity of γ may be the opposite of that of the two other internal states): the transverse measurements in A correspond to some observable that has appropriate matrix elements and parity, electric dipole for instance. We know (see for instance[16]) that phase determination in Fock states is transitive: fixing the phase between a and γ on the one hand, between β and γ on the other, will determine the relative phase of α and β. Under these conditions, in standard quantum mechanics, the macroscopic angular momentum that appears in B can be a consequence of measurements in A of physical quantities that have nothing to to with angular momentum, so that the measurement apparatuses have no reason to take any angular

[e]To avoid this conclusion, one can either give up angular momentum conservation in measurements (making them even more special physical processes than usually thought!), or take the Everett interpretation ("relative state" or "many minds" interpretation) where no transverse polarization ever appears in B, even after the measurements..

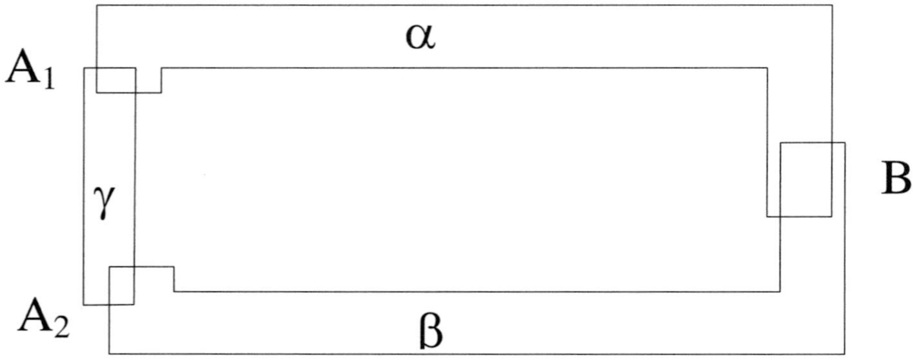

Fig. 4. In a variant of the situation shown in Fig. 3, measurements are performed in region A, at points A_1 and A_2, of the relative phase between a small local condensate (in internal state γ) and two extended condensates (respectively in internal states α and β). By transitivity, this determines the relative phase of the large condensates, resulting by interference in a transverse spin polarization in region B. Even if the measurements with the local A condensate do not involve angular momentum (assumed to have no matrix elements between the overlapping states), this operation can create a macroscopic angular momentum in region B.

momentum recoil at all. Still, they create a large angular momentum in B. Again, if we do not accept the idea of angular momentum appearing from nothing, we must follow EPR and accept that the angular momentum was there from the beginning, even if we had no way to predict its direction[f].

4. Possible Objections

In this text, we have discussed thought experiments, not attempted to propose feasible experiments. We have just assumed that the states that are necessary for the discussion can be produced, and that they are sufficiently robust to undergo a series of measurements, with no other perturbation than the measurements themselves; this may require that the sequence of measurements be sufficiently fast. Of course, one could always object that these double Fock states are fundamentally not physical, for instance because some selection rule forbids them. This would be in contradiction with the generally accepted postulate that all quantum states belonging to the space of states (Hilbert space) of any physical system are accessible. If this postulate is true, there should be no fundamental reason preventing the preparation of a double Fock state, even for a system containing many particles.

A second objection could be that these states may exist but be so fragile that, in practice, it will always be impossible to do experiments with them. In the context of second order phase transitions and spontaneous symmetry breaking, Anderson[27–29] has introduced the notion of spontaneous phase symmetry breaking for superfluid

[f]As above, the only other logical possibility is to choose the other extreme: the Everett interpretation, where no angular momentum exists even after the measurements.

Helium 4 and superconductors. According to this idea, coupled superfluid systems at thermal equilibrium are not in Fock states: as soon as they become superfluid by crossing the second order transition, some unavoidable small perturbation always manages to transform the simple juxtaposition of the two Fock states into a single coherent Fock state containing all the $N_a + N_b$ particles, for which the two quantum states have a well defined relative phase. This assumption is for example implicit in the work of Siggia and Ruckenstein,[15] where the two condensates are considered as having a well defined phase from the beginning[g].

In our discussion, we have assumed neither the existence of a second order phase transition nor even thermal equilibrium, just the availability of the large initial double Fock state. Is there any general mechanism that favours coherent states over double Fock states? Decoherence may actually inroduce this preference. As in ref.,[11] we can introduce the so called coherent "phase states" by:

$$\mid \phi, N > \;=\frac{1}{\left(2^N N!\right)^{1/2}}\left[e^{i\phi/2}\left(a_{u_a,\alpha}\right)^{\dagger}+e^{-i\phi/2}\left(a_{u_b,\beta}\right)^{\dagger}\right]^N \mid \text{vac.} > \tag{21}$$

in terms of which the ket $\mid \Phi >$ of (2) can be written:

$$\mid N_a : u_a, \alpha \; ; \; N_b : u_b, \beta \; > \propto \int_0^{2\pi} d\phi \; e^{i(N_b - N_a)\phi} \mid \phi, N > \tag{22}$$

If M is large, the phase state (21) have a macroscopic transverse orientation in an azimuthal direction defined by ϕ; this orientation is likely to couple to the external environment, as most macroscopic variable do. For instance, if the spin of particles is associated with a magnetic moment, the different phase states create different macroscopic magnetic fields that will affect at least some microscopic particles of the environment, transferring them into states that are practically orthogonal for different values of ϕ. In other words, the basis of phase states is the "preferred basis" for the system coupled to its environment. As a consequence, the coherent superposition (22) spontaneously transforms into a superposition where each component, defined by a very small ϕ domain, is correlated with a different state of the environment. The correlation quickly propagates further and further into the environment, without any limit as long as the Schrödinger equation is obeyed (this is the famous Schrödinger cat paradox). As a result, the observation of interference effects between different ϕ values becomes more and more difficult, in practice impossible. In terms of the the trace of the density operator over the environment, the coherent superposition (22) decays rapidly into an incoherent mixture of different ϕ states. For a general discussion of the observability of macroscopically distinct quantum states, see for instance ref.[30]

[g]It is interesting to note in passing that, unexpectedly, Anderson's spontaneous symmetry breaking concept is so closely related to the old idea of hidden/additional variables in quantum mechanics. A specificity, nevertheless, is that Anderson sees the additional variables as appearing during second order superfluid phase transitions.

Decoherence is unavoidable, but does not really affect our conclusions. It just means that, when the measurements are performed in region A and determine the transverse polarization (in the standard interpretation), they fix at the same time the spin directions in B as well as the state of the local environment. The real issue is not coherence, or the coupling to the environment; it is the emergence of a single macroscopic result, which is considered as an objective fact and a result of the observation in the standard interpretation (but of course not in the Everett interpretation). In the end, decoherence is not an essential issue in our discussion.

A third objection might be size limitations: are there inherent limits to the size of highly populated Fock states and Bose–Einstein condensates? Is there any reason why large sizes should make them extremely sensitive to small perturbations? One could think for instance of thermal fluctuations that may introduce phase fluctuations and put some temperature dependent limit on the size of the coherent system. Other possible mechanisms, such as inhomogeneities of external potentials, might break the condensate into several independent condensates, etc. Generally speaking, we know that ideal condensed gases are extremely sensitive to small perturbations[h], but fortunately also that repulsive interactions between the atoms tend to stabilize condensed systems. They do not only introduce a finite compressibility of the condensate, but also tend to stabilize the macroscopic occupation of a single quantum state.[31] This should increase the robustness of large systems occupying a unique single Fock state, even if extended in space.

Experimentally, Bose–Einstein condensates in dilute gases at very low temperatures provide systems that are very close to being in a highly populated Fock state. Nevertheless, until now experiments have been performed with gas samples that are about the size of a tenth of a millimeter; one can therefore not exclude that new phenomena and unexpected perturbations will appear when much larger condensates are created. In any case, even if the non-local effects that we have discussed are limited to a range of a tenth of a millimeter (or any other macroscopic length), they remain non-local effects on which a perfectly valid EPR type argument can be built!

5. Conclusion

We can summarize the essence of this article by saying that, in some quantum situations where macroscopic systems populate Fock states with well defined populations, the EPR argument becomes significantly stronger than in the historical example with two microscopic particles. The argument speaks eloquently if favour of a pre-existing relative phase of the two states - alternatively, if one prefers, of an interpretation where the phase remains completely undetermined even after the measurements (Everett interpretation) - but certainly not in favour of the orthodox

[h]For instance, condensate in ideal gases tend to localize themselves in tiny regions of space.[32] Nevertheless, this is a pathology introduced by the infinite compressibility of the condensate in an ideal gas; it disappears as soon as the atoms have some mutual repulsion.

point of view where the phase appears during the measurements. If we stick to this orthodox view, surprising non-local effects appear in the macroscopic world. These effects can be expressed in various ways, including considerations on macroscopic angular momentum conservation, but not in terms of violations of Bell type inequalities (this is because the form of the Λ integral in (16), with positive terms inside it, automatically ensures that the Bell inequalities are satisfied). In any case, Bohr's denial of physical reality of the measured system alone becomes much more difficult to accept when this system is macroscopic. Of course, no one can predict what Bohr would have replied to an argument involving macroscopic spin assemblies, and whether or not he would have maintained his position concerning the emergence of a single macroscopic result during the interaction of the measured system with the measurement apparatuses.

Another conclusion is that quantum mechanics is indeed incomplete, not necessarily in the exact sense meant by EPR, but in terms of the postulates related to the measurement: they do not really specify what is the reaction of the measured system on the measurement apparatus ("recoil effect"). Ignoring this reaction was of course completely natural at the time when quantum mechanics was invented: only quantum measurements of microscopic systems were conceivable at that time, so that these effects were totally negligible. But now this is no longer true, so that we need a more complete theory for quantum measurement on a macroscopic system "in which all the assumptions about relative energy and time scales, etc.. are made explicit and if necessary revised".[25] Bose-Eintein condensates in gases seem to be good candidates to explore this question theoretically and experimentally.

Acknowledgments

The author is grateful to W. Mullin, A. Leggett, C. Cohen-Tannoudji and J. Dalibard for useful discussions and comments.

References

1. A. Einstein, B. Podolsky and N. Rosen, *Phys. Rev.* **47**, 777 (1935).
2. N. Bohr, *Phys. Rev.* **48**, 696 (1935).
3. J. Javanainen and Sun Mi Ho, *Phys. Rev. Lett.* **76**, 161 (1996).
4. T. Wong, M. J. Collett and D. F. Walls, *Phys. Rev.***A 54**, R3718 (1996).
5. J. I. Cirac, C. W. Gardiner, M. Naraschewski and P. Zoller, *Phys. Rev.* **A 54**, R3714 (1996).
6. Y. Castin and J. Dalibard, *Phys. Rev.* **A 55**, 4330(1997).
7. K. Mølmer, *Phys. Rev.* **A 55**, 3195 (1997).
8. K. Mølmer, *J. Mod. Opt.* **44**, 1937 (1997).
9. C. Cohen-Tannoudji, *Collège de France 1999-2000 lectures, chap. V et VI* , *http://www.phys.ens.fr/cours/college-de-france/*.
10. Y. Castin and C. Herzog, *C.R. Acad. Sci. série IV*, **2**, 419 (2001).

11. C. J. Pethick and H. Smith, *Bose–Einstein condensates in dilute gases*, (Cambridge University Press, 2002).
12. K. Mølmer, *Phys. Rev.* **A 65**, 021607 (2002).
13. P. Horak and S. M. Barnett,*J. Phys.* **B 32**, 3421 (1999).
14. J. Dunningham, A. Rau and K. Burnett *Science* **307**, 872 (2005).
15. E. Siggia and A. Ruckenstein, *Phys. Rev. Lett.* **44**, 1423 (1980).
16. F. Laloë, *Europ. Phys. J. section D*, **33**, 87 (2005).
17. W. J. Mullin, R. Krotkov and F. Laloë, *Phys. Rev.* **A74**, 023610 (2006).
18. D. Bohm, *Quantum theory*, (Prentice Hall, 1951).
19. D. Bohm and Y. Aharonov, *Phys. Rev.* **108**, 1070 (1957).
20. F. Laloë, *Am. J. Phys.* **69**, 655 (2001);(more recent version *quant-ph/0209123*).
21. M. Jammer, *The philosophy of quantum mechanics*, (Wiley 1974).
22. J. S. Bell, *J. Physique Colloque C2, supplément au no 3*, **42**, 41 (1981); This article is reprinted in Ref. 23.
23. J. S. Bell, *Speakable and unspeakable in quantum mechanics*, (Cambridge University Press, 1987).
24. M. H. Wheeler, K. M. Mertes, J. D. Erwin and D. S. Hall , *Phys. Rev. Lett.* **89**, 090402 (2002).
25. A. J. Leggett and F. Sols, *Found. Phys.* **21**, 353 (1991).
26. A. J. Leggett, *Broken gauge symmetry in a Bose condensate*, in *Bose–Einstein condensation*, A. Griffin, D. W. Snoke and S. Stringari eds., (Cambridge University Press 1995); see in particular pp. 458-459.
27. P. W. Anderson, *Rev. Mod. Phys.* **38**, 298 (1966).
28. P. W. Anderson, *Basic notions in condensed matter physics*, (Benjamin-Cummins 1984).
29. P. W. Anderson, *Measurement in quantum theory and the problem of complex systems*, in *The Lesson of quantum theory* ed. by J. de Boer, E. Dal and O. Ulfbeck, (Elsevier 1986); see section 3.
30. A. J. Leggett, *J. Phys. Cond. Mat.* **14**, R415 (2002).
31. P. Nozières, *Some comments on Bose-Einstein condensation*, in *Bose-Einstein condensation*; edited by A. Griffin, D. W. Snoke and S. Stringari, (Cambridge University Press, 1995).
32. W. E. Lamb and A. Nordsieck,*Phys. Rev.* **59**, 677 (1941).

THE QUANTUM MEASUREMENT PROCESS: LESSONS FROM AN EXACTLY SOLVABLE MODEL

A. E. ALLAHVERDYAN[1)], R. BALIAN[2)] AND Th. M. NIEUWENHUIZEN[3)]

[1)] *Yerevan Physics Institute, Alikhanian Brothers St. 2, Yerevan 375036, Armenia*
[2)] *Service de Physique Théorique, CEA-Saclay, 91191 Gif-sur-Yvette cedex, France*
[3)] *Institute for Theoretical Physics, University of Amsterdam Valckenierstraat 65, 1018 XE Amsterdam, The Netherlands*

The measurement of a spin-$\frac{1}{2}$ is modeled by coupling it to an apparatus, that consists of an Ising magnetic dot coupled to a phonon bath. Features of quantum measurements are derived from the dynamical solution of the measurement, regarded as a process of quantum statistical mechanics. Schrödinger cat terms involving both the system and the apparatus, die out very quickly, while the registration is a process taking the apparatus from its initially metastable state to one of its stable final states. The occurrence of Born probabilities can be inferred at the macroscopic level, by looking at the pointer alone. Apparent non-unitary behavior of the measurement process is explained by the arisal of small many particle correlations, that characterize relaxation.

Keywords: Apparatus; Quantum measurement model; Spin; Registration; Collapse; Schrödinger cats; Decoherence

1. Introduction

As any new student is taught, the result of a quantum measurement process is coded in the collapse postulate and the Born rule. Among theorists, this formal approach has too often led to the opinions that quantum measurements either require a special theory, or that the measurement process itself is a piece of physics that need not be worried about — how different is the life of an experimentalist, whose task it is to perform the measurement!

It is our aim to discuss features of the quantum measurement process along the lines of a rich enough, realistic, but still solvable model. Most of these points are well known in literature, while a few other ones are put forward by the solution of the problem. In particular, it is shown that the collapse is explained by a dynamical approach relying only on the Schrödinger equation and on statistical properties issued from the large size of the apparatus.

The most standard theory of measurements is the von Neumann-Wigner theory, where the apparatus is described in terms of pure states.[1,2] We consider that a sensible approach to the problem should rely on quantum statistical physics. Following other works in the literature ,[3–14] we consider an explicit model for the measurement apparatus.

We shall first outline the model, and then highlight various relevant aspects of the process. The measurement that we describe is ideal (or non-demolishing, or of the first kind) in the sense that the sole perturbation brought to the system by the dynamical process is von Neumann's reduction of the state.

We keep aside all derivations and technical details, one or the other can be found in our earlier papers on this model.[14–18]

2. The model and its solution

For the tested system S, we take the simplest quantum system, that is, a spin $\frac{1}{2}$. Its initial state is represented by the 2×2 density matrix $\hat{r}(0)$. The observable to be measured is its third Pauli matrix $\hat{s}_z$ with eigenvalues s_i equal to ± 1.

The interaction of S with the apparatus A should trigger A from some initial state $\hat{\mathcal{R}}(0)$ into either one of two possible final states $\hat{\mathcal{R}}_\Uparrow$ or $\hat{\mathcal{R}}_\Downarrow$, associated with $s_\uparrow = +1$ or $s_\downarrow = -1$, respectively. (Here $\Uparrow$ and $\Downarrow$ refer to A, while $\uparrow$ and $\downarrow$ refer to S.) These states $\hat{\mathcal{R}}_\Uparrow$ and $\hat{\mathcal{R}}_\Downarrow$ should have no overlap, and should be distinguishable at our scale by observation of some pointer variable. Registration in either one should be permanent and insensitive to weak external perturbations. Symmetry between $\hat{\mathcal{R}}_\Uparrow$ and $\hat{\mathcal{R}}_\Downarrow$ should prevent any bias.

In order to satisfy these conditions, we take for A a system that can undergo a phase transition with broken invariance. Attempting to conciliate mathematical tractability and realistic features, we choose A as the simplest quantum object that displays two phases, namely a sufficiently large Ising system. Though finite, the number of degrees of freedom will be sufficiently large so that the relaxation towards one of the equilibrium states $\hat{\mathcal{R}}_\Uparrow$, $\hat{\mathcal{R}}_\Downarrow$ is irreversible and that the order parameter has weak fluctuations in each possible final state. The transition of A from a metastable state to one of its stable, macroscopic states eliminates the infamous problem of observers ("Wigner's friend", "Mind-body problem"), see e.g. ,[1] from the quantum measurements process: After the measurement, the pointer variable will have a stable value, and can be read off at any moment, or just not, without causing a back-reaction on the already finished measurement process.

Our apparatus A = M + B simulates a *magnetic dot*: the magnetic degrees of freedom M consist of $N \gg 1$ spins with Pauli operators $\hat{\sigma}_a^{(n)}$, $(a = x, y, z)$, while the non-magnetic degrees of freedom such as phonons behave as a thermal bath B. Anisotropic interactions between these spins can generate Ising ferromagnetism below the Curie temperature T_{C}. As pointer variable we take the order parameter, which is the magnetization in the z-direction, represented (within normalization) by the quantum observable

$$\hat{m} = \frac{1}{N} \sum_{n=1}^{N} \hat{\sigma}_z^{(n)} \ . \tag{1}$$

The initial state $\hat{\mathcal{R}}\,(0)$ of A is the metastable paramagnetic state ($\langle \hat{m} \rangle = 0$), prepared by first thermalizing A = M+B at a temperature T_0 above T_{C}, then suddenly cooling

B into equilibrium at the temperature T below T_{C}. We expect the final state of the process to involve for A the two stable ferromagnetic states $\hat{\mathcal{R}}_{\Uparrow}$ or $\hat{\mathcal{R}}_{\Downarrow}$, with broken invariance. The equilibrium temperature T will be imposed to M by the phonon bath through weak coupling between the magnetic and non-magnetic degrees of freedom. Within fluctuations small as $1/\sqrt{N}$, the order parameter Eq. 1 vanishes in $\hat{\mathcal{R}}(0)$ and takes two opposite values in the states $\hat{\mathcal{R}}_{\Uparrow}$ and $\hat{\mathcal{R}}_{\Downarrow}$, $\langle \hat{m} \rangle_i$ equal to $+m_{\mathrm{F}}$ for $i = \Uparrow$ and to $-m_{\mathrm{F}}$ for $i = \Downarrow$. As in real magnetic registration devices, information will be stored by A in the form of the sign of the final magnetization.

2.1. *The Hamiltonian*

The full Hamiltonian can be decomposed into terms associated with the system, with the apparatus and with their coupling:

$$\hat{H} = \hat{H}_{\mathrm{S}} + \hat{H}_{\mathrm{A}} + \hat{H}_{\mathrm{SA}} \ . \tag{2}$$

In an ideal measurement the observable $\hat{s}$ to be measured should not change during the process, so that it should commute with $\hat{H}$. The simplest self-Hamiltonian that ensures this property is the trivial one: $\hat{H}_{\mathrm{S}} = 0$. The coupling between the tested system and the apparatus is a spin-spin coupling,

$$\hat{H}_{\mathrm{SA}} = -g\hat{s}_z \sum_{n=1}^{N} \hat{\sigma}_z^{(n)} = -Ng\hat{s}_z\hat{m} \ . \tag{3}$$

Before the measurement and after it, g will be equal to zero.

The apparatus A consists, as indicated above, of a magnet M and a phonon bath B, and its Hamiltonian can be decomposed into

$$\hat{H}_{\mathrm{A}} = \hat{H}_{\mathrm{M}} + \hat{H}_{\mathrm{B}} + \hat{H}_{\mathrm{MB}} \ . \tag{4}$$

The magnetic part is chosen as the long-range Ising interaction,

$$\hat{H}_{\mathrm{M}} = -\frac{1}{2} JN\hat{m}^2 \ . \tag{5}$$

The magnet-bath interaction, needed to drive the apparatus to equilibrium, is taken as a standard spin-boson Hamiltonian

$$\hat{H}_{\mathrm{MB}} = \sqrt{\gamma} \sum_{n=1}^{N} \left(\hat{\sigma}_x^{(n)} \hat{B}_x^{(n)} + \hat{\sigma}_y^{(n)} \hat{B}_y^{(n)} + \hat{\sigma}_z^{(n)} \hat{B}_z^{(n)} \right) \ , \tag{6}$$

which couples each component $a = x,\ y,\ z$ of each spin $\hat{\sigma}^{(n)}$ with some hermitean linear combination $\hat{B}_a^{(n)}$ of phonon operators. The dimensionless constant $\gamma \ll 1$ characterizes the strength of the thermal coupling between M and B, which is weak.

The bath Hamiltonian $\hat{H}_{\mathrm{B}}$ is a large set of harmonic oscillators. It will be involved in our problem only through its *autocorrelation function* in the equilibrium state at temperature $T = 1/\beta$, defined by

$$\frac{1}{Z_{\mathrm{B}}} \mathrm{tr}_{\mathrm{B}} \left[e^{-\beta \hat{H}_{\mathrm{B}}} \hat{B}_a^{(n)}(t) \, \hat{B}_b^{(p)}(t') \right] = \delta_{n,p}\delta_{a,b} \, K(t - t') \ , \tag{7}$$

$$\hat{B}_a^{(n)}(t) \equiv e^{i\hat{H}_{\mathrm{B}}t/\hbar} \hat{B}_a^{(n)} e^{-i\hat{H}_{\mathrm{B}}t/\hbar}, \tag{8}$$

in terms of the evolution operator of B alone. We choose for our model as Fourier transform

$$\tilde{K}(\omega) = \int_{-\infty}^{+\infty} \mathrm{d}t\; e^{-i\omega t} K(t) \tag{9}$$

of $K(t)$ the simplest expression having the required properties, namely

$$\tilde{K}(\omega) = \frac{\hbar^2}{4} \frac{\omega e^{-|\omega|/\Gamma}}{e^{\beta\hbar\omega} - 1}, \tag{10}$$

known as a quasi-Ohmic spectrum. The temperature dependence accounts for the quantum bosonic nature of the phonons. The Debye cutoff Γ characterizes the largest frequencies of the bath, and is assumed to be larger than all other frequencies entering our problem.

2.2. *Disappearance of Schrödinger cats*

2.2.1. *Dephasing*

The full density operator $\hat{\mathcal{D}}(t)$ of S+A is initially factorized as

$$\hat{\mathcal{D}}(0) = \hat{r}(0) \otimes \hat{\mathcal{R}}(0), \tag{11}$$

with S described by

$$\hat{r}(0) = \begin{pmatrix} r_{\uparrow\uparrow}(0) & r_{\uparrow\downarrow}(0) \\ r_{\downarrow\uparrow}(0) & r_{\downarrow\downarrow}(0) \end{pmatrix}, \tag{12}$$

and A by $\hat{\mathcal{R}}(0)$. $\hat{\mathcal{D}}(t)$ evolves according to the Liouville-von Neumann equation. We have exactly solved this equation, for values of the parameters of the model satisfying

$$N \gg 1, \qquad \hbar\Gamma \gg T \gg \gamma J \gg \frac{J}{N}\left(\frac{g}{\hbar\Gamma}\right)^2, \qquad \hbar\Gamma \gg J > g. \tag{13}$$

Over the very brief time-scale

$$\tau_{\mathrm{red}} = \frac{1}{\sqrt{2N}} \frac{\hbar}{g}, \tag{14}$$

the off-diagonal blocks $\hat{\Pi}_\uparrow \hat{\mathcal{D}}(t) \hat{\Pi}_\downarrow$ and $\hat{\Pi}_\downarrow \hat{\mathcal{D}}(t) \hat{\Pi}_\uparrow$, where $\hat{\Pi}_\uparrow = |\uparrow\rangle\langle\uparrow|$ and $\hat{\Pi}_\downarrow = |\downarrow\rangle\langle\downarrow|$ denote the projection operators on the subspaces $s_\uparrow = +1$ and $s_\downarrow = -1$, respectively, decay to zero. This process takes place on a timescale so short that energy exchange with the apparatus is negligible, so the bath does not play a role in it. Indeed, what occurs is a dephasing process caused by the interaction of the tested spin with the N spins of the apparatus. Just as in a spin-echo setup, due to the phase coherence, the intial state could in principle be retrieved, but this should not occur if we wish the process to be used as a quantum measurement. Recurrences might appear on an extremely long timescale τ_{recur} and, in principle, some mechanism is needed to exclude them. Thus, the so-called "Schrödinger cat" terms, superposing up and down projections, that have so much troubled the understanding of the quantum measurement process, just die out in the initial stage of the process.

2.2.2. *Decoherence*

The definitive nature of this process (irreversibility) is ensured by the large number of degrees of freedom of the apparatus, which prevents recurrences to occur over any reasonable timescale. In the model this is ensured by the bath. It starts to play a role at a time scale

$$\tau_{\rm irrev}^{\rm B} = \left[\frac{2\pi\hbar^2}{N\gamma g^2\Gamma^2}\right]^{1/4}. \tag{15}$$

We choose the parameters of the model, in particular, the coupling γ between M and B, in such a way that

$$\tau_{\rm red} \ll \tau_{\rm irrev} \ll \tau_{\rm recur}, \tag{16}$$

that is,

$$N \gg \frac{\gamma\hbar^2\Gamma^2}{8\pi g^2} \gg \frac{4}{N\pi^4}. \tag{17}$$

These conditions are easy to satisfy for large N.

2.3. *Registration of the measurement*

Only the diagonal blocks $\hat{\Pi}_\uparrow\hat{\mathcal{D}}(t)\hat{\Pi}_\uparrow$ and $\hat{\Pi}_\downarrow\hat{\mathcal{D}}(t)\hat{\Pi}_\downarrow$ are then left, and it takes a much longer time,

$$\tau_{\rm reg} = \frac{\hbar}{\gamma(J-T)} \ln\frac{3m_{\rm F}(J-T)}{g}, \tag{18}$$

for the apparatus to register the measurement through the evolution of these diagonal blocks. We find that the final state of the compound system S+A is represented, at a time $t_{\rm f} > \tau_{\rm reg}$, by von Neumann's reduced density operator

$$\hat{\mathcal{D}}\,(t_{\rm f}) = r_{\uparrow\uparrow}(0)|\uparrow\rangle\langle\uparrow|\otimes\hat{\mathcal{R}}_\Uparrow + r_{\downarrow\downarrow}(0)|\downarrow\rangle\langle\downarrow|\otimes\hat{\mathcal{R}}_\Downarrow\,, \tag{19}$$

where $\hat{\mathcal{R}}_\Uparrow$ and $\hat{\mathcal{R}}_\Downarrow$ represent the two stable ferromagnetic states of the apparatus. The success of the measurement also requires the lifetime of the initial, unstable paramagnetic state $\hat{\mathcal{R}}(0)$ of A to be longer than the duration of the measurement, a condition which is satisfied in the model.[18] In earlier work,[14–17] we considered a quartic interaction instead of the quadratic interaction Eq. 5;[19] since the transition is then of first order, the lifetime of the paramagnetic state is even much longer.

The magnetization m is a macroscopic variable, which for large N behaves continuously with small statistical fluctuations $\sim 1/\sqrt{N}$. The quantity

$$p(m;t_{\rm f}) = {\rm Tr}\,\hat{\mathcal{D}}(t_{\rm f})\,\delta(m-\hat{m}) = p_\Uparrow\,\delta(m-m_{\rm F}) + p_\Downarrow\,\delta(m+m_{\rm F}), \tag{20}$$

derived from Eq. 19, can only be interpreted as the probability distribution for m at the time $t_{\rm f}$. In repeated measurements, the prefactors

$$p_\Uparrow = r_{\uparrow\uparrow}(0), \qquad p_\Downarrow = r_{\downarrow\downarrow}(0), \tag{21}$$

are the frequencies with which we shall observe magnetizations $+m_{\rm F}$ and $-m_{\rm F}$, respectively. Because of this connection, the measurement is called *ideal.* Born's rule is therefore derived through the dynamical analysis of the measurement process instead of being posed as a postulate. Indeed, Born probabilities in the frequency interpretation are obtained straightforwardly from Eq. 20: since this equation refers to the pointer variable alone, which is a macroscopic variable, only the above identification can be given for its meaning. After this has been set, the same meaning must hold for Eq. 19 and for the post-measurement state of the tested system,

$$\hat{r}(t_{\rm f}) = r_{\uparrow\uparrow}(0)|\uparrow\rangle\langle\uparrow| + r_{\downarrow\downarrow}(0)|\downarrow\rangle\langle\downarrow| = \begin{pmatrix} r_{\uparrow\uparrow}(0) & 0 \\ 0 & r_{\downarrow\downarrow}(0) \end{pmatrix}. \tag{22}$$

This expression thus represents a distribution over an ensemble of measured spins in the frequency interpretation, a simpler result than anticipated, see e.g. [20–22] It seems incompatible with, e.g., the Copenhagen interpretation or a many world interpretation.

However, the expression Eq. 19 for the final state of S+A contains additional information: it exhibits a full correlation between the apparatus and the measured spin, and involves no Schrödinger cat term. The disappearance of such terms has resulted from the exact solution of the model, and did not rely on any "collapse postulate" or "projection".

3. Microscopic versus macroscopic aspects of quantum measurement

Several important features of quantum measurements are put forward by the solution of the above model.

The apparatus A should *register* the result in a robust and permanent way, so that it can be read off by any observer. Such a registration, which is often overlooked in the literature on measurements, takes place during the second stage of the process as indicated above. It is needed for practical reasons especially since S is a microscopic object. Moreover, its very existence allows us to disregard the observers in quantum measurements. Once the measurement has been registered, the result becomes objective. The literature which attributes a rôle to the mind who observes S is therefore irrelevant.

Registration also requires an *amplification* within the apparatus of a signal produced by interaction with the microscopic system S. For instance, in a bubble chamber, the apparatus in its initial state involves a liquid, overheated in a metastable phase. In spite of the weakness of the interaction between the particle to be detected and this liquid, amplification and registration of its track can be achieved owing to local transition towards the stable gaseous phase. This stage of the measurement process thus requires an irreversible phenomenon. It is governed by the kinetics of bubble formation under the influence of the particle and implies a dumping of free energy in the surrounding liquid, the dynamics of which governs the size of the bubble. Similar remarks hold for photographic plates, photomultipliers or other types of detectors. In our model the amplification is ensured by the interaction between the magnet M and the phonon bath B, which allows the energy exchange and the entropy increase needed to bring M from the state $\hat{\mathcal{R}}(0)$ to $\hat{\mathcal{R}}_{\Uparrow}$ or $\hat{\mathcal{R}}_{\Downarrow}$.

A measurement process thus looks like a *relaxation process*, but with several complications. On the one hand, the final state of A is not unique, and the dynamical process can have several possible outcomes for A. The apparatus is therefore comparable to a material which has a broken invariance, and can relax towards one equilibrium phase or another, starting from a single metastable phase. This is why we chose a model involving such a phase transition.

All these features, registration, amplification, existence of several possible outcomes, thus require the apparatus to be a macroscopic object, whereas S is microscopic.

4. Measurement, a process of quantum statistical mechanics

The large size of A cannot be dealt with by any other means than statistical mechanics. Of course, we must treat S+A as a compound *quantum* system. The use of statistical mechanics compels us to regard our description of a measurement as relevant to a statistical ensemble of processes, not to an individual process. In particular, our equation Eq. 19 for the final state $\hat{\mathcal{D}}(t_{\mathrm{f}})$ defines expectation values, correlations, possibly with small fluctuations, and this feature is imposed both by quantum mechanics and by statistical mechanics for A. This is why we have described the above solution in terms of density operators, not of pure states. In particular, we saw that the initial, metastable state of A was prepared by controlling a macroscopic parameter, the temperature of B. It is thus represented in the quantum formalism by a *mixed state*, coded in its density operator $\hat{\mathcal{R}}(0)$. (It is impossible in an actual experiment to make a complete quantum preparation of a large object, and the assumption that A might lie initially in a pure state and end up in one among some pure states corresponds to a very unrealistic thought measurement — nevertheless this assumption is frequent in measurement theory, see e.g.[1,2]) Likewise, each of the final states $\hat{\mathcal{R}}_i$, characterized by the value of the pointer variable that will be observed, must again be described by means of a density operator $\hat{\mathcal{R}}_i$, and not by means of pure states as in the von Neumann-Wigner approach, too often followed in the literature. Indeed, the number of state vectors associated with a sharp value of the *macroscopic* pointer variable is huge for any actual measurement: As always for large systems, we must allow for small fluctuations, negligible in relative value, around the mean value $\pm m_{\mathrm{F}}$ of $\hat{m}$.

However, the evolution of A towards one among its possible final states $\hat{\mathcal{R}}_i$ is *triggered by interaction with* S, in a way depending on the initial microscopic state of S and, for an ideal measurement, the outcome should be correlated to the final microscopic state of S, a property exhibited by the form Eq. 19 of the final state $\hat{\mathcal{D}}(t_{\mathrm{f}})$ of S+A. Thus, contrary to theories of standard relaxation processes, the theory of a measurement process requires a simultaneous control of microscopic and macroscopic variables. In order to solve the coupled equations of motion for A and S which involve reduction and registration, we made use of a kind of coarse graining, which is adequate for A, becoming exact in the limit of a large A, but we had to treat the small system S exactly.

Moreover, the final state of S+A keeps *memory* of the initial state of S, at least partly. The very essence of a measurement lies in this feature, whereas memory effects are rarely considered in standard relaxation processes.

5. Irreversibility

A quantum measurement is irreversible for two reasons. On the one hand, the *loss of the off-diagonal blocks*, exhibited in the expression Eq. 19 of the final density of S+A, requires an irreversibility of the process. Even if the initial state were pure, a final state involving only diagonal projections would be a statistical mixture – this irreversibility is associated with the loss of specifically quantum correlations between $\hat{s}_x$ or $\hat{s}_y$ and A, embedded in the off-diagonal blocks. On the other hand, the registration by the apparatus requires an irreversible *relaxation* from the metastable paramagnetic state $\hat{\mathcal{R}}(0)$ towards the stable ferromagnetic states $\hat{\mathcal{R}}_{\Uparrow}$ or $\hat{\mathcal{R}}_{\Downarrow}$, as in the ordinary dynamics of a phase transition.

The irreversibility of the transformation leading from $\hat{\mathcal{D}}(0)$ to $\hat{\mathcal{D}}(t_{\rm f})$ is measured by the entropy balance. The von Neumann entropy of the initial state is split into contributions from S and A, respectively, as

$$S\left[\hat{\mathcal{D}}(0)\right] = -\mathrm{tr}\hat{\mathcal{D}}(0)\ln\hat{\mathcal{D}}(0) = S_{\rm S}\left[\hat{r}(0)\right] + S_{\rm A}\left[\hat{\mathcal{R}}(0)\right] , \tag{23}$$

whereas that of the final state Eq. 19 is

$$S\left[\hat{\mathcal{D}}(t_{\rm f})\right] = S_{\rm S}\left[\sum_i \hat{\Pi}_i \hat{r}(0)\hat{\Pi}_i\right] + \sum_i p_i S_{\rm A}\left[\hat{\mathcal{R}}_i\right] . \tag{24}$$

The increase of entropy from Eq. 23 to Eq. 24 clearly exhibits the two above-mentioned reasons. On the one hand, when the initial density operator $\hat{r}(0)$ involves off-diagonal blocks $\hat{\Pi}_i\hat{r}(0)\hat{\Pi}_j$ $(i \neq j)$, their truncation raises the entropy. On the other hand, a robust registration requires that the possible final states $\hat{\mathcal{R}}_i$ of A are more stable than the initial state $\hat{\mathcal{R}}(0)$, so that also their entropy is larger. The latter effect dominates because the apparatus is large.

An apparatus is a device which allows us to *gain some information* on the state of S by reading on A the outcome $+m_{\rm F}$ or $-m_{\rm F}$. The price we have to pay, for being thus able to determine the matrix elements $r_{\uparrow\uparrow}(0)$ and $r_{\downarrow\downarrow}(0)$ referring to $s_z = +1$ and -1, are a complete *loss of information* about the off-diagonal elements $r_{\uparrow\downarrow}(0)$ and $r_{\downarrow\uparrow}(0)$ of the initial state of S, and a rise in the thermodynamic entropy of the apparatus.

The solution of our model shows that the so-called "measurement problem", to wit, the fact that the final state Eq. 19 does not seem to be related unitarily to the initial state, has the same nature as the celebrated "paradox of irreversibility" in statistical mechanics,[23] with additional quantum features. Here too, it is the large size of the apparatus which produces destructive interferences, thus generating *inaccessible recurrence times*; such times behave as exponentials of the level density, which itself is an exponential of the number of degrees of freedom. As in the solution of the irreversibility paradox,[24] we witness here a *cascade* of correlations involving more and more spins of M, towards which the initial order embedded in the off-diagonal elements of $\hat{r}(0)$ escapes without any possibility of retrieval. Mathematically speaking, such correlations should be included in the final state, but the expression Eq. 19 is physically exact in the sense that many-spin correlations cannot be detected and have no observable implication.

6. Meaning of von Neumann's reduction and of Born's rule

The solution of our model shows that the disappearance of the off-diagonal elements of $\hat{\mathcal{D}}(t)$ is a *real dynamical phenomenon*, involving an irreversible process. Indeed, we found the collapsed final state Eq. 19 by merely working out the unitary Liouville-von Neumann equation, without any other extra ingredient than statistics.

It should be stressed that this disappearance concerns the overall system S+A, and not S or A separately: von Neumann's reduction is a property of the *compound system* S+A, which arises for an ideal measurement. In fact, if we take the trace over the system S, which means that we are no longer interested in S after the time $t_{\rm f}$ but only in the indications of the apparatus as in Eq. 20, off-diagonal blocks of $\hat{\mathcal{D}}(t)$, even if they were present, would drop out. Likewise, tracing over A would yield as marginal density matrix $\hat{r}(t)$ for S simply the diagonal elements of $\hat{r}(0)$, even if Eq. 19 had included off-diagonal blocks.

The elimination of the off-diagonal blocks, not only for the marginal density matrix $\hat{r}$ of S, but also from the overall density matrix $\hat{\mathcal{D}}$ of S+A, contrasts with what happens in usual decoherence processes (for a review on decoherence processes see[2,25]). There, a weak interaction of a system with its environment, which behaves as a thermal bath, destroys off-diagonal blocks in the density matrix of the system, but the back reaction of this system on its environment is usually not considered. Here, the reduction on the timescale (13) is the result of the interaction (3) of S with the pointer M, without intervention of the bath B, and the whole properties of S+A after this process are of interest, including the correlations $\langle \hat{s}_x \hat{m}^k \rangle$ and $\langle \hat{s}_y \hat{m}^k \rangle$ for $k \geq 1$, which, after increasing initially, vanish on the reduction timescale. Because the latter timescale is not related to the bath, it is misleading to regard reduction as a decoherence process.

Born's rule also involves both S and A. As exhibited in the expression Eq. 19 of the final state, it means that the outcome of the measurement, namely $+m_{\rm F}$ associated with $\hat{\mathcal{R}}_{\Uparrow}$, or $-m_{\rm F}$ associated with $\hat{\mathcal{R}}_{\Downarrow}$, is *fully correlated* in the ideal measurement that we consider with the final state $|\uparrow\rangle$ or $|\downarrow\rangle$ of S. We noted moreover that the frequency $p_{\Uparrow}$ of the occurrence of $+m_{\rm F}$ in repeated measurements, exhibited in Eq. 20, is equal to the number $r_{\uparrow\uparrow}(0) = {\rm Tr}\hat{\Pi}_{\uparrow}\hat{r}(0)$. For an ideal measurement it is also equal to the element $r_{\uparrow\uparrow}(t_f)$ of the marginal density operator of S. From the macroscopic character of $p_\Uparrow$ (i.e. being related to the possible values of the pointer), we can thus infer that $r_{\uparrow\uparrow}(0) = r_{\uparrow\uparrow}(t_f)$ can be interpreted as a probability for the microscopic system S to lie in the $+z$-direction in the final state. From $p_\Uparrow$ and $p_\Downarrow$ we also get through $r_{\uparrow\uparrow}(0)$ and $r_{\downarrow\downarrow}(0)$, related by Eq. 21, partial probabilistic information about the state which describes initially the considered statistical ensemble.

When a statistical ensemble of quantum systems is described by a pure state, any sub-ensemble is described by the same pure state. On the contrary, for a mixture, we can split the statistical ensemble into sub-ensembles described by different states, pure or not, provided we have collected the information needed to distinguish which sub-ensemble each system belongs to. This is precisely what happens in the final state Eq. 19 of our ideal measurement. The outcome registered by A provides us with the criterion required to sort the successive experiments into subsets labeled by $\uparrow$ or $\downarrow$. The ensemble for which S + A has the density operator Eq. 19, is thus split into sub-ensembles, in each of which S and A are decorrelated and S lies in the eigenstate $|\uparrow\rangle$ or $|\downarrow\rangle$ of $\hat{s}_z$. It is because the final state of S + A is a mixture, owing to the physical reduction, and because A relaxes towards either one of the states $\mathcal{R}_i$ without Schrödinger cats, that we can use an ideal measurement process as a *preparation* of S, initially in $\hat{r}(0)$, into a new state controlled by the filtering of the outcome of A.[26]

7. Relation to the pre-measurement

In a Stern-Gerlach experiment the pre-measurement stage is related to the region of space where the magnetic field is inhomogenous and the individual particles "decide" either to the upper or to the lower beam. After this, the detection by a remote detector is somewhat trivial if only the position and not the spin is measured. Likewise, in our model, one may look for a stage where it is determined that the tested particle ends up with its spin either up or down. Clearly, this should happen in the dephasing process, which is phase coherent and takes place on the shortest timescale relevant to the measurement, the reduction time $\tau_{\rm red}$. Here the single tested spin interacts with the $N \gg 1$ spins of the apparatus, while the bath is still ineffective.

To produce evidence for this connection with pre-measurement, let us decide to stop the process after the reduction stage, but before the bath sets in, so on a timescale $\tau_{\rm red} \ll \tau \ll \tau_{\rm irrev}$. Clearly, the measurement has not been performed since no registration has taken place. The apparatus is still in the paramagnetic state, but small multiparticle corelations have been developed with the tested spin. Removing the apparatus amounts to trace it out, thus neglecting these anyhow tiny correlations. However, for the tested spin we end up with the mixture Eq. 22, and no recurrence will subsequently occur if the coupling with the apparatus is removed.

Comparing with the initial state Eq. 12, it is seen that the Schrödinger cat terms have indeed already disappeared in this stage. This implies a physics different from the initial state. If, after this stop of the measurement, the spin is measured again along the z-axis with another apparatus, this will still end up as it was described above. But if in the second measurement the spin is measured along the x-axis, the outcome for $\langle \hat{s}_x \rangle$ will not be $r_{\uparrow\downarrow}(0) + r_{\downarrow\uparrow}(0)$, but just zero, and likewise for $\langle \hat{s}_y \rangle$. More precisely, we shall find for repeated measurements of s_x or s_y the values $+1$ or -1 with the same probability, in contrast with measurements on the initial state Eq. 12.

8. Statistical interpretation of quantum mechanics

The very concept of a physical quantity is related in quantum mechanics to the possibility of its measurement. Our model, aimed at understanding measurements as a quantum dynamical process, has compelled us to work in the framework of quantum statistical mechanics, using density operators rather than pure states. The process that we described thus refers to a *statistical ensemble* of measurements on a statistical ensemble of systems, not to a single measurement experiment.

It is often argued ("ignorance interpretation") that a statistical mixture $\hat{\mathcal{D}}$, characterizing only our knowledge of the system, should be interpreted as a collection of "underlying pure states" $|\varphi_k\rangle$ which would have more "physical reality" than $\hat{\mathcal{D}}$. Like a microstate of classical statistical mechanics, each $|\varphi_k\rangle$ would be associated with a particular realization of the ensemble; it would represent an individual system, occurring in the ensemble with a relative frequency q_k ("realist interpretation", for more on this, see e.g. [27]). The probabilities that appear through the pure states $|\varphi_k\rangle$ and through the weights q_k would have two different natures, the former, "purely quantal", being a property of the object, and the latter resulting from our lack of knowledge.

That this is a false idea was stressed by de Muynck.[27] On the one hand, contrary to what happens in classical statistical mechanics, the decomposition into a given mixture $\hat{\mathcal{D}}$ is in general not unique. For instance, it is impossible to distinguish whether the unpolarized state of a spin $\frac{1}{2}$ describes a population of spins pointing (with the same weight) in the $\pm z$-directions, or in the $\pm x$-directions, or isotropically in arbitrary directions. Thus no physical meaning can be given to pure states $|\varphi_k\rangle$ that would underlie $\hat{\mathcal{D}}$, since they cannot be defined unambiguously. On the other hand, a pure state has no more, no less "physical reality" than a mixture, since it is also just a mathematical tool which allows us to predict any expectation value for a statistical ensemble of systems, and to evaluate any probability .[28,29] Indeed, the non-commutativity of the observables which represent the physical quantities in quantum mechanics entails an *irreducibly probabilistic* nature of the theory. Within our quantum approach, we therefore refrain from imagining a more "fundamental" description which would underlie the statistical interpretation and would apply to individual systems.[a]

Anyhow, even if one wished to deal only with pure initial states, each one leading through a unitary transformation to a pure final state, all conclusions drawn from the form Eq. 19 of the final density operator would remain valid in a statistical sense. We are interested only in generic experiments; very unlikely events will never be observed, due to the huge value of the recurrence times. The situation is comparable to that of a classical gas: individual trajectories are reversible, and some of them may exhibit a pathological behavior. However, the consideration of the whole bunch of possible trajectories associated with the physical situation leads to statistical properties that agree with the more feasible theoretical analysis in the language of statistical mechanics – here of density operators.

Note finally that the lack of off-diagonal blocks in the expression Eq. 19 of the final state S+A allows us to use for this state a classical probabilistic description, with classical correlations. In the first stages of the process, the density operator $\hat{\mathcal{D}}(t)$ presents all the singular features of quantum mechanics that arise from the non-commutativity of the physical quantities. The dynamics of the large system S+A modifies, as usual in statistical mechanics, the qualitative properties, letting, for instance, irreversibility emerge from reversible microscopic equations of motion. Moreover, here, we witness the *emergence of standard probabilistic*, scalar-like correlations between S and A in the final state from the quantum description in which $\hat{\mathcal{D}}(t)$ behaves as an operator-like probability distribution describing a statistical ensemble.

Acknowledgement

The authors are grateful for discussion with Willem de Muynck.

References

1. J. A. Wheeler and W. H. Zurek, *Quantum Theory and Measurements* (Princeton University Press, 1983); P. Busch, P. J. Lahti and P. Mittelstaedt, The Quantum Theory of Measurements (Springer, Berlin, 1996).
2. M. Schlosshauer, *Rev. Mod. Phys.* **76** 1267 (2004).
3. K. Hepp, *Helvetica Physica Acta* **45**, 237 (1972).
4. M. Cini, *Nuovo Cimento B* **73**, 27 (1983).
5. B. Gaveau and L.S. Schulman, *J. Stat. Phys.* **58**, 1209 (1990).
6. L. S. Schulman, *Ann. Phys.* **212**, 315 (1991).
7. F. Haake and M. Zukowski, *Phys. Rev. A* **47**, 2506 (1993).
8. A. Venugopalan, *Phys. Rev. A* **61**, 012102 (2000).
9. D. Mozyrsky and V. Privman, *cond-mat/9903403.*
10. A. E. Allahverdyan, R. Balian and Th. M. Nieuwenhuizen, *Phys. Rev. A* **64**, 032108 (2001).
11. D. Braun, F. Haake, and W. T. Strunz, *Phys. Rev. Lett.* **86**, 2913 (2001).
12. G. Sewell, *math-ph/0505032.*

[a]Such a description may exist though, be it not at the quantum level, but Beyond the Quantum.

13. R. Merlin, *Europhys. Lett.* **76**, 541 (2006); *quant-ph/05072508*.
14. A. E. Allahverdyan, R. Balian and Th. M. Nieuwenhuizen, *Europhys. Lett.* **61**, 452 (2003).
15. A. E. Allahverdyan, R. Balian and Th. M. Nieuwenhuizen, *The quantum measurement process: an exactly solvable model*, Atti d. Fond. G. Ronchi, An. LVIII, N. 6, 719 (2003); *cond-mat/0309188*.
16. A. E. Allahverdyan, R. Balian and Th. M. Nieuwenhuizen, *The quantum measurement process in an exactly solvable model*, in AIP Conf. Proc. 750, ed. A. Khrennikov (Am. Inst. Phys., Melville, NY, 2005), pp. 26–34.
17. A. E. Allahverdyan, R. Balian and Th. M. Nieuwenhuizen, *Physica E* **29**, 261 (2005).
18. A. E. Allahverdyan, R. Balian and Th. M. Nieuwenhuizen, *Phase Transitions and Quantum Measurements*, *quant-ph/0508162*. In AIP Conf. Proc. 810: *Quantum Theory: Reconsideration of Foundations-3*, G. Adenier, A. Yu Khrennikov and Th. M. Nieuwenhuizen, eds, (Am. Inst. Phys., Melville, NY, 2006), pp 47-58.
19. For our quadratic interaction the paramagnet is unstable at low temperatures, and this is a drawback for our purposes, since the state will decay without coupling with the tested spin. More precisely, for $p = 2$ and $g = 0$ the instability time for the paramagnetic state can be estimated to be or order $\ln N$; if N is so large that this time is larger than the measurement time $\propto 1/\gamma$, then the model with $p = 2$ is still reasonable as an apparatus.
20. R. T. Cox, *Am. J. Phys.* **14**, 1 (1946).
21. B. de Finetti, *Theory of Probability* (Wiley, New York, 1990).
22. C. M. Caves, C. A. Fuchs, R. Schack, *J. Math. Phys.* **43** (2002) 4537.
23. R. Balian, *From Microphysics to Macrophysics*, in two volumes (Springer, Berlin, 1992).
24. J. E. Mayer and M. G. Mayer, *Statistical Mechanics*, second ed. (Wiley, New York, 1977), pp. 145–154; R. Balian, *Am. J. Phys.* **67**, 1078 (1999).
25. D. Giulini, E. Joos and C. Kiefer, *Decoherence and the Appearance of a Classical World in Quantum Theory* (Springer, Berlin, 1996); and references therein.
26. R. Balian and M. Vénéroni, *Ann. Phys.* **174**, 229 (1987).
27. W. M. de Muynck, *Foundations of Quantum Mechanics, an Empiricist Approach*, (Kluwer, Dordrecht, 2002).
28. N. G. van Kampen, *Physica A* **153**, 97 (1988).
29. R. Balian, *Am. J. Phys.* *57*, 1019 (1989).

PART B

Quantum Mechanics and Quantum Information

POVMs: A SMALL BUT IMPORTANT STEP BEYOND STANDARD QUANTUM MECHANICS

W. M. de MUYNCK

Theoretical Physics, Eindhoven University of Technology
Eindhoven, the Netherlands,
E-mail: w.m.d.muijnck@tue.nl

It is the purpose of the present contribution to demonstrate that the generalization of the concept of a quantum mechanical observable from the Hermitian operator of standard quantum mechanics to a positive operator-valued measure is not a peripheral issue, allegedly to be understood in terms of a trivial nonideality of practical measurement procedures, but that this generalization touches the very core of quantum mechanics, viz. complementarity and violation of the Bell inequalities.

Keywords: Positive operator-valued measure; Complementarity; Bell inequalities.

1. Introduction

I shall refer to the usual quantum mechanical formalism dealt with in quantum mechanics textbooks as the *standard* formalism. In this formalism a quantum mechanical (standard) observable is represented by a Hermitian operator, having a spectral representation consisting of projection operators E_m, which constitute an *orthogonal* $(TrE_mE_{m'} = O,\ m \neq m')$ decomposition of the identity operator,

$$E_m \geq O,\ E_m^2 = E_m,\ \sum_m E_m = I.$$

The projection operators E_m are said to generate a projection-valued measure (PVM). A *generalized* observable is represented by a positive operator-valued measure (POVM), generated by a set of non-negative operators M_m that in general are *not* projection operators and constitute a *non*-orthogonal decomposition of the identity operator, i. e., in general $TrM_mM_{m'} \neq O$, and

$$M_m \geq O,\ \sum_m M_m = I. \tag{1}$$

In generalized quantum mechanics measurement probabilities are given according to

$$p_m = Tr\rho M_m, \tag{2}$$

ρ a density operator, and the set $\{M_m\}$ satisfying (1). In the following generalized and standard observables will be referred to by their POVM and PVM, respectively.

Nowadays it is more and more realized that standard quantum mechanics is not completely adequate for dealing with quantum information, and that it is necessary to consider POVMs. Thus, as is well known, standard observables can only yield information on *diagonal* elements of ρ. Since in general the operators M_m of a POVM need not commute with each other, POVMs may yield information on *off-diagonal* elements of ρ, too. Particularly interesting is the existence of *complete* POVMs allowing to reconstruct the density operator from the set of probabilities (2) obtained in a measurement of one single *generalized* observable. By means of 'quantum tomography' state reconstruction can be achieved also in standard quantum mechanics; however, in general this method requires measurement of a large number of standard observables.

POVMs are obtained in a natural way when applying quantum mechanics to the interaction of object and measuring instrument. Thus, let $\rho^{(o)}$ and $\rho^{(a)}$ be the initial density operators of object and measuring instrument, respectively, and let $\rho^{(oa)}_{fin} = U\rho^{(o)} \otimes \rho^{(a)}U^\dagger$, $U = e^{-\frac{i}{\hbar}HT}$ be the final state of the interaction. If in this latter state a measurement is performed of the pointer observable $\{E^{(a)}_m\}$, then the measurement probabilities are found according to

$$p_m = Tr_{oa}\, \rho^{(oa)}_{fin} E^{(a)}_m = Tr_o \rho^{(o)} M_m, \tag{3}$$

with M_m given by

$$M_m = Tr_a \rho^{(a)} U^\dagger E^{(a)}_m U. \tag{4}$$

From expression (4) it follows that there is no reason to expect that M_m should be a projection operator; and in general it isn't.

The generalization of the mathematical formalism by means of the introduction of POVMs entails a considerable extension of the domain of application of quantum mechanics. In particular, it is possible that the set of operators $\{M_m\}$ of a POVM spans the whole of Hilbert-Schmidt space, in which case we have a *complete* measurement. Such *complete* measurements are experimentally feasible,[1–3] and may have considerable practical importance because of their richer informational content.

As can be seen from (3) and (4), it is the interaction of object and measuring instrument which is at the basis of the notion of a POVM. In quantum mechanics textbooks measurement is generally treated in an axiomatic way, and a detailed description of it is virtually absent. Bohr was one of the few to take the problem seriously, but he was mainly interested in the *macroscopic* phase of the measurement, which, according to him, was to be described by *classical* mechanics. However, not the macroscopic but rather the *microscopic* phase of the measurement, in which the microscopic information is transferred from the microscopic object to the measuring instrument, is crucial for obtaining *quantum* information. This phase should be described by quantum mechanics.

The influence of measurement has been of the utmost importance in the early days of quantum mechanics. In particular has it been a crucial feature at the inception of the notion of *complementarity*. In section 2 it will be shown that POVMs indeed play a crucial role there. By hindsight it can be concluded that much confusion could have been prevented if at that time it would have been realized that the standard formalism of quantum mechanics (as laid down, for instance, in von Neumann's authoritative book[4]) is just a preliminary step towards a more general formalism. As it is evident now, the standard formalism is not even able to yield a proper description of the so-called thought experiments, at that time being at the heart of our understanding of quantum mechanics.[2]

In this contribution the importance of the generalized formalism of POVMs is demonstrated by means of two examples, the first one elucidating Bohr's concept of complementarity in the sense of mutual disturbance of the measurement results of jointly measured incompatible standard observables (section 2). As a second example, in section 3 a *generalized* Aspect experiment is discussed using the POVM formalism, thus demonstrating that violation of the Bell inequalities can be seen as a consequence of complementarity rather than as being associated with nonlocality.

2. Complementarity

2.1. *The Summhammer, Rauch, Tuppinger experiment*

In this section I will now discuss as an example of the double-slit experiment a neutron interference experiment performed by Summhammer, Rauch and Tuppinger.[5] Other examples can be found in many different areas of experimental physics.[2] Let us first consider the two limiting cases which can be treated by means of standard

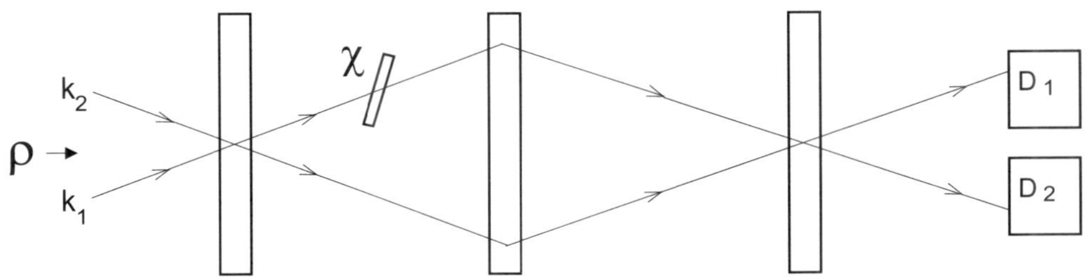

Fig. 1. Pure interference measurement.

quantum mechanics, viz. what I shall denote as the 'pure interference measurement' (cf. figure 1) and the 'pure path measurement' (cf. figure 2). In the figures a neutron interferometer is schematically represented by three vertical slabs in which Bragg reflection takes place and interference of the different paths can be realized after a possible phase shift χ has been applied in one of the paths. Detectors D_1 and D_2 can be placed either behind the third slab (figure 1), or behind the second one, in which case the experiment reduces to a 'which-path' measurement (in figure 2 this is realized by putting an ideal absorber in one path, while adding the measurement

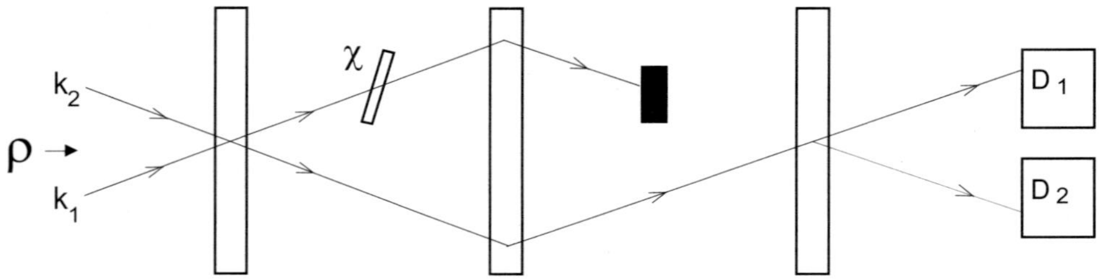

Fig. 2. Pure path measurement.

frequencies of detectors D_1 and D_2 to obtain the probability p_+ the neutron was in the second path; $p_- = 1 - p_+$ is the probability that the neutron was in the path of the ideal absorber).

The observables measured in the measurement arrangements of figures 1 and 2 are standard observables which can easily be found on the basis of elementary considerations.[6] With ρ the initial (incoming) state we find:
pure interference measurement: $p_n = Tr\rho Q_n$, $n = 1, 2$, $\{Q_1, Q_2\}$ the PVM of the standard interference observable;
pure path measurement: $p_m = Tr\rho P_m$, $m = +, -$, $\{P_+, P_-\}$ the PVM of the standard path observable.
It will not be necessary to display these observables explicitly; it is sufficient to know that the operators Q_n and P_m are projection operators, defining PVMs of standard quantum mechanics.

In the experiment performed by Summhammer, Rauch and Tuppinger[5] an absorber (transmissivity a) is inserted in one of the paths (figure 3). In the limits $a = 1$

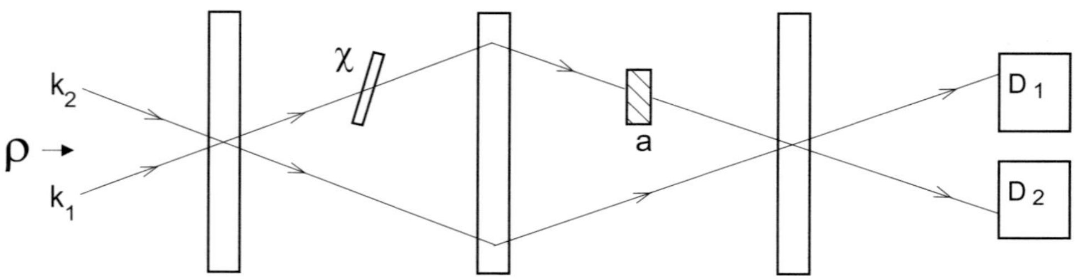

Fig. 3. Neutron interference measurement by Summhammer, Rauch and Tuppinger.

and $a = 0$ the Summhammer, Rauch, Tuppinger experiment reduces to the pure interference and the pure path measurement, respectively. The interesting point is that in between these limits the experiment is no longer described by a PVM, but by a POVM. Indeed, we find for $0 \leq a \leq 1$: $p_i = Tr\rho M_i, i = 1, 2, 3$, in which $i = 3$ refers to those events in which the neutron is absorbed by the absorber. It is easily seen[6] that the operators M_i are given according to

$$\begin{cases} M_1 = \frac{1}{2}[P_+ + aP_- + \sqrt{a}(Q_1 - Q_2)], \\ M_2 = \frac{1}{2}[P_+ + aP_- - \sqrt{a}(Q_1 - Q_2)], \\ M_3 = (1-a)P_-. \end{cases} \tag{5}$$

In the following way the measurement represented by the POVM $\{M_1, M_2, M_3\}$, M_i given by (5), can be interpreted as a joint nonideal measurement of the interference and path observables defined above. Define the bivariate POVM (R_{mn}) as follows:

$$(R_{mn}) := \begin{pmatrix} M_1 & M_2 \\ \frac{1}{2}M_3 & \frac{1}{2}M_3 \end{pmatrix}. \tag{6}$$

Then the two marginals, $\{\sum_m R_{mn}, n = 1, 2\}$ and $\{\sum_n R_{mn}, m = +, -\}$, are easily found. It directly follows from (5) that these marginals are related to the PVMs $\{Q_1, Q_2\}$ and $\{P_+, P_-\}$, respectively, according to

$$\begin{pmatrix} \sum_m R_{m1} \\ \sum_m R_{m2} \end{pmatrix} = \frac{1}{2} \begin{pmatrix} 1+\sqrt{a} & 1-\sqrt{a} \\ 1-\sqrt{a} & 1+\sqrt{a} \end{pmatrix} \begin{pmatrix} Q_1 \\ Q_2 \end{pmatrix}, \tag{7}$$

$$\begin{pmatrix} \sum_n R_{+n} \\ \sum_n R_{-n} \end{pmatrix} = \begin{pmatrix} 1 & a \\ 0 & 1-a \end{pmatrix} \begin{pmatrix} P_+ \\ P_- \end{pmatrix}. \tag{8}$$

The important feature of (7) and (8) is that one marginal contains information only on the standard interference observable, whereas the other marginal only refers to the standard path observable. Actually, the bivariate POVM (6) was construed so as to realize this.

Equations (7) and (8) are applications of a general definition of a *nonideal* measurement,[7] to the effect that a POVM $\{M_i\}$ is said to represent a nonideal measurement of POVM $\{N_j\}$ if

$$M_i = \sum_j \lambda_{ij} N_j, \; \lambda_{ij} \geq 0, \; \sum_i \lambda_{ij} = 1.$$

This expression compares the measurement procedures of POVMs $\{M_i\}$ and $\{N_j\}$, to the effect that the first can be interpreted as a nonideal or inaccurate version of the second, the nonideality matrix (λ_{ij}) representing the nonideality. A convenient measure of this nonideality is the average row entropy of the nonideality matrix,

$$J_{(\lambda)} := -\frac{1}{N} \sum_{ij} \lambda_{ij} \ln \frac{\lambda_{ij}}{\sum_{j'} \lambda_{ij'}}, \; N \text{ the dimension of matrix } (\lambda_{ij}). \tag{9}$$

As is seen from (7) and (8) the measurement procedure depicted in figure 3 gives rise to two nonideality matrices, to be denoted by $(\lambda_{mm'})$ and $(\mu_{nn'})$, respectively. Under variation of the parameter a the nonideality matrices $(\lambda_{mm'})$ and $(\mu_{nn'})$ are seen to exhibit a behavior that is very reminiscent of the idea of complementarity as presented for the first time by Bohr in his Como lecture:[8] in one limit ($a = 1$) ideal information is obtained on the standard interference observable, whereas no information at all is obtained on the standard path observable; in the other limit ($a = 0$) the roles of the standard interference and path observables are interchanged. For values $0 < a < 1$ we have intermediate situations in which information is obtained on the probability distributions of *both* standard observables, to the effect that information on one observable gets less accurate as the information on the other one gets more ideal. By changing the measurement arrangement so as to also obtain information on another (incompatible) standard observable, the information on the first observable gets blurred to a certain extent.

2.2. *The Martens inequality*

Complementary behavior as discussed above is a rather common feature of quantum mechanical measurement; many other examples can be given.[2] Using the nonideality measure $J_{(\lambda)}$ (9) it is possible to give a general account of this complementarity.[7] Let a bivariate POVM (R_{mn}) satisfy

$$\begin{array}{l}\sum_n R_{mn} = \sum_{m'} \lambda_{mm'} P_{m'},\ \lambda_{mm'} \geq 0,\ \sum_m \lambda_{mm'} = 1,\\ \sum_m R_{mn} = \sum_{n'} \mu_{nn'} Q_{n'},\ \mu_{nn'} \geq 0,\ \sum_n \mu_{nn'} = 1,\end{array} \tag{10}$$

in which $\{P_m\}$ and $\{Q_n\}$ are maximal PVMs. Then the corresponding nonideality

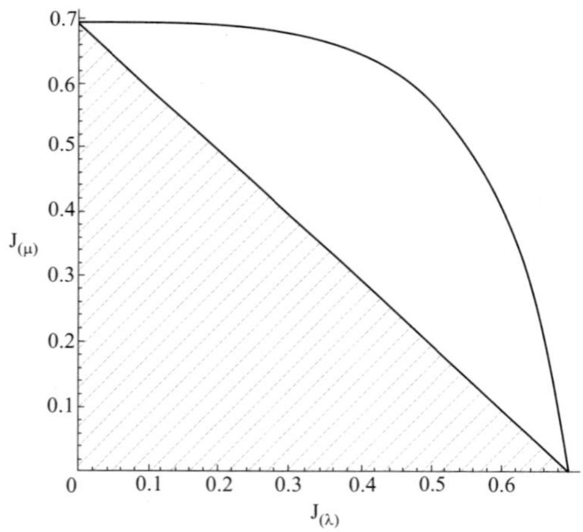

Fig. 4. Parametric plot of $J_{(\lambda)}$ and $J_{(\mu)}$ for the Summhammer, Rauch,Tuppinger experiment.

measures $J_{(\lambda)}$ and $J_{(\mu)}$ satisfy the *Martens inequality*[a]

$$J_{(\lambda)} + J_{(\mu)} \geq -\ln\{\max_{mn} Tr P_m Q_n\}.$$

[a]For nonmaximal PVMs this expression has to be slightly generalized.[7]

For the Summhammer, Rauch,Tuppinger experiment we obtain

$$J_{(\lambda)} = \tfrac{1}{2}[(1+a)\ln(1+a) - a\ln a],$$
$$J_{(\mu)} = \tfrac{1}{2}[2\ln 2 - (1+\sqrt{a})\ln(1+\sqrt{a}) - (1-\sqrt{a})\ln(1-\sqrt{a})].$$

In figure 4 a parametric plot is given of these quantities. The shaded area contains values of $J_{(\lambda)}$ and $J_{(\mu)}$ forbidden by the Martens inequality. This latter inequality, hence, does represent the Bohr-Heisenberg idea of complementarity in the sense of a mutual disturbance of the information obtained in a joint nonideal measurement of two standard observables.

2.3. *Martens inequality versus Heisenberg uncertainty relation*

It should be stressed that the Martens inequality is a general feature of quantum mechanical *measurement procedures* satisfying (10). In particular it is independent of the density operator ρ. This feature distinguishes the Martens inequality from the Heisenberg uncerainty relation

$$\Delta A \Delta B \geq \frac{1}{2}|Tr\rho[A, B]_-|, \tag{11}$$

A and B standard observables. It is interesting to remember that for a long time it has been the Heisenberg uncerainty relation (11) that was supposed to describe complementarity in the sense of mutual disturbance in a joint nonideal measurement of two standard observables. This has been questioned by Ballentine[9] to the effect that the Heisenberg inequalities (11) do not refer to *joint* measurement of incompatible observables at all, since they can be tested by *separate ideal* measurements of the two standard observables in question. According to Ballentine even "Quantum mechanics has nothing to say about joint measurement of incompatible observables."

As far as *standard* quantum mechanics is concerned, Ballentine is certainly right. However, as is seen from section 2.2, the *generalized* quantum mechanics of POVMs is able to deal with joint nonideal measurement of incompatible observables. The Martens inequality, rather than the Heisenberg inequality, is representing the concomitant complementarity. Although Bohr and Heisenberg had a perfect intuition as regards the physics going on in a double-slit experiment, they were not able to give a comprehensive treatment of it, due to the fact that they did not have at their disposal the generalized formalism of POVMs. As a consequence they were restricted to a discussion of the limiting cases only (in our example $a = 0$ and $a = 1$). They unjustifiedly thought[10] that the Heisenberg inequality (11) was the mathematical expression of their intuition on the intermediate cases (in our example corresponding to $0 < a < 1$). However, rather than the Heisenberg inequality it is the Martens inequality, derived from the generalized formalism, which serves this purpose. It seems that, due to a too one-sided preoccupation with measurement, Bohr and Heisenberg overlooked the possibility that not only *measurement* but also *preparation* might yield a contribution to complementarity, the Heisenberg uncertainty relations referring to the latter contribution.

3. Bell Inequalities

As is well known, the Bell inequalities cannot be derived from *standard* quantum mechanics; they were derived by Bell[11] from a local hidden-variables theory. As a consequence, it is generally believed that violation of the Bell inequalities by the standard Aspect experiments[12,13] is a consequence of *nonlocality*. In this section it will be demonstrated that our understanding of the Bell inequalities, too, can considerably be enhanced by applying the generalized formalism.[2]

3.1. *Generalized Aspect experiment*

For this purpose the following experiment is considered, to be referred to as the *generalized Aspect experiment* (cf. figure 5). In the experiment each photon of a

Fig. 5. Generalized Aspect experiment.

correlated photon pair impinges on a semi-transparent mirror (transmissivities γ_1 and γ_2, respectively). In the paths of the transmitted and reflected photon beams of photon i ($i = 1, 2$) polarization measurements are performed in directions θ_i and θ'_i, respectively. Using ideal detectors D_1, D'_1, D_2 and D'_2, a measurement result (occurrence or nonoccurrence of a click in each of the four detectors) is obtained for each individual photon pair. The four (standard) Aspect experiments[12,13] are special cases of the present experiment, satisfying $(\gamma_1, \gamma_2) = (1, 1)$, $(1, 0)$, $(0, 1)$ or $(0, 0)$, respectively (in each of these experiments the two detectors which are certain *not* to click have been omitted).

The generalized Aspect experiment can be analyzed, analogously to the discussion in section 2, in terms of complementarity in the sense of mutual disturbance in a joint nonideal measurement of incompatible standard observables. Let us first consider the measurement performed in one arm of the interferometer ($i = 1$ or 2). In agreement with definition (10) this measurement can be interpreted as a joint nonideal measurement of the (standard) polarization observables (PVMs) $\{E_+^{\theta_i}, E_-^{\theta_i}\}$ and $\{E_+^{\theta'_i}, E_-^{\theta'_i}\}$ in directions θ_i and θ'_i, respectively. Thus, by expressing, for a given i, the joint detection probabilities of photon detectors D_i and D'_i as $p_{m_in_i} = Tr\rho R^{\gamma_i}_{m_in_i}$, the bivariate POVM $(R^{\gamma_i}_{m_in_i})$ is defined according to

$$(R^{\gamma_i}_{m_in_i}) = \begin{pmatrix} O & \gamma_i E^{\theta_i}_+ \\ (1-\gamma_i)E^{\theta'_i}_+ & \gamma_i E^{\theta_i}_- + (1-\gamma_i)E^{\theta'_i}_- \end{pmatrix}, \; i = 1, 2. \tag{12}$$

The marginals of $(R^{\gamma_i}_{m_in_i})$ are found as

$$\begin{pmatrix} \sum_{n_i} R^{\gamma_i}_{+n_i} \\ \sum_{n_i} R^{\gamma_i}_{-n_i} \end{pmatrix} = \begin{pmatrix} \gamma_i & 0 \\ 1-\gamma_i & 1 \end{pmatrix} \begin{pmatrix} E^{\theta_i}_+ \\ E^{\theta_i}_- \end{pmatrix} \text{ (detector } D_i), \tag{13}$$

$$\begin{pmatrix} \sum_{m_i} R^{\gamma_i}_{m_i+} \\ \sum_{m_i} R^{\gamma_i}_{m_i-} \end{pmatrix} = \begin{pmatrix} 1-\gamma_i & 0 \\ \gamma_i & 1 \end{pmatrix} \begin{pmatrix} E^{\theta'_i}_+ \\ E^{\theta'_i}_- \end{pmatrix} \text{ (detector } D'_i). \tag{14}$$

As functions of γ_i, $0 \leq \gamma_i \leq 1$ the nonideality matrices in (13) and (14) show complementary behavior completely analogous to that of (7) and (8), and illustrated by figure 4. From this complementarity it can be concluded that, for instance, a measurement result for the polarization in direction θ_i, obtained in the generalized Aspect experiment ($0 < \gamma_i < 1$), can be different from one obtained if an *ideal* measurement of this standard observable ($\gamma_i = 1$) would have been performed. Moreover, it follows that this difference is a consequence of changing the measurement arrangement from $\gamma_i = 1$ to $\gamma_i < 1$, or vice versa. It is also seen that for $\gamma_i = 1$ the marginal (14) is given by $\{O, I\}$, which is a completely uninformative observable, not yielding any information on the state of photon i. This justifies the omission, referred to above, of the corresponding detector in the standard Aspect experiment.

The generalized Aspect experiment depicted in figure 5 can be interpreted in an analogous way as a joint nonideal measurement of the *four* standard observables $\{E^{\theta_1}_+, E^{\theta_1}_-\}$, $\{E^{\theta'_1}_+, E^{\theta'_1}_-\}$, $\{E^{\theta_2}_+, E^{\theta_2}_-\}$, and $\{E^{\theta'_2}_+, E^{\theta'_2}_-\}$, a quadrivariate POVM being obtained as the direct product of the bivariate POVMs given in (12):

$$R^{\gamma_1\gamma_2}_{m_1n_1m_2n_2} = R^{\gamma_1}_{m_1n_1} R^{\gamma_2}_{m_2n_2}. \tag{15}$$

From this expression it is evident that there is no disturbing influence on the marginals in one arm of the interferometer by changing the measurement arrangement in the other arm. Since observables referring to different objects do commute with each other, this is not unexpected. Complementarity in the sense of mutual disturbance of measurement results is effective in both of the arms *separately*, disturbance being caused in each arm by changing the measurement arrangement in that very arm.

3.2. *Complementarity and nonlocality as alternative explanations of violation of the Bell inequalities*

The interesting outcome of the present discussion of the generalized Aspect experiment is the existence of a *quadrivariate* probability distribution

$$p^{\gamma_1\gamma_2}_{m_1n_1m_2n_2} = Tr\rho R^{\gamma_1}_{m_1n_1} R^{\gamma_2}_{m_2n_2} \tag{16}$$

for the experimentally obtained measurement results. According to a theorem proven by Fine,[14] and by Rastall,[15] the existence of this quadrivariate probability distribution implies that the Bell inequalities are satisfied by the four bivariate marginals $p^{\gamma_1\gamma_2}_{m_1m_2}$, $p^{\gamma_1\gamma_2}_{m_1n_2}$, $p^{\gamma_1\gamma_2}_{n_1m_2}$ and $p^{\gamma_1\gamma_2}_{n_1n_2}$ which can be derived from (16) for fixed (γ_1, γ_2).

It should be noted that this holds true *also* for each of the *standard* Aspect experiments, corresponding to one of the limiting cases $(\gamma_1, \gamma_2) = (1, 1)$, etc.. Evidently, violation of the Bell inequalities by these experiments must be caused by the fact that no quadrivariate probability distribution exists from which the four bivariate probabilities $p^{\gamma_1=1\gamma_2=1}_{n_1n_2}$, $p^{\gamma_1=1\gamma_2=0}_{n_1m_2}$, $p^{\gamma_1=0\gamma_2=1}_{m_1n_2}$, and $p^{\gamma_1=0\gamma_2=0}_{m_1m_2}$ can be derived as marginals. So, the question to be answered is, why such a quadrivariate probability distribution does not exist, even though for each of the standard Aspect experiments separately one is given by (16).

A step towards answering this question is the observation that the quadrivariate probability distributions $p^{\gamma_1\gamma_2}_{m_1n_1m_2n_2}$ given by (16) are different for different (γ_1, γ_2): they depend on the measurement arrangement, and so, in general, do their marginals. Hence, changing the measurement arrangement from one standard Aspect experiment to another yields a disturbance of the measurement probabilities, preventing the Bell inequalities from being derivable from the existence of a single quadrivariate probability distribution.

In accepting this explanation we may choose between two different disturbing mechanisms, viz. nonlocality or complementarity. In the first case it is assumed that the probabilities in one arm of the interferometer are influenced in a nonlocal way by changing the measurement arrangement in the *other* arm. This is the explanation that is generally accepted. The alternative explanation, based on complementarity, takes into account the disturbing influence of a change of the measurement arrangement performed in an arm of the interferometer on the measurement probabilities measured in that *same* arm, as expressed by (13) and (14).

In deciding which of the alternatives, nonlocality or complementarity, to accept, it is important to realize that, if four standard observables (PVMs) $\{E^1_i\}$, $\{F^1_j\}$, $\{E^2_k\}$, and $\{F^2_\ell\}$ are mutually compatible, a quadrivariate probability distribution, viz. $Tr\rho E^1_i F^1_j E^2_k F^2_\ell$, exists even in the standard formalism. Hence, *in*compatibility is a necessary condition for violation of the Bell inequalities. But, since observables referring to causally disjoint regions of space-time do commute, only observables referring to the *same* region can be incompatible. Hence, *in*compatibility is a *local* affair, as, consequently, is violation of the Bell inequalities. It seems that complementarity can yield a *local* explanation of violation of the Bell inequalities, based on mutual disturbance in the joint nonideal measurements of incompatible standard observables carried out separately in each arm of the interferometer. Such an explanation could not be given on the basis of the standard formalism since, as demonstrated in section 2, that formalism is not able to deal with this kind of complementarity. Dependence on the measurement arrangement is only evident when considering the bivariate probabilities $Tr\rho R^{\gamma_i}_{m_in_i}$, derived from (12), which do not exist in the standard formalism.

References

1. W. M. de Muynck, *Journ. Phys. A: Math. Gen.* **31**, 431 (1998).
2. Willem M. de Muynck, *Foundations of quantum mechanics, an empiricist approach, Fundamental theories of physics* **127**, (Kluwer Academic Publishers, Dordrecht, Boston, London, 2002).
3. A. E. Allahverdyan, R. Balian, and Th. M. Nieuwenhuizen, *Phys. Rev. Lett.* **92**, 120402 (2004).
4. J. von Neumann, *Mathematische Grundlagen der Quantenmechanik*, (Springer, Berlin, 1932); or, *Mathematical foundations of quantum mechanics*, (Princeton Univ. Press, 1955).
5. J. Summhammer, H. Rauch, and D. Tuppinger, *Phys. Rev. A* **36**, 4447 (1987).
6. W. M. de Muynck and H. Martens, *Phys. Rev. A* **42**, 5079 (1990).
7. H. Martens and W. de Muynck, *Found. of Phys.* **20**, 255, 357 (1990).
8. N. Bohr, Como Lecture, *The quantum postulate and the recent development of atomic theory* in: N. Bohr, *Collected Works*, (J. Kalckar, ed. North–Holland, Amsterdam, 1985), Vol. 6, pp. 113–136.
9. L. E. Ballentine, *Rev. Mod. Phys.* **42**, 358 (1970).
10. W. M. de Muynck, *Found. of Phys.* **30**, 205 (2000).
11. J. S. Bell, *Physics* **1**, 195 (1964).
12. A. Aspect, P. Grangier, and G. Roger, *Phys. Rev. Lett* **47**, 460 (1981).
13. A. Aspect, J. Dalibard, and G. Roger, *Phys. Rev. Lett.* **49**, 1804 (1982).
14. A. Fine, *Journ. Math. Phys.* **23**, 1306 (1982); *Phys. Rev. Lett.* **48**, 291 (1982).
15. P. Rastall, *Found. of Phys.* **13**, 555 (1983).

STATE REDUCTION BY MEASUREMENTS WITH A NULL RESULT

G. NIENHUIS

Huygens Laboratorium, Universiteit Leiden,
Postbus 9504, 2300 RA Leiden, The Netherlands
E-mail: nienhuis@molphys.leidenuniv.nl

We review two topics of quantum optics that shed new light on the effect of state reduction by a quantum measurement. One topic is the observation of quantum jumps switching on and off the fluorescence of a trapped atomic ion. The other one is the spontaneous decay of a single atom, described by the method of quantum trajectories. This method is based on the decomposition of the density matrix of an open system into an ensemble of time-dependent pure state vectors. Here we consider single histories of the spontaneously emitting atom. It is shown that in both cases the evolution is affected by a detection with a null result.

Keywords: Null measurement; State reduction; Quantum jumps; Quantum trajectories.

1. Introduction

The probabilistic nature of quantum mechanics implies that identical systems that are described by identical state vectors can produce different measurement outcomes. This raises the natural question: what is the cause of these differences? Interpretations of quantum mechanics mainly differ in their answer to this question. Here we distinguish only two possible answers, which gives a rough but broadly accepted first classification in the possible points of view.

The modest and wise answer to the question for the unknown causes of differences in measurement results is simply: *These differences are caused by something unknown.* This reflects the view of some of the greatest among the founders of quantum mechanics, such as Schrödinger, Einstein, De Broglie and Bohm. It regards the probabilistic nature of quantum mechanics as arising from uncertainty about the actual state of the system that is described by a state vector. The differences in measurement results are viewed as reflecting hidden differences in two systems with the same state vector. The existence of unknown causes for the quantum indeterminacy that is postulated here would imply that quantum mechanics is incomplete, which leaves room for the exploration of various kinds of hidden variables. This view allows one to maintain that the outcome of a measurement reflects a property that the system already had before the measurement. In this sense, this view may be termed realistic.

The main alternative answer to the question as to the causes of different measurement outcomes is best summarized by the reply: *There is nothing that causes these differences.* This reflects the view of Bohr and Heisenberg, and forms the basis of what is now termed the Copenhagen interpretation of quantum mechanics. It restores the completeness of quantum mechanics, but the price is non-negligible. When one holds this view, one can no longer insist that the outcome of a measurement reflects a property that the system had before and independent of the measurement. After all, the same system having the same properties could have produced a different outcome. Therefore the measurement result is fundamentally undetermined prior to the measurement. According to this interpretation, the probabilistic nature of quantum mechanics does not arise from a lack of knowledge. Uncertainty is replaced by indeterminacy. Presently, this view of quantum mechanics is the most widely accepted among physicists, even though differences remain in the precise wording.

The central problem in these considerations is the role of the measurement. In textbook quantum mechanics, it is normally stated that an observable quantity corresponds to a Hermitian operator $\hat{Q}$ on state space, with eigenvectors $|\phi_n\rangle$ and eigenvalues q_n, so that $\hat{Q}|\phi_n\rangle = q_n|\phi_n\rangle$. When the state of the system is described by the normalized state vector $|\psi\rangle$, a measurement produces the eigenvalue q_n with probability $p_n = |\langle\psi|\phi_n\rangle|^2$, provided that the eigenstates $|\phi_n\rangle$ are normalized. After a measurement with this outcome, the state of the system is described by the eigenvector $|\phi_n\rangle$, which replaces the original state $|\psi\rangle$. Here it is tacitly assumed that a measurement is instantaneous, and has no time duration.

In the present paper we consider two simple cases of detection of a photon emitted by an atom where the state of the atom changes when no photon is observed. The first case is the intermittent fluorescence of a strong atomic transition, where one of the states of the transition is weakly coupled to another non-emitting state. This has been introduced as a prototype case for the macroscopic observability of quantum jumps.[1–3] Such a jump corresponds to the switch-off or the switch-on of the fluorescence, and may be viewed as a transition to or from the non-emitting state. The second case is a single history of an atom that is brought in a superposition state of the ground state and a decaying excited state. This is probably the simplest case of a quantum trajectory, which is a single pure-state realization of the history of an open quantum system.[4] In this case, the the trajectories contain a discontinuous jump to the ground state, that is registered by the detection of an emitted photon. A common feature of both cases is that a time-dependent state-vector projection results from the non-detection of a photon.

2. Macroscopic Fluorescence Switching

2.1. *Evolution of density matrix*

Macroscopically observable quantum jumps can arise in three-state systems with one weak and one strong transition. A simple example is a V configuration, sketched in Fig. 1. The ground state g is coupled to two excited states e and s by driving

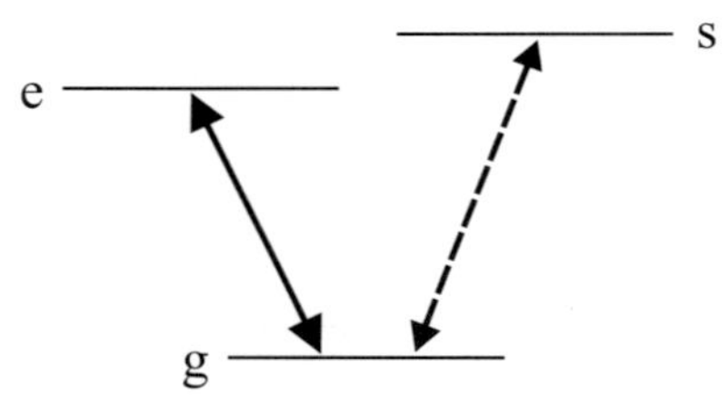

Fig. 1. Three-level scheme, with strong transition (solid line) and weak transition (dashed line).

fields, and spontaneous emission from these states produces fluorescence radiation. The coupling between states g and e is strong, and the natural lifetime of state e is of the order of nanoseconds. The fluorescence from this transition is directly observable. However, the transition between the states g and s is weak, and the state s is metastable, with a macroscopic lifetime, of the order of seconds, say. This can be the case when the transition $g \rightarrow s$ is forbidden. Emission by state s is rare, and the fluorescent photons contribute negligibly to the fluorescence intensity. Then what is the effect of the presence of this weakly coupled state s on the fluorescence on the strong transition?

The answer must be based on the behavior of the density matrix $\bar{\rho}$ of the three-state atom. In the absence of the coupling to state s, we have effectively a two-state atom, and when the driving field is sufficiently strong, so that the transition is saturated, the states g and e have the same population of $\bar{\rho}_{gg} \approx \bar{\rho}_{ee} \approx 1/2$. The fluorescent intensity expressed in the number of photons per unit time is then $\Gamma_e \bar{\rho}_{ee} \approx \Gamma_e/2$, with Γ_e the spontaneous decay rate of state e. When the coupling to state s is switched on, and when this transition is also saturated, in the steady state the populations of all three states will be the same, so that they are about equal to $1/3$. Notice that the weakness of the transition concerns both spontaneous and stimulated transitions. For a weak transition the time needed to reach the steady state is long, but the ratio between the steady-state populations $\bar{\rho}_{gg}$ and $\bar{\rho}_{ss}$ is no different than for a strong transition driven by light with the same intensity. In fact, the ratio between spontaneous and stimulated transitions does not depend on the strength of the atomic transition matrix elements, but only on the intensity of the driving field, according to Einstein's discussion of the radiative transition coefficients. The situation is characterized by a separation of time scales, a short time scale, of the order of Γ_e^{-1}, and a long time scale, on the order of Γ_s^{-1}. When both driving fields are switched on simultaneously, on the short time scale the weak coupling between the states g and s is not yet effective, and the population ρ_{ee} of state e rapidly attains the value $1/2$. The buildup of the state s only occurs on the long time scale. Its population ρ_{ss} grows from zero to its steady-state value $1/3$, while the population ρ_{ee} of state e decreases from $1/2$ to $1/3$.

Since we are accustomed to identify the fluorescence intensity to the product of a spontaneous decay rate Γ and a population, this suggests that the effect of the weakly coupled state s is simply that the fluorescent intensity would be $\Gamma_e/3$, which is less than the value $\Gamma_e/2$ in the absence of the coupling to state s. One would therefore expect that the weakly-coupled state reveals itself by a weaker intensity of the steady-state fluorescence. In contrast, Cook and Kimble[5] argue that a transition to the state s switches off the fluorescence on the transition $e \to g$, and that (spontaneous or stimulated) decay of the state s switches the fluorescence on again. The result is a fluorescence signal that is randomly interrupted by dark periods, in analogy to a random telegraph signal. This raises the question what is wrong with the suggestion mentioned above. The main point is that the density-matrix description gives an average of many pure-state histories of the system. In most cases the difference between the average of a signal as calculated by using the density matrix and an actual history of the signal is small. When fluctuations of the signal are large, this is no longer true, and then we need different techniques to predict the nature of single histories.

2.2. *Intensity correlation and waiting-time distribution*

A fundamental quantity in the theoretical description is the intensity distribution $f(t)$, defined as the time-dependent photon emission rate following an earlier emission at time 0. It is given by

$$f(t) = \Gamma_e \rho_{ee}(t) \, , \tag{1}$$

in terms of the matrix element for the state e of the time-dependent density matrix $\rho(t)$ for the three-level system, with the initial condition that at time zero a fluorescence photon was detected. This implies that the atom was in the ground state g, so that $\rho(0) = |g\rangle\langle g|$. The function $f(t)$ can be evaluated by solving the optical Bloch equations, which are the evolution equations for the density matrix of the three-state system. For complete saturation, the function $f(t)$ approaches the value $\Gamma_1/2$ on the short time scale, and then it decays slowly to the lower value $\Gamma_1/3$.

It requires some arguing and intuition to conclude from this that the fluorescence displays dark periods.[6,7] It is more convincing to use an alternative time-dependent function $w(t)$, which is defined as the probability distribution of the time intervals one has to wait for the first next photon emission after an earlier emission. This waiting-time distribution can be evaluated by starting from a simplified version of the optical Bloch equations, where the gain term of the ground-state population by spontaneous emission from the upper states is omitted.[8,9] This gain term corresponds to the creation of a fluorescent photon, so that the resulting simplified density matrix is subjected to the condition that no photon emission took place since the initial time g. The waiting-time distribution $w(t)$ is just the trace of the conditional density matrix. According to its definition, $w(t)$ is normalized, in the sense that

$$\int_0^\infty dt\ w(t) = 1\ . \tag{2}$$

We may view $f(t)$ as the probability density to have a photon emission at time t. Since this emission can be the first, or the second, ..., we can express the relation between f and w in the form of the Dyson equation

$$f(t) = w(t) + \int_0^t dt'\ f(t')w(t-t')\ . \tag{3}$$

The first term on the right-hand side corresponds to the history where the emission at time t was the first one since time zero. The second term describes the histories with emissions in between, the last one at time t'. This relation (3) can be converted into an algebraic relation

$$\hat{f}(v) = \hat{w}(v) + \hat{f}(v)\hat{w}(v) \tag{4}$$

for the Laplace transforms, which are defined in the standard way by

$$\hat{f}(v) = \int_0^\infty dt\ f(t)e^{-vt}\ ,\ \hat{w}(v) = \int_0^\infty dt\ w(t)e^{-vt}\ . \tag{5}$$

This relation can be used to show that the slow decrease of $f(t)$ on the long time scale gives rise to a weak long-time tail on the waiting-time distribution $w(t)$.[9] This tail has an exponential behavior characterized by the slow rate $R_{\rm on}$, which is the effective switching rate from the weakly coupled state s to the strongly coupled pair of states g and e. For short times the behavior of $w(t) \approx w_{\rm short}(t)$ is indistinguishable from the case of the system consisting only of the strongly coupled states g and e. The long-time behavior of $w(t)$ has the form[9]

$$w_{\rm long}(t) = \frac{R_{\rm off}R_{\rm on}}{\Gamma_e\bar{\rho}_{ee}} \exp(-R_{\rm on}t)\ , \tag{6}$$

with $R_{\rm off}$ the effective switching rate from the strongly coupled pair of states to the weakly coupled state s.

2.3. *Single histories*

The interpretation of this expression (6) for the long-time tail is now obvious. After each photon emission the next photon arrives in most cases within a short time, determined by the two-state waiting-time distribution $w_{\rm short}(t)$. The average waiting time for such a rapid successor is $(\Gamma_e\bar{\rho}_{ee})^{-1}$. However, there is a small probability $R_{\rm off}/(\Gamma_e\bar{\rho}_{ee})$ that the atom jumps to the state s, so that one has to wait a long time, on average $(R_{\rm on})^{-1}$, for the arrival of the next photon. When both transitions are saturated, the two-state steady state value of the population of state e is $\bar{\rho}_{ee} = 1/2$. The fluorescence at rate $\Gamma_e/2$ is interrupted at random instants by a dark period, which occurs at a rate $R_{\rm off}$. The switch-on that ends the dark period occurs at the rate $R_{\rm on}$. On average, all three states have the same population 1/3, and the ratio of the switching rates is $R_{on}/R_{off} = 2$, since during a bright period the atom spends half of the time in the lower state g, from where the coupling to the dark state s can only arise. Hence, the average duration of the bright period is twice as long as the duration of a dark period. The long-time decrease of the intensity correlation from $\Gamma_e/2$ to $\Gamma_e/3$ simply arises from the fact that one third of the time, the atom passes through a dark period. A typical single history of bright and dark periods is sketched in Fig. 2.

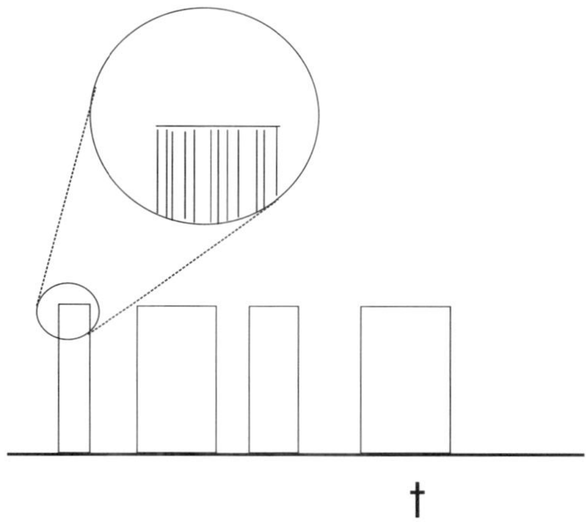

Fig. 2. Single history of bright and dark periods of fluorescence on the long time scale. On the short time scale, the fluorescence consists of single photon counts

The switching on and off of the fluorescence occurs at random instants whose precise values are not predicted by quantum mechanics. Therefore quantum mechanics does not describe a single run, only the statistics of many runs. The fluorescence during a bright period is easily observable, and it is basically a macroscopic signal. As viewed from the perspective of quantum measurement theory, observation of the fluorescence intensity is equivalent to measuring whether the atom is in the dark state s or not. In accordance with the view that the outcome of a measurement cannot be regarded as a property that the system had before the measurement, we would have to conclude that the observation that the fluorescence is off projects the atom in the state s. What happens when no one is looking remains undetermined. This shows that a macroscopic observation with a null result can be a full-fledged quantum measurement. The fact that the fluorescence is either off or on depends on this specific observation of the fluorescence intensity. It is the continuous observation on the long time scale that projects the atom either in the dark or in the bright state. Other observations, such as the phase properties of the fluorescence light, would have a different natural basis of natural states.

3. Spontaneous Decay as a Quantum Trajectory

We consider now a two-state atom, with ground state g, and excited state e. When the atom is prepared in the excited state at time 0, it will emit a photon by spontaneous emission. The atomic initial state is then denoted as $|\psi(0)\rangle = |e\rangle$. The exponential decay law states that $N(t) = N(0)\exp(-\Gamma t)$, with N the population of the excited state e, and Γ the rate of spontaneous decay. However, this law does not hold for a single history of a single initially excited atom. When we consider a single atom, initially in e, that is constantly monitored by a (supposedly ideal) photo-detector, a photon will be observed at some time t_0. Then as observers we will conclude that the atom remained in the excited state e up to time T, and that it was in the ground state g for times $t > T$. This implies that $N(t) = 1$ for $t < T$, and $N(t) = 0$ for $t > T$. The exponential decay law is reproduced only after averaging over many runs of the experiment, where the probability distribution over the decay time T is

$$w(T) = \Gamma\exp(-\Gamma T)\ . \tag{7}$$

This is illustrated in Fig. 3.

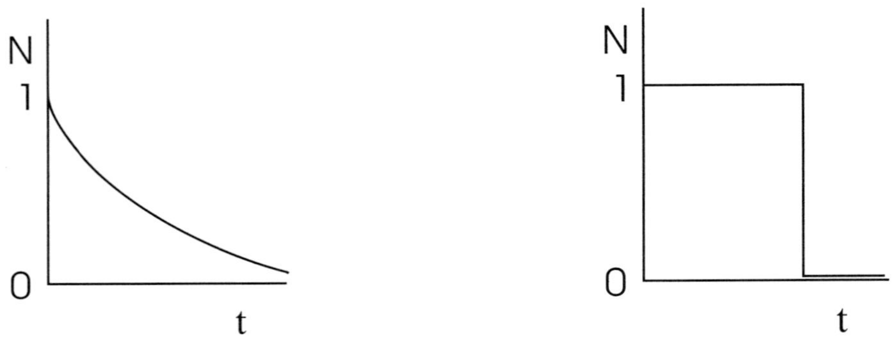

Fig. 3. Exponential decay (left) of excited state population is average over many single runs (right).

Now suppose that the atom is initially in a superposition of the excited state and the ground state, so that

$$|\psi(0)\rangle = a_e|e\rangle + a_g|g\rangle\ , \tag{8}$$

with the normalization condition $|a_e|^2 + |a_g^2| = 1$. We define the state vector $|\phi_0(t)\rangle$ as describing the state of the atom under the condition that no emission occurred until time t. Obviously, at time 0 this state must be the initial state, so that $|\phi_0(0)\rangle = |\psi(0)\rangle$. In analogy to the case of an initial excited state one might think that the state does not change as long as there is no emission. After all, we have seen that the excited state does not change as long as there is no emission, and the ground state cannot change. Still, this conclusion is false. In the state $|\psi(0)\rangle$, there is a non-zero probability $|a_g|^2$ that the atom is in the ground state. So with this probability, the atom will never emit a photon. Conversely, when no photon has been seen after many lifetimes, we can be pretty sure that the atom is in the ground state. This shows that $|\phi_0(\infty)\rangle = |g\rangle$. This shows that the time dependence of $|\phi_0(t)\rangle$ is not completely trivial. In particular, how can we explain and describe the decay of this state to the ground state $|g\rangle$?

The answer is found by first considering the evolution of an initially excited atom. The initial atomic state $|\psi(0)\rangle = |e\rangle$ gives rise to a time-dependent state of the atom and the radiation field, which we schematically represent as

$$|\Psi(t)\rangle = e^{-\Gamma t/2}|e\rangle|0\rangle + |g\rangle \int d\vec{k}\ B(\vec{k},t)|\vec{k}\rangle\ , \tag{9}$$

where $|\vec{k}\rangle$ indicates the state of the electromagnetic field with one photon with wave vector $\vec{k}$, while $|0\rangle$ is the vacuum state of the field. Expressions for the amplitudes $B(\vec{k},t)$ that determine the time-dependent distribution over the photon momenta can be directly obtained,[10] but we do not need them here. For simplicity, we adopted the Born-Markov approximation and the interaction representation.

Now we turn to the initial state (8). From the superposition principle, we know that the corresponding time-dependent state is

$$|\Psi(t)\rangle = |\psi_0(t)\rangle|0\rangle + a_e|g\rangle \int d\vec{k}\ B(\vec{k},t)|\vec{k}\rangle\ , \tag{10}$$

where the atomic state $|\psi_0(t)\rangle$ that multiplies the zero-photon state is

$$|\psi_0(t)\rangle = a_e e^{-\Gamma t/2}|e\rangle + a_g|g\rangle\ . \tag{11}$$

Since this state $|\Psi(t)\rangle$ does not factorize in an atomic part and a field part, the state is entangled.

The strength of the state $|\psi_0(t)\rangle$ is

$$P_0(t) = |a_e|^2 e^{-\Gamma t} + |a_g|^2\ . \tag{12}$$

Its physical significance is the probability that no photon was emitted before time t. Its time derivative determines the waiting-time distribution for the emission of a photon by the relation to $w(t) = -dP_0(t)/dt$, so that

$$w(T) = \Gamma|a_e|^2 e^{-\Gamma T}\ . \tag{13}$$

In the special case that the atom is initially in the excited state ($|a_e|^2 = 1$), this reduces to the expression (7). The time integral $W(t) = \int_0^t dT\ w(T) = 1 - P_0(t)$ of $w(T)$ is the probability of a photon emission before the time t, which is complementary to the no-photon probability P_0. In the limit $t \to \infty$ this gives the probability that a photon is emitted at any time, which is equal to the initial population $|a_e|^2$ of the excited state, as it should be. This confirms that there is a finite probability $|a_g|^2 = 1 - |a_e|^2$ that no photon will be emitted at any time.

Now we can answer the question raised above for the expression for the conditional state $|\phi_0(t)\rangle$ of the atom when no photon is emitted. It is just the normalized version of $|\psi_0(t)\rangle$. We can express it as

$$|\phi_0(t)\rangle = b_e(t)|e\rangle + b_g(t)|g\rangle \ , \tag{14}$$

with

$$b_e(t) = a_e e^{-\Gamma t/2}/\sqrt{P_0(t)} \ , \ b_g(t) = a_g/\sqrt{P_0(t)} \ . \tag{15}$$

From the evolution equation for the state vector $|\psi_0(t)\rangle$ we find evolution equations for the amplitudes b_g and b_e in the form[11]

$$\frac{db_e}{dt} = -\frac{\Gamma}{2} b_e |b_g|^2 \ , \ \frac{db_g}{dt} = \frac{\Gamma}{2} b_g |b_e|^2 \ . \tag{16}$$

These equations determine the physical (normalized) conditional state vector $|\phi_0(t)\rangle$. It is important to notice that these equations are non-linear. This reflects the fact that the conditional state $|\phi_0(t)\rangle$ does not obey the superposition principle: it is not the superposition of the states corresponding to the initial states $|e\rangle$ and $|g\rangle$, with amplitudes a_e and a_g.

In Fig. 4 we plot the decaying probability $|b_e|^2$ that the atom is in the excited state during the period that no photon was emitted. The plot has the shape of an inverted tanh curve.

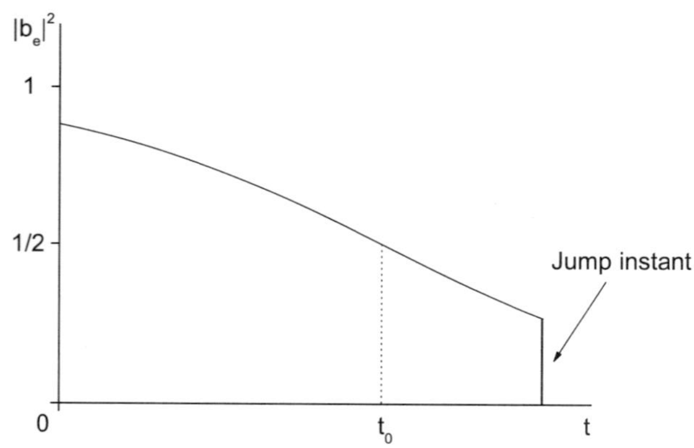

Fig. 4. Single run of the excited-state population. Up to the instant of the decay, the population decays with an tanh behavior. The time t_0 is the time that the population passes 1/2.

In this case the projection of the atomic state unto the ground state proceeds continuously during the period that no photon is emitted. This indicates that the probability that the atom is actually in the ground state increases gradually when one waits in vain for the photodetector to click. When the detector does click, the projection to the ground state is instantaneous, as indicated by the jump instant in Fig. 4.

This description of spontaneous emission of a single atom is a simple example of the methods of quantum trajectories. This is a way to describe the density matrix representing the state of an open system as a distribution over state vectors of the system. Here the open system is the atom, and the environment (the bath) is the quantized electromagnetic field. The mixture of states represented by the density matrix $\rho(t)$ is the statistical average over pure states represented by the state vector $|\phi(t)\rangle$. When the average is indicated by pointy brackets, this relation is simply $\rho(t) =< |\phi(t)\rangle\langle\phi(t)| >$. This is the basic idea of the Monte Carlo wave function method.[4,12] In the present case, the quantum trajectories are simply the normalized states $|\phi_0(t)\rangle$, that are interrupted by a quantum jump to the ground state, as sketched in Fig. 4. Hence, for a jump instant T, the pure-state history of the atom is given by $|\phi_0(t)\rangle = |g\rangle$ for $t > T$. Hence, the single histories differ only in the instant T of the jump. The distribution over these jump instants is provided by the waiting-time distribution (13). Averaging the expression for $|b_e(t)|^2$ over these jump instants reproduces the exponential decay law $< |b_e(t)|^2 >= |a_e|^2 \exp(-\Gamma t)$. It has been noted[11,13] that the evolution of the single histories $|\phi_0(t)\rangle$ before the emission is identical to the results of the neoclassical radiation history,[14] which is based on the idea that the decay of an excited atom results from classical radiation reaction. Now we see that the neoclassical description is just incomplete: what is missing is the quantum jump to the ground state. When the population of the excited state is small, the probability for a jump is small, and the long-time tail of decay as described by the tanh law becomes indistinguishable from the exponential decay law.

4. Discussion

We have discussed quantum jumps that arise for two simple configurations of a single atom. The first case occurs when one of the states of a driven transition is weakly coupled to a metastable state. In this case, the fluorescence radiation is switched off and on at random instants, reflecting the transition of the atom to and from the metastable state to the strongly coupled pair of states. When the fluorescent intensity is constantly monitored, the observance of no fluorescence must be viewed as a quantum measurement of the state of the atom, which projects it onto the metastable state. The second case is the spontaneous emission of a single two-state atom that is initially in a superposition of the ground state and the excited state. In this case, the observation that the atom emits no photon leads to a continuous decrease of the probability that the atom is in the excited state. When no photon has appeared after several lifetimes, it becomes unlikely that a photon will ever show up. This is a continuous version of the state projection that accompanies a quantum measurement. In both cases, a central concept in the description is the distribution $w(T)$ of times that one has to wait for a photon to appear. Also in both cases, the state of the atom becomes determined by a measurement which gives a null result: no photons are observed. In these cases the state projection occurs as time goes by. Note that in the standard description of a quantum measurement time does not enter. It is simply stated that the state is projected onto the eigenstate of the observed quantity when the corresponding eigenvalue is measured. Our examples demonstrate that simple atomic and optical physics produce cases where this simple picture is not adequate.

References

1. W. Nagourney, J. Sandberg and H. Dehmelt, *Phys. Rev, Lett.* **56**, 2797 (1986).
2. Th. Sauter, W. Neuhauser, R. Blatt and P. Toschek, *Phys. Rev. Lett.* **57**, 1696.
3. J. C. Bergquist, R. G. Hulet, W. M. Itano and D. J. Wineland, *Phys. Rev. Lett.***57**, 1699.
4. R. Dum, P. Zoller and H. Ritsch, *Phys. Rev. A* **45**, 4879 (1992).
5. R. J. Cook and H. J. Kimble, *Phys. Rev. Lett.* **54**, 1023 (1985).
6. J. Javanainen, *Phys. Rev. A* **33**, 2121 (1986).
7. A. Schenzle, R. G. DeVoe and R. G. Brewer, *Phys. Rev. A* **33**, 2127 (1986).
8. C. Cohen-Tannoudji and J. Dalibard, *Europhys. Lett.* **1**, 441 (1986); P. Zoller, M. Marte and D. F. Walls, *Phys. Rev. A* **35**, 198 (1987).
9. G. Nienhuis, *Phys. Rev. A* **35**, 4639 (1987).
10. H. J. Kimble and L. Mandel, *Phys. Rev. A* **13**, 2123 (1976).
11. D. Bouwmeester and G. Nienhuis, *Quantum Semiclass. Opt.* **8**, 277 (196).
12. J. Dalibard, Y. Castin and K. Molmer, *Phys. Rev. Lett.* **68**, 580 (1992).
13. D. Bouwmeester, R. J. C. Spreeuw, G. Nienhuis and J. P. Woerdman, *Phys. Rev. A* **49**, 4170 (1994).
14. C. R. Stroud, Jr. and E. T. Jaynes, *Phys. Rev. A* **1**, 106 (1970)

SOLVING OPEN QUESTIONS IN THE BOSE–EINSTEIN CONDENSATION OF AN IDEAL GAS VIA A HYBRID MIXTURE OF LASER AND STATISTICAL PHYSICS

M. KIM, A. SVIDZINSKY and M. O. SCULLY

Institute for Quantum Studies and Dept. of Physics, Texas A&M Univ., Texas 77843
Applied Physics and Materials Science Group, Eng. Quad.,Princeton Univ., Princeton 08544

Fluctuations in the Bose-Einstein condensate (BEC) remain a rich field of study even in the ideal gas limit. We here present the laser master equation approach to the problem in the spirit of Eugene P. Wigner who said: "With classical thermodynamics, one can calculate almost everything crudely; with kinetic theory, one can calculate fewer things, but more accurately; and with statistical mechanics, one can calculate almost nothing exactly."The combination of kinetic theory plus statistical mechanics proves to be a powerful combination for the calculation of essentially exact BEC equilibrium results.

Keywords: Bose–Einstein condensation; Statistical physics; Optics; Ideal Bose gas; Laser master equation.

1. Introduction

Historically, Einstein was the first to demonstrate the existence of the "Bose" condensate.[1,2] After Bose[3] followed an extreme "photon as a particle" path to the famous Planck distribution for the average number of thermal quanta (having frequency ν at temperature T) given by

$$\bar{n}_\nu = \frac{1}{e^{\frac{\hbar\nu}{k_BT}} - 1}, \tag{1}$$

Einstein showed atoms can display a similar behavior, he found that

$$\bar{n}_\epsilon = \frac{1}{e^{(\epsilon-\mu)/k_BT} - 1}, \tag{2}$$

where ϵ is the kinetic energy of the atom, and μ is the chemical potential ($\mu = 0$ for photons).

Then, in a stroke of genius, Einstein[2] went on to show that there is a critical temperature below which a macroscopic number of atoms would occupy the lowest energy state of the potential (e.g., a square well) holding the atoms. His simple analysis yields the average number of atoms in the ground state $\bar{n}_0$

$$\bar{n}_0 = N\left[1 - \left(\frac{T}{T_c}\right)^\alpha\right], \qquad \alpha = \begin{cases} 3/2 & \text{square potential} \\ 3 & \text{harmonic potential} \end{cases} \tag{3}$$

in terms of the temperature T, critical temperature T_c, and the number of atoms in the trap N.

Lively debate followed. Uhlenbeck began the bosonic bickering. He focused on the region near T_c, where things are most interesting and challenged Einstein's analysis.[4] Indeed for a finite number of atoms there is no sharp "critical point" and he worried that there didn't seem to be any way to have a phase transition without a cusp. For today's mesoscopic condensates, with $N \lesssim 10^3$, the problem is of experimental relevance. In Appendix A we derive $\bar{n}_0(T)$ for a mesoscopic number of particles. We show, in particular, that the cusp appears if we disregard the chemical potential μ, while replacement of summation by an integral practically does not change the answer.

Perhaps, one would think that all problems concerning the ideal Bose gas would have been solved by Einstein (or Uhlenbeck or . . .) in the 20's or 30's. Not so. Consider the simple problem of the squared variance $\Delta n_0^2 \equiv \langle (n_0 - \bar{n}_0)^2 \rangle$ of the number of particles in the ground state. In the late 70's, Ziff, Uhlenbeck, and Kac noted that there was a problem with the usual treatment of fluctuations:[5]

> "[When] the grand canonical properties for the ideal Bose gas are derived, it turns out that some of them differ from the corresponding canonical properties – even in the bulk limit! . . . The grand canonical ensemble . . . *loses its validity* for the ideal Bose gas in the condensed region."

As late as the end of the 20th century, it was noted that: "The grand canonical fluctuation catastrophe"[5] has been discussed by generations of physicists who have not solved the problem, see Fig. 1.

Indeed, the canonical partition function for a Bose gas of N particles at temperature T

$$Z(T,N) = \frac{\beta}{2\pi} \sum_{\{n\}} \int_{-i\pi}^{i\pi} d\mu \, \exp\left[-\mu\left(N - \sum_s n_s\right)\right] e^{-\beta E_{\{n\}}}, \tag{4}$$

has not been so well studied as one might have thought. Herzog and Olshanii in the late 90's say:[10]

> "To our knowledge there is no simple analytical expression for the canonical partition function in [the] case of N bosons in a 3-D trap!"

Motivated by the above (and by suggestions by Willis Lamb), we reconsidered and extended our previous work on the laser-phase transition analogy to include BEC. We found a new approach to the problem on N bosons, in thermal equilibrium below T_c. We emphasize that the present work provides another example in which steady state (detailed balance) solutions to nonequilibrium equations provides a useful tool for the study of systems in thermal equilibrium.

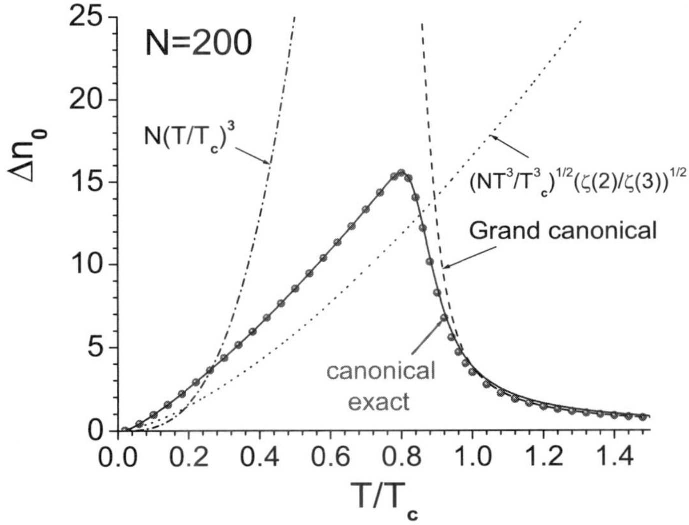

Fig. 1. Variance $\Delta n_0 = \sqrt{\langle n_0^2 \rangle - \langle n_0 \rangle^2}$ of the condensate particle number as a function of temperature for an ideal Bose gas of $N = 200$ atoms in an isotropic harmonic trap. Solid line is a solution of the condensate (laser-like) master equation. Large dots are the exact numerical results obtained in the canonical ensemble.[7] Dot line is a plot of $\Delta n_0 = \sqrt{\frac{\zeta(2)N}{\zeta(3)} \left(\frac{T}{T_c}\right)^3}$ which is obtained in the thermodynamic limit.[8] Dash line for Δn_0 is the grand canonical answer $\sqrt{\bar{n}_0(\bar{n}_0 + 1)}$ which gives the catastrophically large fluctuations below T_c. Dash-dot line is a plot of $\Delta n_0 = N - \langle n_0 \rangle = N \left(\frac{T}{T_c}\right)^3$, which is proposed by D. ter Haar[9] in the low temperature regime (adapted to a harmonic trap). This had the correct zero limit as $T \to 0$, but is not right for higher temperatures.

2. Laser Master Equation Analysis of BEC Statistics

Thus we seek a nonequilibrium equation of motion for the ground state of an ideal gas in a 3-D harmonic trap coupled to the thermal reservoir, as is shown elsewhere.[11] We find that the density matrix obeys the following master equation

$$\begin{aligned}\dot{\rho}_{n_0,n_0} &= -K_{n_0}(n_0+1)\rho_{n_0,n_0} + K_{n_0-1}n_0\rho_{n_0-1,n_0-1} \\ &\quad -H_{n_0}n_0\rho_{n_0,n_0} + H_{n_0+1}(n_0+1)\rho_{n_0+1,n_0+1}, \end{aligned} \tag{5}$$

where the cooling and heating coefficients K_{n_0} and H_{n_0} are given by

$$K_{n_0} = \sum_k 2\pi W_k g_k^2 \langle \eta_k + 1 \rangle \langle n_k \rangle_{n_0}, \tag{6}$$

and

$$H_{n_0} = \sum_k 2\pi W_k g_k^2 \langle \eta_k \rangle \langle n_k + 1 \rangle_{n_0}. \tag{7}$$

Here W_k is the heat bath density of states, $\langle \eta_k \rangle$ is the average occupation number of the kth heat bath oscillator and the $\langle n_k \rangle_{n_0}$ is the average number of atoms in the kth excited state, given n_0 atoms in the condensate. The coefficient K_{n_0} denotes the cooling rate from the excited states to the ground state, and similarly H_{n_0} is the heating rate for the ground state.

2.1. *Low temperature approximation*

At very low temperatures the number of non-condensed atoms is very small, we one can therefore approximate $\langle n_k \rangle_{n_0} + 1$ by 1 in Eq. (7). Then the heating coefficient is a constant proportional to the total average number of thermal excitations in the reservoir at all energies corresponding to the energy levels of the trap

$$H_{n_0} \simeq \kappa \sum_k \langle \eta(\epsilon_k) \rangle = \kappa \sum_{l,m,n} \frac{1}{e^{(\hbar\Omega/k_B T)(l+m+n)} - 1}. \tag{8}$$

This is the same sum given in Appendix A, namely

$$H_{n_0} \simeq \kappa (k_B T/\hbar\Omega)^3 \zeta(3) = \kappa N (T/T_c)^3. \tag{9}$$

In addition, at low enough temperatures, the average occupations in the reservoir are small and $\eta_k + 1 \simeq 1$. Therefore, the cooling term (6) is governed by the total number of excited state bosons,

$$K_{n_0} \simeq \kappa \sum_k \langle n_k \rangle_{n_0} = \kappa (N - n_0). \tag{10}$$

By writing the equation of motion for $\langle n_0 \rangle$ from Eq. (5), using H_{n_0} and K_{n_0} from Eqs. (9) and (10), we find

$$\begin{aligned} \frac{1}{\kappa}\dot{p}_{n_0} = & -\left[(N+1)(n_0+1) - (n_0+1)^2\right] p_{n_0} + [(N+1)n_0 - n_0^2] p_{n_0-1} \\ & - N\left(\frac{T}{T_c}\right)^3 [n_0 p_{n_0} - (n_0+1)p_{n_0+1}]. \end{aligned} \tag{11}$$

The resulting steady state distribution for the number of condensed atoms is given by

$$p_{n_0} = \frac{1}{Z_N} \frac{\mathcal{H}^{N-n_0}}{(N-n_0)!}, \tag{12}$$

where

$$\mathcal{H} = N \left(\frac{T}{T_c}\right)^3, \tag{13}$$

$Z_N = 1/p_N$ is the partition function, where, from the normalization condition $\sum_{n_0} p_{n_0} = 1$ we have

$$Z_N = e^{\mathcal{H}} \Gamma(N+1, \mathcal{H})/N!, \tag{14}$$

in which $\Gamma(\alpha, x) = \int_x^\infty t^{\alpha-1}e^{-t}dt$ is an incomplete gamma-function.

The mean value and the variance can be calculated from the distribution (12) for an arbitrary finite number of atoms in the Bose gas,

$$\langle n_0 \rangle = N - \mathcal{H} + \mathcal{H}^{N+1}/Z_N N!, \tag{15}$$

$$\Delta n_0^2 \equiv \langle n_0^2 \rangle - \langle n_0 \rangle^2 = \mathcal{H}\left(1 - (\langle n_0 \rangle + 1)\mathcal{H}^N/Z_N N!\right). \tag{16}$$

Figure Fig. 2.1 shows the first four central moments for the ideal Bose gas in an isotropic harmonic trap as calculated via the solution of the condensate master equation in the low temperature approximation. The approximations (9), (10) and, therefore, the results (15), (16) are clearly valid at low temperatures, i.e., in the weak trap limit, $T \ll \varepsilon_1$, where ε_1 is an energy gap between the first excited and the ground levels of a single-particle spectrum in the trap. However, in the case of a harmonic trap the results (15), (16) show qualitatively correct behavior for all temperatures, including $T \gg \varepsilon_1$ and $T \sim T_c$.[12]

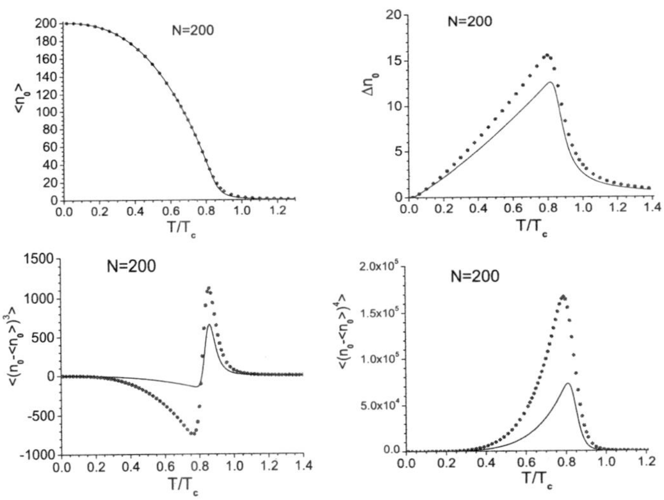

Fig. 2. The first four central moments for the ideal Bose gas in an isotropic harmonic trap with $N = 200$ atoms as calculated via the solution of the condensate master equation in the low temperature approximation, Eq. (12). The dots are "exact" numerical result obtained from the canonical ensemble.[7]

2.2. *Quasithermal approximation*

In the previous section we saw that $< n_0 >$ is well described by the "low temperature" approximation, even at $T = T_c$. However the higher moments are only qualitatively given by this theory. Next we extend the previous, low temperature, results by making a very reasonable approximation for the average non-condensate occupation numbers in the cooling and heating coefficients at higher temperatures, namely

$$\langle n_k \rangle_{n_0} = \eta_k \sum_{k>0} \langle n_k \rangle_{n_0} / \sum_{k'} \eta_{k'} = \frac{(N - \bar{n}_0)}{(e^{\varepsilon_k/T} - 1)\mathcal{H}} \,. \tag{17}$$

The cooling and heating coefficients in the quasithermal approximation of Eq. (17) are now given by

$$K_{n_0} = (N - n_0)(1 + \eta), \qquad H_{n_0} = \mathcal{H} + (N - n_0)\eta, \tag{18}$$

where

$$\mathcal{H} = \sum_{k>0} \frac{1}{(e^{\varepsilon_k/T} - 1)}. \tag{19}$$

Compared with the low temperature approximation (10) and (9), these coefficients acquire an additional parameter η given by

$$\eta = \frac{1}{N - n_0} \sum_{k>0} \langle \eta_k \rangle \langle n_k \rangle_{n_0} = \frac{1}{\mathcal{H}} \sum_{k>0} \frac{1}{(e^{\varepsilon_k/T} - 1)^2}. \tag{20}$$

Now, for arbitrary temperatures, the condensate master equation (6) contains two parameters, $\mathcal{H}$ and η,

$$\begin{aligned} \dot{p}_{n_0} &= -\kappa\{(1+\eta)[(N - n_0)(n_0 + 1)p_{n_0} - (N - n_0 + 1)n_0 p_{n_0 - 1}] \\ &\quad + [\mathcal{H} + (N - n_0)\eta] n_0 p_{n_0} - [\mathcal{H} + (N - n_0 - 1)\eta](n_0 + 1)p_{n_0+1}\}. \end{aligned} \tag{21}$$

The steady-state solution of Eq. (21) is

$$p_{n_0} = \frac{1}{Z_N} \frac{(N - n_0 + \mathcal{H}/\eta - 1)!}{(\mathcal{H}/\eta - 1)!(N - n_0)!} \left(\frac{\eta}{1+\eta}\right)^{N - n_0}, \tag{22}$$

and the canonical partition function $Z_N = 1/p_N$ is

$$Z_N = \sum_{n_0=0}^{N} \binom{N - n_0 + \mathcal{H}/\eta - 1}{N - n_0} \left(\frac{\eta}{1+\eta}\right)^{N - n_0}. \tag{23}$$

The average number of atoms condensed in the ground state of the trap is now found to be

$$\langle n_0 \rangle = N - \mathcal{H} + p_0 \eta (N + \mathcal{H}/\eta) \tag{24}$$

and the squared variance can be also calculated analytically yielding

$$\Delta n_0^2 = (1+\eta)\mathcal{H} - p_0(\eta N + \mathcal{H})\Big(N - \mathcal{H} + 1 + \eta\Big) - p_0^2(\eta N + \mathcal{H})^2, \tag{25}$$

where

$$p_0 = \frac{1}{Z_N} \frac{(N + \mathcal{H}/\eta - 1)!}{N!(\mathcal{H}/\eta - 1)!} \left(\frac{\eta}{1+\eta}\right)^{N} \tag{26}$$

is the probability that there are no atoms in the condensate.

Similarly we find the higher order central moments $< (n_0 - \bar{n}_0)^m >$. The first four central moments for the Bose gas in a harmonic trap with $N = 200$ atoms are presented in Fig. 3 in the quasithermal approximation.

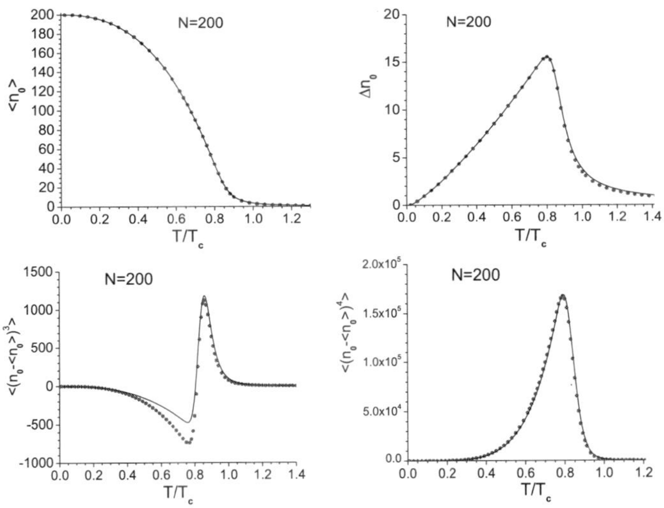

Fig. 3. The first four central moments for the ideal Bose gas in an isotropic harmonic trap with $N = 200$ atoms as calculated via the quasithermal approximation. Dots are "exact" numerical result obtained in the canonical ensemble.[7]

3. A Very Accurate Hybrid Approach

In section 2 we gave a good treatment of the first two moments. However the third moment was not so well treated (see Fig. 3). Here we correct that shortcoming and, in so doing, point the way to solving for the moments of the weakly interacting Bose gas.

3.1. *Cumulants of BEC fluctuation for an ideal Bose gas*

The canonical-ensemble density matrix $\hat{\rho}$ is given by

$$\rho_{\mathbf{k}}(n_{\mathbf{k}}) = \exp(-n_{\mathbf{k}}\varepsilon_{\mathbf{k}}/T)(1 - \exp(-\varepsilon_{\mathbf{k}}/T)). \tag{27}$$

Please note that the statistical distribution of the number of excited atoms, $n = \sum_{\mathbf{k}\neq 0} n_{\mathbf{k}}$ is a simple "mirror" image of the distribution of the number of condensed atoms,

$$\rho(n) = \rho_0(n_0 = N - n). \tag{28}$$

A convenient way to find and to describe this is to use the characteristic function

$$\Theta_n(u) = Tr\{e^{iu\hat{n}}\hat{\rho}\}, \tag{29}$$

which upon taking the Fourier transform of $\Theta_n(u)$ yields the probability distribution

$$\rho(n) = \frac{1}{2\pi}\int_{-\pi}^{\pi} e^{-iun}\Theta_n(u)du. \tag{30}$$

Taylor expansions of $\Theta_n(u)$ and $\log\Theta_n(u)$ give the (non-central) moments and cumulants, or semi-invariants:

$$\Theta_n(u) = \sum_{m=0}^{\infty}\alpha_m\frac{u^m}{m!}, \qquad \alpha_m \equiv \langle n^m\rangle = \frac{d^m}{du^m}\Theta_n(u)|_{u=0}, \tag{31}$$

$$\log\Theta_n(u) = \sum_{m=1}^{\infty}\kappa_m\frac{(iu)^m}{m!}, \qquad \kappa_m = \frac{d^m}{d(iu)^m}\log\Theta_n(u)|_{u=0}, \qquad \Theta_n(u=0) = 1. \tag{32}$$

The cumulants κ_r, initial moments α_m, and central moments $\mu_m \equiv \langle(n-\bar{n})^m\rangle$ are related to each other by the simple binomial formulas[13,14] via the mean number of the non-condensed atoms $\bar{n} = N - \bar{n}_0$,

$$\mu_r = \sum_{k=0}^{r}(-1)^k\binom{r}{k}\alpha_{r-k}\bar{n}^k, \qquad \alpha_r = \sum_{k=0}^{r}\binom{r}{k}\mu_{r-k}\bar{n}^k,$$

$$\bar{n} = \kappa_1, \quad \langle(n-\bar{n})^2\rangle \equiv \mu_2 = \kappa_2, \quad \langle(n-\bar{n})^3\rangle \equiv \mu_3 = \kappa_3, \langle(n-\bar{n})^4\rangle \equiv \mu_4 = \kappa_4 + 3\kappa_2^2,$$

$$\langle(n-\bar{n})^5\rangle \equiv \mu_5 = \kappa_5 + 10\kappa_2\kappa_3, \qquad \langle(n-\bar{n})^6\rangle \equiv \mu_6 = \kappa_6 + 15\kappa_2(\kappa_4+\kappa_2^2) + 10\kappa_3^2, \quad \ldots \tag{33}$$

Instead of calculation of the central moments, $\mu_m = \langle(n-\bar{n})^m\rangle$, it is more convenient, in particular for the analysis of the non-Gaussian properties, to solve for the cumulants κ_m, which are related to the moments by simple binomial expressions, the first six are

$$\kappa_1 = \bar{n}, \qquad \kappa_2 = \mu_2, \qquad \kappa_3 = \mu_3, \qquad \kappa_4 = \mu_4 - 3\mu_2^2,$$

$$\kappa_5 = \mu_5 - 10\mu_2\mu_3, \qquad \kappa_6 = \mu_6 - 15\mu_2(\mu_4 - 2\mu_2^2). \tag{34}$$

The essence of the BEC fluctuations are given in terms of the "generating cumulants" $\tilde{\kappa}_m$ which are related to the cumulants κ_m by the combinatorial formulas in Eq. (37),

$$\kappa_1 = \tilde{\kappa}_1, \quad \kappa_2 = \tilde{\kappa}_2 + \tilde{\kappa}_1, \quad \kappa_3 = \tilde{\kappa}_3 + 3\tilde{\kappa}_2 + \tilde{\kappa}_1, \quad \kappa_4 = \tilde{\kappa}_4 + 6\tilde{\kappa}_3 + 7\tilde{\kappa}_2 + \tilde{\kappa}_1, \quad \ldots \tag{35}$$

The main advantage of the cumulant analysis of the probability distribution $\rho(n)$ is the simple fact that the cumulant of a sum of independent stochastic variables is equal to a sum of the partial cumulants, $\kappa_r = \sum_{\mathbf{k}\neq 0} \kappa_r^{(\mathbf{k})}$. This is a consequence of the fact that $\log \Theta_n(u) = \log \Pi_{\mathbf{k}\neq 0} \Theta_{n_{\mathbf{k}}}(u) = \sum_{\mathbf{k}\neq 0} \log \Theta_{n_{\mathbf{k}}}(u)$. The characteristic function can be easily calculated from the equilibrium density matrix as follows

$$\Theta_{n_{\mathbf{k}}}(u) = Tr\{e^{iu\hat{n}_{\mathbf{k}}}\hat{\rho}_{\mathbf{k}}\} = Tr\{e^{iu\hat{n}_{\mathbf{k}}}e^{-\varepsilon_{\mathbf{k}}\hat{n}_{\mathbf{k}}/T}\}\left(1 - e^{-\varepsilon_{\mathbf{k}}/T}\right) = \frac{z_{\mathbf{k}} - 1}{z_{\mathbf{k}} - z}. \quad (36)$$

Here we have introduced the exponential function of the single-particle energy spectrum $\varepsilon_{\mathbf{k}}$, namely $z_{\mathbf{k}} = \exp(\varepsilon_{\mathbf{k}}/T)$, and a variable $z = \exp(iu)$ which has the character of a "fugacity". As a result, we obtain an explicit formula for the characteristic function and all cumulants of the number of excited (and, according to the equation $n_0 = N - n$, also the condensed) atoms in the ideal Bose gas in an arbitrary trap as:[15]

$$\log \Theta_n(u) = \sum_{\mathbf{k}\neq 0} \log\left(\frac{z_{\mathbf{k}} - 1}{z_{\mathbf{k}} - z}\right) = \sum_{m=1}^{\infty} \tilde{\kappa}_m \frac{(e^{iu} - 1)^m}{m!} = \sum_{r=1}^{\infty} \kappa_r \frac{(iu)^r}{r!},$$

$$\tilde{\kappa}_m = (m-1)! \sum_{\mathbf{k}\neq 0} \left(e^{\varepsilon_{\mathbf{k}}/T} - 1\right)^{-m}; \qquad \kappa_r = \sum_{m=1}^{r} \sigma_r^{(m)} \tilde{\kappa}_m. \quad (37)$$

Here we use the Stirling numbers of the 2nd kind,[13]

$$\sigma_r^{(m)} = \frac{1}{m!}\sum_{k=0}^{m}(-1)^{m-k}\binom{m}{k}k^r, \quad \left(e^x - 1\right)^k = k!\sum_{n=k}^{\infty}\sigma_n^{(k)}\frac{x^n}{n!}, \quad (38)$$

that yield a simple expression for the cumulants κ_r via the generating cumulants $\tilde{\kappa}_m$. The first four moments are summarized in Fig. 4.

3.2. *Hybrid approach to condensate fluctuations*

We now show how to combine ideas from the canonical ensemble quasiparticle formalism[15] of section 3.1 (which works well at temperatures not too close to T_c when $\sqrt{\mu_2} \ll \bar{n}_0$) with the physics of the master equation approach,[16] in order to obtain essentially perfect quantitative agreement with the exact numerical solution of the canonical partition function at *all* temperatures for the fluctuation statistics of the Bose gas. Such a hybrid technique was presented in.[17]

The central tool used in the ideal gas analysis of[16] was the laser-like master equation Eq. (6) for the probability p_{n_0} of finding n_0 atoms in the condensate, given that there are N total particles

$$\frac{1}{\kappa}\dot{p}_{n_0} = -K_{n_0}(n_0+1)p_{n_0} + K_{n_0-1}n_0 p_{n_0-1} - H_{n_0}n_0 p_{n_0} + H_{n_0+1}(n_0+1)p_{n_0+1}, \quad (39)$$

where κ is an uninteresting rate constant, H_{n_0} and K_{n_0} are heating and cooling coefficients. In equilibrium the rates of any two opposite processes are equal to each other, e.g., $K_{n_0}(n_0+1)p_{n_0} = H_{n_0+1}(n_0+1)p_{n_0+1}$. The detailed balance condition yields

$$\frac{p_{n_0+1}}{p_{n_0}} = \frac{K_{n_0}}{H_{n_0+1}}. \quad (40)$$

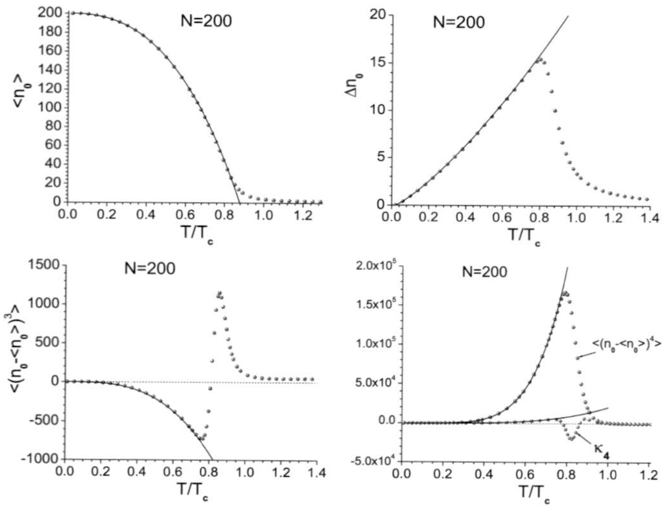

Fig. 4. Average condensate particle number $< n_0 >$, its variance $\Delta n_0 = \sqrt{< (n_0 - \bar{n}_0)^2 >}$, third and fourth central moments $< (n_0 - \bar{n}_0)^m >$ $(m = 3, 4)$ and fourth cumulant κ_4 as a function of temperature for an ideal gas of $N = 200$ particles in a harmonic trap. Solid lines show the result obtained from Eq. (37). Dots are "exact" numerical simulation in the canonical ensemble.

Since the occupation number of the ground state cannot be larger than N there is a canonical ensemble constraint $p_{N+1} = 0$ and, hence, $K_N = 0$. In contrast to p_{n_0}, the ratio p_{n_0+1}/p_{n_0} as a function of n_0 shows simple monotonic behavior. We approximate K_{n_0} and H_{n_0} by a few terms of the Taylor expansion near the point $n_0 = N$

$$K_{n_0} = (N - n_0)(1 + \eta) + \alpha(N - n_0)^2, \tag{41}$$

$$H_{n_0} = \mathcal{H} + (N - n_0)\eta + \alpha(N - n_0)^2. \tag{42}$$

Parameters $\mathcal{H}$, η and α are independent of n_0; they are functions of the occupation of the excited levels. We derive them below by matching the first three central moments in the low temperature limit with the result of the previous section. We note that the detailed balance equation (40) is the Padé approximation[18] of the function p_{n_0+1}/p_{n_0}. Padé summation has proven to be useful in many applications, including condensed-matter problems and quantum field theory.

Eqs. (40)-(42) yield an analytical expression for the condensate distribution function

$$p_{n_0} = \frac{1}{\mathcal{Z}_N} \frac{(N - n_0 - 1 + x_1)!(N - n_0 - 1 + x_2)!}{(N - n_0)!(N - n_0 + (1 + \eta)/\alpha)!}, \tag{43}$$

where $x_{1,2} = (\eta \pm \sqrt{\eta^2 - 4\alpha\mathcal{H}})/2\alpha$ and $\mathcal{Z}_N$ is the normalization constant determined by $\sum_{n_0=0}^{N} p_{n_0} = 1$. In the particular case $\eta = \alpha = 0$ Eq. (43) reduces to Eq. (12) obtained in the low temperature approximation.

Using the distribution function (43) we find that, in the validity range of[15] (at low enough T), the first three central moments $\mu_m \equiv < (n_0 - \bar{n}_0)^m >$ are

$$\bar{n}_0 = N - \mathcal{H}, \qquad \mu_2 = (1+\eta)\mathcal{H} + \alpha\mathcal{H}^2, \tag{44}$$

$$\mu_3 = -\mathcal{H}(1 + \eta + \alpha\mathcal{H})(1 + 2\eta + 4\alpha\mathcal{H}). \tag{45}$$

Eqs. (44), (45) thus yield

$$\mathcal{H} = N - \bar{n}_0, \qquad \eta = \frac{1}{2}\left(\frac{\mu_3}{\mu_2} - 3 + \frac{4\mu_2}{\mathcal{H}}\right), \tag{46}$$

$$\alpha = \frac{1}{\mathcal{H}}\left(\frac{1}{2} - \frac{\mu_2}{\mathcal{H}} - \frac{\mu_3}{2\mu_2}\right). \tag{47}$$

Substitute for $\bar{n}_0$, μ_2 and μ_3 in Eqs. (46), (47) their expressions obtained in the previous section yields the unknown parameters $\mathcal{H}$, η and α. The distribution function (43) together with Eqs. (46), (47) provides complete knowledge of the condensate statistics at all T.

Fig. 5 shows the average condensate particle number $\bar{n}_0$, its variance, third and fourth central moments μ_m and fourth cumulant κ_4 as a function of T for an ideal gas of $N = 200$ particles in a harmonic trap. Solid lines are the result of the hybrid approach (we call it CNB5) which is in remarkable agreement with the "exact" dots at all temperatures both for μ_m and κ_4. Central moments and cumulants higher than fourth order are not shown here, but they are also remarkably accurate at all temperatures. Results of the previous section[15] are given by dashed lines which are accurate only at sufficiently low T. Deviation of higher order cumulants ($m = 3, 4, \ldots$) from zero indicates that the fluctuations are not Gaussian.

Clearly the hybrid method describes the condensate statistics at all T with flying colors.

Acknowledgments

We wish to thank Michael Fisher and Franck Laloe for technical discussions and stimulating interactions on the present problem. MOS is also indebted to the Lorentz center of Leiden University for an excellent meeting on the present and related topics. He also gratefully acknowledges the support of the Office of Naval Research (Award No. N00014-03-1-0385) and the Robert A. Welch Foundation (Grant No. A-1261).

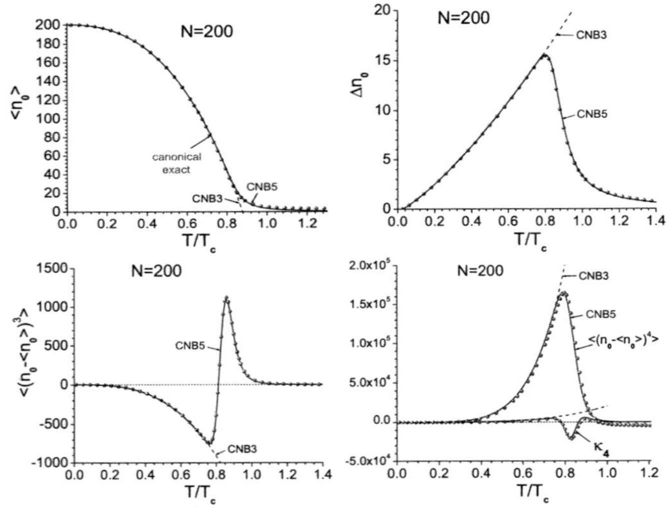

Fig. 5. Average condensate particle number $< n_0 >$, its variance $\Delta n_0 = \sqrt{< (n_0 - \bar{n}_0)^2 >}$, third and fourth central moments $< (n_0 - \bar{n}_0)^m >$ $(m = 3, 4)$ and fourth cumulant κ_4 as a function of temperature for an ideal gas of $N = 200$ particles in a harmonic trap. Solid lines (CNB5) show the result of the hybrid approach[15,17] yields dashed lines (CNB3). Dots are "exact" numerical simulation in the canonical ensemble.

Appendix A. Mean number of condensate particles: cusp vs smooth crossover

Here we find the mean number of condensate particles $\bar{n}_0$ for a three dimensional (3D) isotropic harmonic trap in different approximations. We use the grand canonical ensemble and write the equation that the total number of particles in the trap N is equal to the average particle number for the chemical potential μ:

$$N = \sum_{k=0}^{\infty} \bar{n}_{\mathbf{k}} = \sum_{\mathbf{k}=0}^{\infty} \frac{1}{\exp[\beta(\varepsilon_{\mathbf{k}} - \mu)] - 1}, \tag{A.1}$$

where for the 3D isotropic harmonic trap $\varepsilon_{\mathbf{k}} = \hbar\Omega(k_x + k_y + k_z)$, Ω is the trap frequency and $\beta = 1/k_B T$. Following Eq. (A.1), we can relate the chemical potential μ to $\bar{n}_0$ as $1 + 1/\bar{n}_0 = \exp(-\beta\mu)$. Thus, we have

$$N = \sum_{\mathbf{k}=0}^{\infty} \langle n_{\mathbf{k}} \rangle = \sum_{\mathbf{k}=0}^{\infty} \frac{1}{(1 + 1/\bar{n}_0)\exp(\beta\varepsilon_{\mathbf{k}}) - 1}. \tag{A.2}$$

For the 3D isotropic harmonic trap the degeneracy of the nth energy level is $(n+2)(n+1)/2$ and we obtain

$$N = \bar{n}_0 + \frac{1}{2}\sum_{n=1}^{\infty}\frac{(n+2)(n+1)}{(1+1/\bar{n}_0)\exp(\beta n\hbar\Omega)-1}. \tag{A.3}$$

Eq. (A.3) is a self-consistent equation for $\bar{n}_0$. For large $\bar{n}_0$ one can neglect the $1/\bar{n}_0$ as it appears under the sum in (A.3); this is equivalent to the approximation $\mu = 0$. However, for a mesoscopic number of particles (e.g., a few hundred) such approximation is not accurate. In Fig. A1 we plot solution of Eq. (A.3) for $\bar{n}_0$ as a function of temperature for $N = 200$ and compare it with the answer obtained assuming $\mu = 0$. If we keep $\mu \neq 0$ the solution shows a smooth crossover near the critical temperature T_c, while for $\mu = 0$ the curve has a cusp.

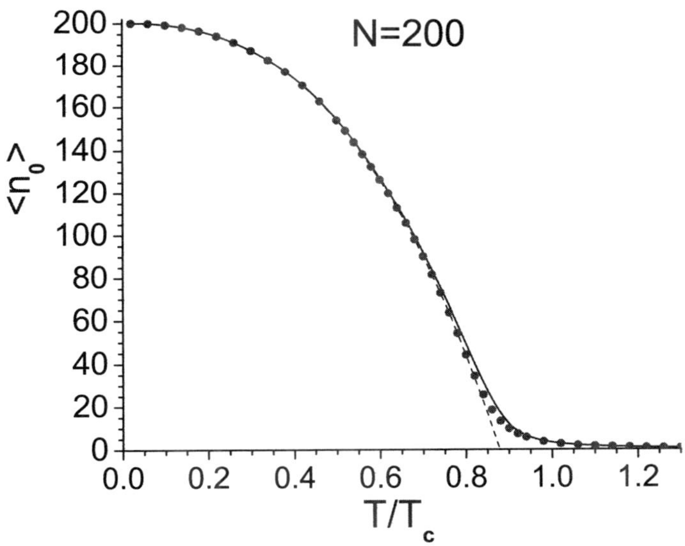

Fig. A1. The average condensate particle number versus temperature for $N = 200$ particles in an isotropic harmonic trap. Solid line is solution of Eq. (A.3), while the dash line shows the answer for $\mu = 0$. Dots are obtained numerically in the canonical ensemble.[7]

Next we replace the sum by an integral in Eq. (A.3)

$$N = \bar{n}_0 + \frac{1}{2}\int_1^{\infty}\frac{(x+2)(x+1)}{(1+1/\bar{n}_0)\exp(x\beta\hbar\Omega)-1}dx \tag{A.4}$$

and solve this integral equation for $\bar{n}_0$ numerically. For $N = 200$ we find that the answer is very close to the solution of Eq. (A.3) at all T and also shows a smooth crossover near T_c. Thus, replacement of the sum by an integral does not yield a cusp in $\bar{n}_0(T)$. The cusp appear only if we disregard the chemical potential μ.

In the thermodynamic limit $N \ggg 1$ one can write

$$\int_1^\infty \frac{(x+2)(x+1)}{(1+1/\bar{n}_0)\exp(x\beta\hbar\Omega)-1}dx \approx \int_0^\infty \frac{x^2}{\exp(x\beta\hbar\Omega)-1}dx = 2\left(\frac{k_B T}{\hbar\Omega}\right)^3 \zeta(3), \tag{A.5}$$

then Eq. (A.4) yields the following analytical formula for $\bar{n}_0$

$$\bar{n}_0(T) = N\left[1-\left(\frac{T}{T_c}\right)^3\right], \tag{A.6}$$

where

$$k_B T_c = \hbar\Omega\left(\frac{N}{\zeta(3)}\right)^{1/3} \tag{A.7}$$

is the temperature of BEC transition in the thermodynamic limit. Eq. (A.6) shows a cusp at $T = T_c$ and is accurate only in the thermodynamic limit in which $\mu = 0$ is a good approximation.

For a mesoscopic number of particles Eq. (A.6) is inaccurate. To improve the accuracy one can treat Eq. (A.3) in the following way[19]

$$N - \bar{n}_0 = \frac{1}{2\left(\frac{1}{\bar{n}_0}+1\right)}\sum_{n=1}^{\infty}\frac{(n+2)(n+1)}{\exp(\beta n\hbar\Omega)-\frac{\bar{n}_0}{\bar{n}_0+1}}. \tag{A.8}$$

For $\bar{n}_0 \gg 1$, the term $\bar{n}_0/(\bar{n}_0+1)$ inside the summation can be approximated by 1. Then we obtain a quadratic equation for the mean number of particles in the ground state

$$N - \bar{n}_0 = \frac{\mathcal{H}}{\frac{1}{\bar{n}_0}+1},$$

whose solution is

$$\bar{n}_0 = \frac{1}{2}\left(N-\mathcal{H}-1+\sqrt{(N-\mathcal{H}-1)^2+4N}\right), \tag{A.9}$$

here

$$\mathcal{H} = \frac{1}{2}\sum_{n=1}^{\infty}\frac{(n+2)(n+1)}{\exp(\beta n\hbar\Omega)-1}.$$

Analytical expression (A.9) shows a smooth crossover near T_c for a mesoscopic number of particles N. Figure A2 compares Eq. (A.9) obtained in the grand canonical ensemble (solid line) for $N = 200$ with the numerical calculation of $\bar{n}_0(T)$ from the exact recursion relations in the canonical ensemble (dots).[7] One can see that for the average particle number both ensembles yield very close answers. Dash line shows the plot of the solution (A.6), which is valid only for a large number of particles N.

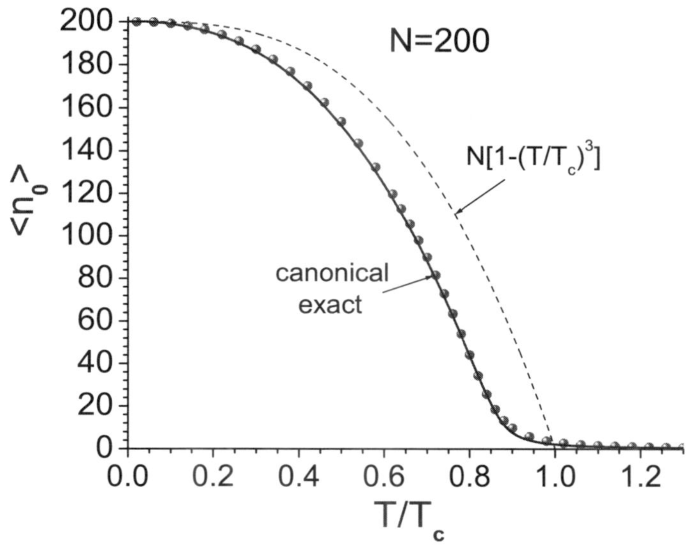

Fig. A2. The average condensate particle number versus temperature for $N = 200$ particles in an isotropic harmonic trap. Solid line is Eq. (A.9), while the dash line shows the thermodynamic limit formula (A.6). "Exact" dots are obtained numerically in the canonical ensemble.[7]

References

1. A. Einstein, *Sitzungsberichte der preussischen Akademie der Wissenschaften*, XXII. Gesamtsitzung, 261 (1924).
2. A. Einstein, *Sitzungsberichte der preussischen Akademie der Wissenschaften*, I. Sitzung der physikalisch-mathematischen Klasse, 3 (1925).
3. S. N. Bose, *Z. Phys.* **26**, 178 (1924).
4. "Some Strangeness in the proportion: A Centennial Symposium to Celebrate the achievements of Albert Einstein", ed. H. Woolf, (Addison Wesley, p. 524, 1981).
5. R. M. Ziff, G. E. Uhlenbeck, and M. Kac, *Phys. Rep.* **32**, 169 (1977).
6. M. Holthaus, E. Kalinowski, and K. Kirsten, *Ann. Phys. (N.Y.)* **270**, 198 (1998).
7. M. Wilkens and C. Weiss, *J. Mod. Opt.* **44**, 1801 (1997).
8. H. D. Politzer, *Phys. Rev. A* **54**, 5048 (1996).
9. D. ter Haar, *Lectures on selected topics in statistical mechanics* (Elsevier, Amsterdam, 1977), Chapter 4; I. Fujiwara, D. ter Haar and H. Wergeland, *J. Stat. Phys.* **2**, 329 (1970).
10. C. Herzog and M. Olshanii, *Phys. Rev. A* **55**, 3254 (1997).
11. For a recent review see V. V. Kocharovsky, Vl. V. Kocharovsky, M. Holthaus, C. H. R. Ooi, A. A. Svidzinsky, W. Ketterle and M. O. Scully, *Ad. in AMO Physics* **53**, 291 (2006).
12. M. O. Scully, *Phys. Rev. Lett.* **82**, 3927 (1999).
13. *Handbook of Mathematical Functions*, edited by M. Abramowitz and I. A. Stegun (Dover, New York, 1972).

14. T. N. Thiele, *Theory of Observations* (London, 1903); C. W. Gardiner, *Handbook of Stochastic Methods* (Springer, Heidelberg, 1983).
15. V. V. Kocharovsky, Vl. V. Kocharovsky, and M. O. Scully, *Phys. Rev. Lett.* **84**, 2306 (2000); *Phys. Rev. A* **61**, 053606 (2000).
16. V. V. Kocharovsky, M. O. Scully, S. Y. Zhu and M. S. Zubairy, *Phys. Rev. A* **61**, 023609 (2000).
17. A. A. Svidzinsky and M.O. Scully, *Phys. Rev. Lett.* **97**, 190402 (2006).
18. G. A. Baker, Jr. and P. Graves-Morris, Padé Approximants. (New York: Cambridge University Press, 1996).
19. A. N. Jordan, C. H. R. Ooi, and A. A. Svidzinsky, *Phys. Rev. A* **74**, 032506 (2006).

TWIN–PHOTON LIGHT SCATTERING AND CAUSALITY

G. PUENTES*, A. AIELLO, and J. P. WOERDMAN

Huygens Laboratory, Leiden University,
P.O. Box 9504, 2300 RA Leiden, The Netherlands.
**E-mail: graciana@molphys.leidenuniv.nl*

We present some of our results on multi-mode scattering of entangled photon pairs. We describe these scattering processes in terms of trace-preserving and non-trace-preserving local quantum maps. We show that non-trace preserving local maps can lead to apparent violations of causality, when the two-photon states are post-selected by coincidence measurements.

Keywords: Entanglement; Causality; Quantum maps.

1. Introduction

Quantum non-locality has played a crucial role in the foundations of quantum mechanics ever since it was theoretically discovered by Einstein, Podolsky, and Rosen (EPR) in 1935.[1] Bell's findings (1964) that these quantum non-local correlations, also referred to as entanglement, could not be explained in terms of classical local hidden variable models[2] triggered a prolific experimental activity, starting by Aspect *et al.* (1982),[3] who experimentally verified a violation of Bell's inequalities for the first time. These successful experiments have formed, in more recent years, the basis of quantum information science. Within this context it appears to be important to characterize entanglement and its robustness in different kinds of conditions.

In this spirit, we present some of our recent experimental results on the effects that different types of scattering processes can have on the degree of polarization-entanglement of twin-photon pairs. First, we briefly present the set-up used in our twin-photon scattering experiments. Second, we discuss some of the constraints imposed by special relativity (i.e., causality condition) on the possibly allowed experiments. Third, we motivate the description of scattering processes as trace-preserving and non-trace-preserving quantum maps.[4] Specifically, we show that non-trace-preserving maps can lead to an apparent violation of causality and that this can be explained in terms of post-selection during the quantum state reconstruction procedure. Finally we draw our conclusions.

2. Our experiments

In our experiments we want to analyze the effect that a given multi-mode scattering process can have over the polarization degrees of freedom of entangled photons. The pairs of photons (A-B) are initially created in the polarization singlet state by degenerate type II spontaneous parametric down conversion (SPDC),[5,6] where a pump-photon from a Krypton-Ion laser at 413.1 nm is split in two twin-photons of half energy and double wavelength. Then one of the two photons (A) propagates through a local scattering medium (i.e., a scattering medium acting on only one photon). The different scattering media we analyzed range from milk to multi-mode polymer fibers.[7] The polarization density matrix (ρ_{AB}) of scattered two-photon states are then reconstructed via a standard quantum tomographic procedure[8] (see Fig. 1). ¿From this reconstructed density matrix we extract the entanglement content (i.e., the tangle[9]) and the degree of purity (i.e., the linear entropy[10]). The measured data is then displayed in a tangle vs linear entropy plane[11,12] (see Fig. 3 below).

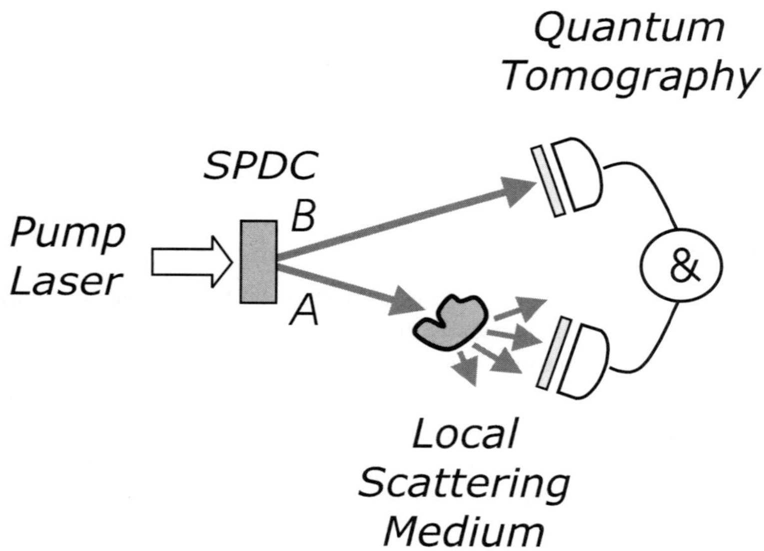

Fig. 1. Scheme of our experimental set-up. A high-frequency pump photon is split in two lower-frequency twin-photons (A-B) by SPDC. Photon A undergoes a local scattering process. The polarization density matrix of the scattered photon-pair is then reconstructed via a quantum tomographic procedure.

In the next sections we will show what are the restrictions on any experiment aiming at quantifying entanglement, and we will discuss whether these restrictions can apparently be violated by using local scattering media and quantum tomographic detection.

3. Causality condition

In any kind of Bell-type measurement there are two logical loopholes that have to be closed in order to demonstrate that entanglement is a truly non-local feature of quantum mechanics, which can not be explained in terms of local hidden variable models. These are the detection loophole and the causality loophole (and it should be noted that they have not been closed at the same time in any experiment up to date). The first loophole refers to the detection efficiency and is grounded on the fact that all experiments so far detect only a small subset of all pairs created.[13] Closure of the second loophole demands that the measurement processes of the two observers A and B (Alice and Bob) are space-like separated events, so that they cannot signal each other.[14] There is thus no way to infer from the result of a local measurement on one wing of the experiment, which measurement has been performed on the other wing. This causality condition is also referred to as 'no-signalling condition'. This idea is schematized in Fig. 2. Consider a pair of photons initially created in a polarization correlated state (for instance by SPDC), which propagate in the forwards direction of the space-time diagram. Each of these photons is then detected at time t_D. Each detection event is determined by an independent choice of the polarizer setting (θ_i, $i = A, B$) symbolized by a circle. The choices of the polarizers settings have to be space-like separated enough so that their forward cone of events do not intersect before time t_D, which is the time where each photon is absorbed (a click on a detector). Moreover, each individual detection event has to be registered on both sides independently and compared only after the whole measurement procedure is finished. This does not exclude the existence of correlations between A and B, since they could result from common causes in the overlap region of their backward cone. The overlap regions, where the two cones intersect, corresponds to systems causally related. Note that in Fig.1 the arrow of time stops after the detection happens at time t_D, since the photons are irreversibly absorbed during detection. After the irreversible detection process all that remains is classical information (photon counts). This fact shows dramatically how information is always at the boundary between the quantum and the classical world.[15] Once recorded, this classical information can be cleverly manipulated; for instance only a portion of the measured counts can be selected in order to display the desired quantum correlations. Such a selective procedure (named post-selection), involves only classical communication between Alice and Bob and is present in any type of coincidence measurement procedure. In particular it is present in standard quantum state reconstruction tomographic procedures.

4. Scattering processes as trace-preserving and non-trace-preserving quantum maps

The mathematical description of a multiple-scattering processes involving a quantum object (such as a photon) is somewhat cumbersome. The appropriate formulation depends among other things on a correct specification of the detection mecha-

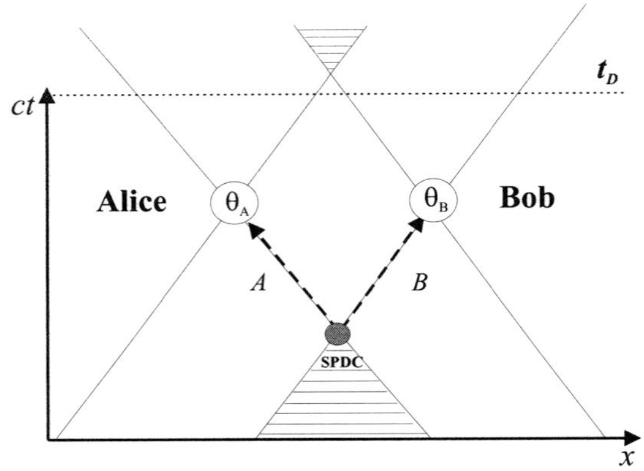

Fig. 2. A pair of twin-photons (A-B) produced by SPDC propagates in the forward direction of the space-time diagram. The cone of events for each photon is determined by a choice of each polarizer setting $\theta_{A,B}$. In order to satisfy the no-signalling condition the forward cones of photon A and B should not intersect before the detection time t_D.

nism. In its most pedestrian form the state of a scattered photon can be described as a pure superposition of transverse-spatial modes with different probability amplitudes (which depend on the scatterer). If the detection system could resolve each individual transverse-mode, the state would remain pure, and the whole evolution would be described by a single unitary operator. In a more realistic scenario, the polarization state of the scattered photon is detected in a multi-transverse-mode fashion, so that the spatial mode information is averaged (or traced over) upon detection, and the state of the system is reduced to a statistical mixture (i.e., a density matrix).[16,17] Thus, we can effectively describe the system as if it was open, where the role of the environment is played by the unobserved internal degrees of freedom (i.e., the transverse-spatial modes of the photons). In this case the evolution of the system cannot be described by a *single* unitary operator. It has to be described in terms of a *set* of unitary operators. This set of unitary operators is usually referred to as "super-operator" or quantum map.[4] A quantum map is a quantum operation that relates input and output density matrices. As much as the evolution operator for closed systems is considered physical only if it is unitary, a quantum operation $\mathcal{E}$ is a physical map that transforms the input density operators ρ^{in} into the output density operators $\rho^{out} \equiv \mathcal{E}(\rho^{in})$ if it satisfies:[4]

- $0 \leq \mathrm{Tr}\{\mathcal{E}(\rho^{in})\} \leq 1$,
- $\mathcal{E}$ is a convex-linear map such that:

$$\mathcal{E}(\sum_i p_i \rho_i) = \sum_i p_i \mathcal{E}(\rho_i) \qquad (p_i \geq 0),$$

- $\mathcal{E}$ is a completely positive map.

The first condition states that $0 \leq \mathrm{Tr}\{\rho_{out}\} \leq 1$. When no irreversible processes such as measurements (i.e., projections) or dissipation (i.e., anisotropic losses) are involved the map satisfies the condition $\mathrm{Tr}\{\mathcal{E}(\rho^{in})\} = 1$, and it is called a trace-preserving map (also referred to as a deterministic map). In the context of polarization optics a particular kind of irreversible process is given by polarization dependent losses. This is in fact the case for dichroic media such as polarizers, which transmit an arbitrary polarization while absorbing the rest. Such type of anisotropic losses must be described by a non-trace-preserving map such that $0 \leq \mathrm{Tr}\{\mathcal{E}(\rho^{in})\} < 1$. We will next see that these types of maps give rise to interesting questions. The second condition requires the map to be linear and to preserve probabilities, and finally the third one guarantees that ρ^{out} is positive semi-definite, so that it represents a legal density matrix.

5. Non-trace-preserving maps and the causality condition

In the context of quantum optics, a common way of proving entanglement between bipartite systems is by measuring the two-photon density matrix ρ_{AB} by means of quantum tomography and then extracting from it a given entanglement measure such as the concurrence or the tangle.[9] One simply infers from the causality condition that if a pair of photons are space-like separated in an initial state ρ_{AB}^{in}, they cannot communicate before detection and thus they cannot affect each other states before detection. This implies that if we act locally on only one of the two photons (say photon A via a local map $\mathcal{E}_A$), and we measure the state of photon B after the action on photon A took place, we should not see any change in the state of subsystem B. This condition is so important that it strongly restricts the possible outcomes we can measure for the two-photon scattered state ρ_{AB}^{out}. In particular, we have experimentally demonstrated and numerically verified that a two-photon scattered state generated by a local scattering process, in the experimental configuration shown in Fig 1, can only have a very particular shape; namely, it can only belong to a generalized class of Werner state (see Fig. 3 (a)). This statement is quite, but not completely, general. In fact, it is only true when the scattering system applied on A has no selective losses, in other words when the action upon subsystem A can be described in terms of a trace-preserving map. On the other hand, we have found that when selective losses are allowed, as in the case of dichroic scattering media, a class of sub-Werner states is obtained for the two-photon scattered state ρ_{AB}^{out}[18] (see Fig. 3 (b)).

What happens then when the scattering system involves selective absorption? So far, we know that this type of scattering media cannot be described by a trace preserving map. But what is the consequence of that? As we will see, non-trace preserving maps can lead to apparent violations of causality. Consider as a trivial example the case of a polarization singlet input state $\rho_{AB}^{in} = |\psi^-\rangle\langle\psi^-|$ where $|\psi^-\rangle = (|HV\rangle - |VH\rangle)/\sqrt{2}$. This state contains maximal information about the bipartite correlations, but minimal information about the state of each subsystem. So if we

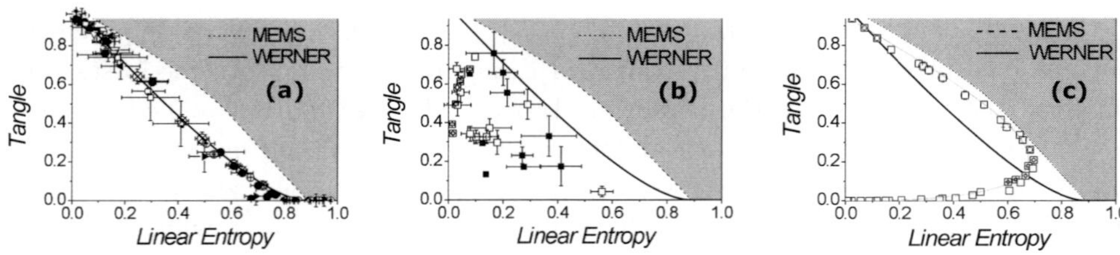

Fig. 3. (a) Generalized-Werner states generated by trace-preserving maps (b) Sub-Werner states generated by non-trace-preserving maps and (c) MEMS states generated by non-trace-preserving maps.

measure the reduced input state of B ($\rho_B^{in} = \mathrm{Tr}_A\{\rho_{AB}^{in}\}$) or the reduced input state of A ($\rho_A^{in} = \mathrm{Tr}_B\{\rho_{AB}^{in}\}$) we obtain in both cases a maximally mixed state (proportional to the 2×2 identity). Let's imagine now that we place a polarizer (we assume for simplicity that it is oriented in the H direction) in the path of photon A, this is a typical dichroic system acting locally on subsystem A. If we then measure the bi-photon output state in a coincidence-count circuit, we will obtain $\rho_{AB}^{out} = |H\rangle\langle H|_A \otimes |V\rangle\langle V|_B$, which is a separable pure state. If we now trace over photon A, we obtain that the output state of B is also fully polarized in the V direction $\mathrm{Tr}_A\{\rho_{AB}^{out}\} = \rho_B^{out} = |V\rangle\langle V|$, so clearly the output state of B obtained in this way is not equivalent to (ρ_B^{in}), so we could claim that photon B was affected by only acting on A, thus violating causality. The reason why the state of B has apparently changed without acting on it, is because there has been classical communication between Alice and Bob (of course the communication was *after* detection so there is no violation of causality). Such is the case in a coincidence-count type of measurement. The state of B has changed only after the tomographic procedure, which only involves local operations and classical communication.

Besides, on a more sophisticated level, non-trace preserving maps have proved to be useful for maximally-entangled-mixed state engineering.[19] Consider as an example the local map proposed in[19] to create maximally entangled mixed states (MEMS) (see Fig. 3 (c)). MEMS are of interest for realistic quantum information applications.[11,12] This map can be implemented by using a medium with anisotropic losses (i.e., dichroic medium) acting locally on one photon of the pair. A medium with anisotropic losses naturally performs a kind of post-selective measurement, since it selectively transmits (or absorbs) a portion of the total number of photons that passes through it.[20] These selective losses introduce a non-trivial renormalization to the measured bipartite output density matrix ρ_{AB}^{out}, in contrast to the trivial isotropic losses given for instance by isotropic scattering, where only the total intensity diminishes (in terms of maps, isotropic or random scattering can be described by a trace preserving map up to an overall renormalization constant). The reconstructed density matrix of the bipartite output state ρ_{AB}^{out} contains all the information about the correlations between systems A and B, which was obtained by performing single photon operations and classical communication. Note that any classical communication between distant observers can be considered as a kind of long range interaction;[15] in this way it is apparently possible to affect the state of B at a distance. But this does *not* violate the no-signalling condition, since the communication or the 'interaction' occurs after detection, by post-selection of the counts.

6. Conclusion

In conclusion, we have shown how a scattering process involving twin photons can be described as a quantum map. In the case of non-trace-preserving maps, as it is the case for scattering systems with polarization dependent losses, the tomographically reconstructed bi-photon density matrices can lead to an apparent violation of the no-signalling condition. This apparent surprise can be overcome when one includes post-selective detection, which is a procedure that involves classical communication between distant observers.

Acknowledgments

This project is part of the program of FOM and is also supported by the EU under the IST-ATESIT contract.

References

1. A. Einstein, B. Podolsky and N. Rosen, *Phys. Rev.* **47**, 777 (1935).
2. J. S. Bell, *Physics* (Long Island City, N.Y.) **1**, p. 195 (1965).
3. A. Aspect, J. Dalibard and G. Roger, *Phys. Rev. Lett.* **49**, p. 1804 (1982).
4. M. A. Nielsen and I. L. Chuang, *Quantum computation and quantum information* (Cambridge Univ. Press, 2000).
5. P. G. Kwiat, K. Mattle, H. Weinfurter, A. Zeilinger, A. V. Sergienko and Y. Shih, *Phys. Rev. Lett.* **75**, p. 4337 (1995).
6. G. Puentes, D. Voigt, A. Aiello and J. P. Woerdman, *Opt. Lett.* **31**, p. 2057 (2006).
7. G. Puentes, D. Voigt, A. Aiello and J. P. Woerdman, *Opt. Lett.* **30**, p. 3216 (2005).
8. D. F. V. James, P. G. Kwiat, W. J. Munro and A. G. White, *Phys. Rev. A* **64**, p. 052312 (2001).
9. W. K. Wootters, *Phys. Rev. Lett.* **80**, p. 2245 (1998).
10. S. Bose and V. Vedral, *Phys. Rev. A* **61**, p. 040101(R) (2000).
11. N. A. Peters, J. B. Altepeter, D. A. Branning, E. R. Jeffrey, T. C. Wei and P. G. Kwiat, *Phys. Rev. Lett.* **92**, p. 133601 (2004).
12. M. Barbieri, F. D. Martini, G. D. Nepi and P. Mataloni, *Phys. Rev. Lett.* **92**, p. 177901 (2004).
13. P. Pearle, *Phys. Rev. D* **2**, p. 1418 (1970).
14. G. Weihs, T. Jennewein, C. Simon, H. Weinfurter and A. Zeilinger, *Phys. Rev. Lett.* **81**, p. 5039 (1998).
15. A. Peres and D. R. Terno, *Rev. Mod. Phys.* **76**, p. 93 (2004).
16. A. Aiello and J. P. Woerdman, *Phys. Rev. A* **70**, p. 023808 (2004).
17. J. L. van Velsen and C. W. J. Beenakker, *Phys. Rev. A* **70**, p. 032325 (2004).
18. G. Puentes, A. Aiello, D. Voigt and J. P. Woerdman, *submitted to Phys. Rev. A* (2006).

19. A. Aiello, G. Puentes, D. Voigt and J. P. Woerdman, *submitted to Phys. Rev. Lett.* (2006).
20. N. Brunner, A. Acin, D. Collins, N. Gisin and V. Scarani, *Phys. Rev. Lett.* **91**, p. 180402 (2003).

SIMULTANEOUS MEASUREMENT OF NON-COMMUTING OBSERVABLES

G. AQUINO and B. MEHMANI*

Institute for Theoretical Physics, University of Amsterdam,
Valckenierstraat 65, 1018 XE, Amsterdam, The Netherlands
**E-mail: bmehmani@science.uva.nl*

It is shown that the full unknown state of a spin-$\frac{1}{2}$ system, S, described by its density matrix, can be determined with a simultaneous measurement with the help of another system A, called assistant, whose initial state is known. The idea is to let S and A interact with each other in a known way during a proper interaction time τ, and then to measure simultaneously two observables, one of S and one of A and then to determine their averages and their correlation. One thus determines the three unknown components of the polarization vector of S by means of repeated experiments using a unique stetting. In this way one can measure all the non-commutative observables of S at the same time, which may seem prohibited in quantum mechanics.

Keywords: State determination; Quantum measurement; Two-level system; Coherent state.

1. Introduction

The determination of the unknown state of a quantum system is one of the most important issues in the field of quantum information.[1,2] This determination involves a measurement process in which a macroscopic system, apparatus, is coupled to the quantum system; during this process the state of both the apparatus and the system is modified.[3] For instance, as currently described in many textbooks, the z-component of the polarization vector of a spin-$\frac{1}{2}$system, S, can statistically be determined by means of a repeated Stern-Gerlach experiment. In this process, the x- and y-component of the polarization vector are destroyed as a consequence of the non-commutation of the spin operator in the transversal directions. Other experimental settings seem therefore necessary to measure the unknown polarization vector of S. Its three components are represented by incompatible observables, the Pauli operators, and their direct determination requires three macroscopic apparatuses, differing through a change of orientation of the magnets and detectors. Likewise, the state of any two-level system, represented by a 2×2 density matrix $\hat{\rho}$ can be fully determined only through measurement of three linearly independent observables which do not commute and cannot be simultaneously measured.
Nevertheless, we will prove that the whole unknown density matrix of such a system S, in particular the full polarization vector of a spin-$\frac{1}{2}$ system, can be determined

indirectly by means of a set of measurements performed simultaneously on S and an auxiliary system, A, which we term the assistant. The strategy is the following: initially S is in the unknown state that we wish to determine, while the assistant A is in some known state. During some time lapse S and A interact in a known fashion. Their joint state is modified, it involves correlations and keeps memory of the initial state of S. A simultaneous measurement of one of the observables of S and A is then performed. Repeating this process provides then three statistical data: the expectation values of these observables and their correlation. We will show that one can infer the three components of the initial polarization vector of S from the three data.

There are two approaches to this problem. Either using another two-level assistant, for solid state applications[5] or using an electromagnetic field as an assistant, for quantum optics applications.

After defining the general problem, we illustrate, in section 3, the first approach, i.e. using another spin-$\frac{1}{2}$ system as an assistant in a known pure state to determine the initial state of S. Then, in section 4, we show that it is also possible to use an electromagnetic field in a coherent state to determine the whole elements of the unknown density matrix of S.

2. Statement of the Problem

The idea of mapping the state $\hat{\rho}$ of an unknown spin-$\frac{1}{2}$ system, S, onto a single observable of S+A system by using another system in a known state $\hat{R}$, A, was first proposed by D'Ariano.[4] It was explicitly implemented in a dynamic form in.[5]

The state of the composite system $S+A$, which is tested, is

$$\hat{\mathcal{R}}_\tau = \hat{U}\hat{\mathcal{R}}_0\hat{U}^\dagger, \tag{1}$$

where the initial state of $S+A$ is $\hat{\mathcal{R}}_0 = \hat{R} \otimes \hat{\rho}$ and the evolution operator is $\hat{U} = e^{-i\hat{H}\tau}$. Therefore, the dynamics of the system yields the required mixing of $\hat{\rho}$ and $\hat{R}$ and the simplest possible non-degenerate observable of the composite system S+A, $\hat{\Omega}$, can be chosen as a factorized quantity

$$\hat{\Omega} = \hat{\omega} \otimes \hat{o}, \tag{2}$$

where the observable $\hat{\omega}$ and $\hat{o}$ pertain to S and A respectively. Then the spectral decomposition of $\hat{\omega}$ and $\hat{o}$ can be used to construct the projection operator $\hat{P}_\alpha$ of $\hat{\Omega}$

$$\hat{\omega} = \sum_{i=1}^{m} \omega_i \hat{\pi}_i, \qquad \hat{o} = \sum_{a=1}^{n} o_a \hat{p}_a, \tag{3}$$

where $\hat{p}_a$ and $\hat{\pi}_i$ are eigen projectors of the observables $\hat{o}$ and $\hat{\omega}$ respectively. Therefore, projection operator $\hat{P}_\alpha$ with $\alpha \equiv (i, a)$ takes the form

$$\hat{P}_\alpha \equiv \hat{P}_{ia} = \hat{\pi}_i \otimes \hat{p}_a. \tag{4}$$

Repeated measurements of $\hat{\Omega}$ which means repeated simultaneous measurements of $\hat{\omega}$ and $\hat{o}$, determines the joint probabilities to observe ω_i for S and o_a for A

$$P_\alpha \equiv P_{ia} = \mathrm{Tr}[\hat{\mathcal{R}}_\tau(\hat{\pi}_i \otimes \hat{p}_a)], \tag{5}$$

where $\hat{\mathcal{R}}_\tau$ is defined in (Eq. 1). In fact the numbers P_α are the diagonal elements of $\hat{U}^\dagger(\hat{\rho} \otimes \hat{R})\hat{U}$ in the factorized basis which diagonalizes $\hat{\omega}$ and $\hat{o}$.
The whole elements of the density matrix $\hat{\rho}$ can be determined by the mapping $\hat{\rho} \rightarrow P_\alpha$. Like the idea of finding a universal observable,[4] if $\hat{H}$ couples S and A properly, this mapping will be expected to be invertible for $n \geq m$. We shall see that even simple interactions can achieve this condition. For given observables $\hat{\omega}$ of S and $\hat{o}$ of A and for a known initial state $\hat{R}$, the precision of this procedure relies on the ratio between the experimental uncertainty of P_α and the resulting uncertainty on $\hat{\rho}$, which can be characterized by the determinant, Δ, of the transformation (5). For $\Delta = 0$ it is impossible to determine $\hat{\rho}$ from P_α. This means, the system is unstable with respect to small errors made during experimental determination of P_α or equivalently $(\langle\hat{\sigma}_z\rangle, \langle\hat{s}_z\rangle, \langle\hat{s}_z\hat{\sigma}_z\rangle)$. Therefore, the Hamiltonian $\hat{H}$ and time interval τ should be chosen so as to maximize $|\Delta|$ over all possible unitary transformations.

3. Spin-$\frac{1}{2}$ Assistant in a Known Pure State

In this section we illustrate the above ideas by studying a two-level system S, namely, an spin-$\frac{1}{2}$ system in a known pure state. The density matrix of a spin-$\frac{1}{2}$ can be represented with the help of Pauli matrices. The determination of $\hat{\rho}$ corresponds to determination of the elements of the polarization vector, $\vec{\rho}$. we let S and A interact during the time interval τ. The observables $\hat{\omega}$ and $\hat{o}$ to be measured are the z-components of spin of S and A and are determined by $\hat{\sigma}_z$ and $\hat{s}_z$ respectively. The projection operators are

$$\hat{\pi}_i = \frac{1}{2}(\hat{1} + \hat{\sigma}_z), \quad \hat{p}_a = \frac{1}{2}(\hat{1} + \hat{s}_z), \tag{6}$$

for i and a equal to ± 1. Experiments will determine the four joint probabilities $P_\alpha = P_{++}, P_{+-}, P_{-+}, P_{--}$. These probabilities are related to the three real parameters $\vec{\rho}$ of $\hat{\rho}$ by inserting (Eq. 6) and $\hat{R} = \frac{1}{2}(\hat{1} + \hat{s}_z)$ into (Eq. 5).

$$P_\alpha = u_\alpha + \vec{v}_\alpha \cdot \vec{\rho}, \tag{7}$$

where

$$u_\alpha = \frac{1}{2}[\hat{U}(\hat{1} \otimes \hat{R})\hat{U}^\dagger]_{\alpha,\alpha}, \quad \vec{v}_\alpha = \frac{1}{2}[\hat{U}(\vec{\sigma} \otimes \hat{R})\hat{U}^\dagger]_{\alpha,\alpha}, \tag{8}$$

with $\alpha = \{ia\} = \{++, +-, -+, --\}$ and matrix elements have been represented in the standard representation of the Pauli matrices, $\vec{\sigma}$ and $\vec{s}$. The probabilities P_α should be positive and normalized for any the density matrix, $\hat{\rho}$, such that $\vec{\rho}^2 \leq 1$. These conditions imply that

$$u_\alpha \geq | v_\alpha |, \qquad \sum_\alpha u_\alpha = 1, \qquad \sum_\alpha \vec{v}_\alpha = 0. \tag{9}$$

The determinant of transformation $\hat{\rho} \to P_\alpha$ can be either

$$\vec{v}_{++} \cdot (\vec{v}_{+-} \times \vec{v}_{+-}), \tag{10}$$

or any other permutations of three of the vectors $\vec{v}_{++}$, $\vec{v}_{+-}$, $\vec{v}_{-+}$ and $\vec{v}_{--}$. Therefore, the determinant of the transformation is four times the volume of the parallelepiped made by three of these vectors. For example,

$$\Delta = 4\vec{v}_{++} \cdot (\vec{v}_{+-} \times \vec{v}_{+-}). \tag{11}$$

If the unitary evolution operator $\hat{U}$ is such that vectors $\vec{v}_\alpha$ are not coplanar, the transformation (Eq. 7) is invertible and one can determine $\vec{\rho}$ from the set of P_α. Alternatively, $\hat{\rho}$ is deduced from $< \hat{\sigma}_z >$, $< \hat{s}_z >$ and $< \hat{s}_z\hat{\sigma}_z >$ at time τ. The notation $\hat{s}_z\hat{\sigma}_z$ is used for simplicity instead of $\hat{s}_z \otimes \hat{\sigma}_z$ which lives in the common Hilbert space of S and A. $\hat{\sigma}_z$, $\hat{s}_z$ and $\hat{s}_z\hat{\sigma}_z$ can be simultaneously measured and are in one to one correspondence with the set of probabilities P_α.

We first look for the upper bound of the determinant of transformation (Eq. 7), $|\Delta|$ implied by the conditions (Eq. 9). First we note that $|\Delta|$ increases with $|\vec{v}_\alpha|$ for each α. We therefore maximize Δ^2 under the constraints

$$\sum_\alpha |\vec{v}_\alpha| = 1, \quad \sum_\alpha \vec{v}_\alpha = 0. \tag{12}$$

This yields a symmetric solution for all these vectors

$$u_\alpha = |\vec{v}_\alpha| = \frac{1}{4}, \quad \cos(\vec{v}_\alpha, \vec{v}_\beta) = \frac{\vec{v}_\alpha \cdot \vec{v}_\beta}{|\vec{v}_\alpha||\vec{v}_\beta|} = -\frac{1}{3}. \tag{13}$$

By definition this means that vectors $\vec{v}_\alpha$ form a regular tetrahedron. These solutions are not unique and they follow from one another by rotating in the space of the spins and permutations of the indexes α. Therefore, the corresponding determinant for the upper bound is

$$|\Delta| = \frac{1}{12\sqrt{3}}. \tag{14}$$

Having a non-zero determinant for the proposed procedure, ensures its feasibility. One simple choice for vectors $\vec{v}_\alpha$ is

$$\begin{aligned} \vec{v}_{++} &= \frac{1}{4\sqrt{3}}(1,1,1), & \vec{v}_{+-} &= \frac{1}{4\sqrt{3}}(-1,1,-1), \\ \vec{v}_{-+} &= \frac{1}{4\sqrt{3}}(1,-1,-1), & \vec{v}_{--} &= \frac{1}{4\sqrt{3}}(-1,-1,1). \end{aligned} \tag{15}$$

This yields a specially simple form for the density matrix of S:

$$\rho_1 = \sqrt{3} < \hat{\sigma}_z >, \quad \rho_2 = \sqrt{3} < \hat{s}_z >, \quad \rho_3 = \sqrt{3} < \hat{s}_z\hat{\sigma}_z >, \tag{16}$$

which gives directly the whole elements of the density matrix, $\hat{\rho}$, in terms of the expectation values and the correlation of the commuting observables $\hat{\sigma}_z$ and $\hat{s}_z$ in the final state.

Next step is to find out the interaction Hamiltonian and the interaction time τ

which give such a description of the tested system, S.
This correspondence can be achieved under the action of the Hamiltonian

$$\hat{H} = \frac{1}{\sqrt{2}}\hat{\sigma}_x(\hat{s}_x \cos\phi + \hat{s}_z \sin\phi) + \frac{1}{2}[(\hat{s}_y - \hat{s}_x)\sin\phi + \hat{s}_z \cos\phi], \tag{17}$$

where 2ϕ is the angle between $\vec{v}_{++}$ and the z-axis, that is, $\cos\phi = \frac{1}{\sqrt{3}}$. Noting that $\hat{H}^2 = \sin^2\chi$, where χ satisfies $\cos\chi = 1/2\cos\phi$, and taking as duration of the evolution $\tau = \chi/\sin\chi$, we obtain $\hat{U} \equiv exp(-i\hat{H}\tau) = \cos\chi - i\hat{H}$. The simpler form

$$\hat{H} = \frac{1}{\sqrt{2}}\hat{\sigma}_x\hat{s}_x + \frac{1}{2}\left(\hat{s}_y \sin\phi + \hat{s}_z\right) \tag{18}$$

of $\hat{H}$ can be obtained by a rotation of $\vec{\hat{s}}$ and also achieves an optimal mapping $\hat{\rho} \to P_\alpha$, provided $\hat{s}_z \to \hat{s}_x \sin\phi + \hat{s}_z \cos\phi$ both in the measured projections $\hat{p}_a = 1/2(1 \pm \hat{s}_z)$ and in the initial state $\hat{R} = \hat{p}_+$. The first term in (Eq. 18) describes, in the spin language, an Ising coupling, while the second term represents a transverse magnetic field acting on the assistant A.

4. Assistant System as a Coherent State of Light

In this section we discuss the possibility of using light as an assistant to determine the elements of the density matrix of a spin-$\frac{1}{2}$ system. We show that in case of using an electromagnetic field in coherent state, one can determine the state of S from the commutative measurements on S and A.
To describe this physical situation we choose the Jaynes-Cummings model,[67] a widely accepted model describing the interaction of matter (two-level atom or spin-$\frac{1}{2}$ system) and a single mode of radiation. This model is exactly solvable but still rather non-trivial and it finds direct experimental realization in quantum optics. The Hamiltonian reads:[6]

$$\hat{H} = \hat{H}_A + \hat{H}_S + \hat{H}_{SA} = \hbar\omega\hat{a}^\dagger\hat{a} + \frac{1}{2}\hbar\omega\hat{\sigma}_z + \hbar\gamma(\hat{\sigma}_+\hat{a} + \hat{\sigma}_-\hat{a}^\dagger) \tag{19}$$

where $\hat{a}^\dagger$ and $\hat{a}$ are the standard photon creation and annihilation operators of the field (the assistant A), with commutation relation $[\hat{a}, \hat{a}^\dagger] = 1$, $\hat{\sigma}_i$ are the standard Pauli matrices for the spin of the two level system S. The total Hamiltonian (19) is the sum of the Hamiltonian of the field $\hat{H}_A$, the Hamiltonian $\hat{H}_S$ of the two level system S and the interaction Hamiltonian $\hat{H}_{SA}$ which can be written as $\hbar\gamma\hat{V}$, with γ the coupling constant. It can be easily checked that the interaction operator, $\hat{V}$, and the total number of excitations, $\hat{N}$:

$$\hat{V} = \hat{\sigma}_+\hat{a} + \hat{\sigma}_-\hat{a}^\dagger, \quad \hat{N} = \hat{a}^\dagger\hat{a} + \hat{\sigma}_+\hat{\sigma}_-. \tag{20}$$

are two constants of motion. $\hat{V}$ and $\hat{N}$ also commute with each other since $\hat{V}^2 = \hat{N}$.

One can then calculate exactly the relevant observables at time t using the Heisenberg equations of motion:

$$\begin{aligned}\dot{\hat{a}} &= -i\omega\hat{a} - i\gamma\hat{\sigma}_-,\\ \dot{\hat{\sigma}}_- &= -i\omega\hat{\sigma}_- + i\gamma\hat{\sigma}_z\hat{a},\\ \dot{\hat{\sigma}}_z &= 2i\gamma(\hat{a}^\dagger\hat{\sigma}_- - \hat{\sigma}_+\hat{a}).\end{aligned} \tag{21}$$

We will briefly outline the result. The exact solution of the above set of equations reads:

$$\begin{aligned}\hat{a}(t) &= e^{i(\gamma\hat{V}-\omega)t}\left[\left(\cos\gamma\hat{K}t - \frac{i\hat{V}\sin\gamma\hat{K}t}{\hat{K}}\right)\hat{a}(0) - \frac{i\sin\gamma\hat{K}t}{\hat{K}}\hat{\sigma}_-(0)\right],\\ \hat{\sigma}_-(t) &= e^{i(\gamma\hat{V}-\omega)t}\left[\left(\cos\gamma\hat{K}t + \frac{i\hat{V}\sin\gamma\hat{K}t}{\hat{K}}\right)\hat{\sigma}_-(0) - \frac{i\sin\gamma\hat{K}t}{\hat{K}}\hat{a}(0)\right],\end{aligned} \tag{22}$$

where $\hat{K} = \sqrt{\hat{N}+1} = \sqrt{\hat{V}^2+1}$. Note that $\hat{K}$ and $\hat{V}$ commute so that their mutual ordering is irrelevant.
We study the case in which the electromagnetic field is in a coherent state, a condition that coincides with the common experimental situation of a resonant laser mode interacting with a spin-$\frac{1}{2}$ system. Assuming initial factorization between S and A, the density matrix of the total system S+A at time $t=0$, is:

$$\hat{\rho}(t=0) = \frac{1}{2}(1 + \langle\hat{\sigma}_i\rangle\hat{\sigma}_i) \otimes \sum_{k=0}^{\infty}\sum_{m=0}^{\infty}\frac{e^{-|\alpha|^2}}{\sqrt{k!}\sqrt{m!}}\alpha^k\alpha^{*m}|k\rangle\langle m| \tag{23}$$

where $|k\rangle$ denotes the ket for the photon quanta and $|\alpha|^2$ is the average photon number in the coherent state.

We consider as possible triplets of commuting observables the following:

$$\begin{aligned}&\hat{\sigma}_z \ , \ \hat{a}^\dagger\hat{a} \ , \ \hat{\sigma}_z\hat{a}^\dagger\hat{a}\\ &\hat{\sigma}_z \ , \ \hat{a}^\dagger \pm \hat{a} \ , \ \hat{\sigma}_z(\hat{a}^\dagger \pm \hat{a})\\ &\hat{\sigma}_x \ , \ \hat{a}^\dagger\hat{a} \ , \ \hat{\sigma}_x\hat{a}^\dagger\hat{a}\end{aligned} \tag{24}$$

but others combinations are possible. A direct evaluation shows that only the third choice produces a set of independent equations relating the measurements of the chosen commutating observables at time $t > 0$, after turning on the interaction, and the state of the system S at time $t = 0$. Only in this case then, a reconstruction of the state of the spin-$\frac{1}{2}$ system at time $t = 0$ is possible . We report here the details of the calculation only for this relevant choice of observables. We have to calculate the expectation values of these three observables at time t. In order to perform this calculation, we will make use of the eigenfunctions of the $\hat{V}$ operator, which are:

$$|\phi_n^\pm\rangle = \frac{|n-1\rangle|+\rangle \pm |n\rangle|-\rangle}{\sqrt{2}} \quad (n \geq 1)\ , \qquad |\phi_0^\pm\rangle = |\phi_0\rangle = |0\rangle|-\rangle \qquad (n=0), \tag{25}$$

and the operators $\hat{V}$ and $\hat{K}$, when applied to these functions, evaluate to:

$$\hat{V}|\phi_n^\pm\rangle = \pm\sqrt{n}\,|\phi_n^\pm\rangle, \qquad \hat{K}|\phi_n^\pm\rangle = \sqrt{n+1}\,|\phi_n^\pm\rangle. \tag{26}$$

Let us introduce the following notation:

$$
\begin{aligned}
\hat{f}(\hat{V}) = e^{i(\gamma\hat{V}-\omega)t} &\Rightarrow f_n^{\pm} = \langle\phi_n^{\pm}|f(\hat{V})|\phi_n^{\pm}\rangle = e^{i(\pm\gamma\sqrt{n}-\omega)t} \to \bar{f}_n^{\pm} = (f_n^{\pm})^* \\
\hat{S}(\hat{K}) = \frac{i\sin\gamma\hat{K}t}{\hat{K}} &\Rightarrow S_n = \langle\phi_n^{\pm}|S(\hat{K})|\phi_n^{\pm}\rangle = \frac{i\sin\gamma\sqrt{n+1}t}{\sqrt{n+1}} \to \bar{S}_n = S_n^* \\
\hat{g}(\hat{V},\hat{K}) = \cos\gamma\hat{K}t + \hat{V}\hat{S}(\hat{K}) &\Rightarrow g_n^{\pm} = \langle\phi_n^{\pm}|\hat{g}(\hat{V},\hat{K})|\phi_n^{\pm}\rangle \\
&\qquad = \cos\gamma\sqrt{n+1}t \pm \sqrt{n}S_n \to \bar{g}_n^{\pm} = (g_n^{\pm})^*
\end{aligned}
$$

where, for the sake of compactness, a bar in some cases is used instead of the asterisk to indicate the complex conjugation operation. In this notation it holds that:

$$
\begin{aligned}
\hat{\sigma}_+(t) &= (\hat{\sigma}_+ g^\dagger(\hat{V},\hat{K}) - \hat{a}^\dagger S^\dagger(\hat{K}))f^\dagger(\hat{V}) \\
\hat{a}^\dagger\hat{a}(t) &= (\hat{a}^\dagger g(\hat{V},\hat{K}) - \hat{\sigma}_+ S^\dagger(\hat{K}))(g^\dagger(\hat{V},\hat{K})\hat{a} - S(\hat{K})\hat{\sigma}_-) \\
\hat{a}^\dagger\hat{a}\hat{\sigma}_+(t) &= (\hat{a}^\dagger g(\hat{V},\hat{K}) - \hat{\sigma}_+ S^\dagger(\hat{K}))(g^\dagger(\hat{V},\hat{K})\hat{a} - S(\hat{K})\hat{\sigma}_-)(\hat{\sigma}_+ g^\dagger(\hat{V},\hat{K}) \\
&\quad - \hat{a}^\dagger S^\dagger(\hat{K}))f^\dagger(\hat{V})
\end{aligned} \tag{27}
$$

For a generic observable $\hat{O}$ we define:

$$
\langle\hat{O}(t)\rangle = Tr[\hat{\rho}(t=0)\hat{O}(t)] = \textstyle\sum_{n=0}^{\infty}\sum_{i=\pm}\langle\phi_n^i|\hat{\rho}(t=0)\hat{O}(t)|\phi_n^i\rangle
$$

We can then proceed to evaluate the expectation values at a generic time t of the chosen triplet of commuting observables. As a first step we have:

$$
\begin{aligned}
\langle\hat{\sigma}_+(t)\rangle &= \sum_{n=0}^{\infty}\frac{e^{-|\alpha|^2}|\alpha|^{2n}}{n!}\left[\left(\frac{\bar{f}_n^+\bar{g}_n^+ - \bar{f}_n^-\bar{g}_n^-}{2\alpha}\sqrt{n} - \bar{S}_n\frac{n-|\alpha|^2}{\alpha}\frac{\bar{f}_n^+ - \bar{f}_n^-}{2}\right)\lambda_1\right. \\
&\quad \left. - \bar{S}_n\frac{\bar{f}_n^+ - \bar{f}_n^-}{2}\frac{\alpha^*\sqrt{n}}{\alpha}\lambda_2 + \left(\frac{\bar{f}_n^+\bar{g}_n^+ + \bar{f}_n^-\bar{g}_n^-}{2} - \sqrt{n}\bar{S}_n\frac{\bar{f}_n^+ - \bar{f}_n^-}{2}\right)\lambda_2^* - \bar{S}_n\frac{\bar{f}_n^+ - \bar{f}_n^-}{2}\alpha^*\right] \\
\langle\hat{a}^\dagger\hat{a}(t)\rangle &= \sum_{n=0}^{\infty}\frac{e^{-|\alpha|^2}|\alpha|^{2n}}{n!}\left[\left(|g_n^+|^2(n-|\alpha|^2) - S_n\sqrt{n}\frac{g_n^+ - g_n^-}{2} + |S_n|^2\right)\lambda_1\right. \\
&\quad \left. + S_n\frac{g_n^+ + g_n^-}{2}(\alpha\lambda_2^* - \alpha^*\lambda_2) + |\alpha|^2|g_n^+|^2\right] \\
\langle\hat{a}^\dagger\hat{a}\hat{\sigma}_+(t)\rangle &= \sum_{n=0}^{\infty}\frac{e^{-|\alpha|^2}|\alpha|^{2n}}{n!}\left[\frac{n}{\alpha}\left(|g_n^+|^2(\sqrt{n}\frac{\bar{f}_n^+\bar{g}_n^+ - \bar{f}_n^-\bar{g}_n^+}{2} - S_n\frac{\bar{f}_n^+\bar{g}_n^+ - \bar{f}_n^-\bar{g}_n^+}{2}\frac{g_n^+ - g_n^-}{2} + \right.\right. \\
&\quad S_n\frac{f_n^+ - f_n^-}{2}(n - \frac{|\alpha|^2(n+1)}{n}(1-\frac{1}{\lambda_1})) + S_n\frac{\bar{f}_n^+\bar{g}_n^+ + \bar{f}_n^-\bar{g}_n^-}{2}\frac{g_n^+ + g_n^-}{2}\frac{|\alpha|^2}{n}(1-\frac{1}{\lambda_1}) \\
&\quad +|S_n|^2(\frac{\bar{f}_n^+\bar{g}_n^+ - \bar{f}_n^-\bar{g}_n^-}{2\sqrt{n}} + \frac{\bar{f}_n^+ + \bar{f}_n^-}{2}S_n - \frac{n+1}{\sqrt{n}}\frac{g_n^- + g_n^+}{2}\frac{\bar{f}_n^+ - \bar{f}_n^-}{2} + 2\sqrt{n}\frac{g_n^+ - g_n^-}{2}\cdot \\
&\quad \left.\frac{\bar{f}_n^+ + \bar{f}_n^-}{2} - \frac{|\alpha|^2}{\sqrt{n}}\frac{g_n^- + g_n^+}{2}\frac{\bar{f}_n^+ - \bar{f}_n^-}{2})\right)\lambda_1 + \left(n\frac{\bar{f}_n^+\bar{g}_n^+ + \bar{f}_n^-\bar{g}_n^-}{2}|g_n^+|^2 + S_n\frac{\bar{f}_n^+ - \bar{f}_n^-}{2}\cdot\right. \\
&\quad |g_n^+|^2\sqrt{n} + 2S_n\frac{\bar{f}_n^+\bar{g}_n^+ + \bar{f}_n^-\bar{g}_n^-}{2}\frac{\bar{g}_n^+ - \bar{g}_n^-}{2} - |S_n|^2(2\frac{g_n^- - g_n^+}{2}\frac{\bar{f}_n^+ - \bar{f}_n^-}{2} + \frac{n+1}{n}\cdot \\
&\quad \left.\frac{g_n^- + g_n^+}{2}\frac{\bar{f}_n^+ + \bar{f}_n^-}{2} + \frac{\bar{f}_n^+\bar{g}_n^+ + \bar{f}_n^-\bar{g}_n^-}{2n} - \frac{\bar{f}_n^+ - \bar{f}_n^-}{2\sqrt{n}}S_n)\right)\lambda_2^* + n\frac{\alpha^*}{\alpha}\left(|S_n|^2\frac{g_n^- + g_n^+}{2}\cdot\right. \\
&\quad \left.\left.\frac{\bar{f}_n^- + \bar{f}_n^+}{2} + S_n(|g_n^+|^2\frac{\bar{f}_n^+ - \bar{f}_n^-}{2}\frac{n+1}{\sqrt{n}} - \frac{\bar{f}_n^+\bar{g}_n^+ - \bar{f}_n^-\bar{g}_n^-}{2}\frac{g_n^+ + g_n^-}{2\sqrt{n}})\right)\lambda_2\right]
\end{aligned} \tag{29}
$$

where we have also defined:

$$\lambda_1 = \frac{1+\langle\hat{\sigma}_z(0)\rangle}{2}, \qquad \lambda_2 = \langle\hat{\sigma}_+(0)\rangle, \qquad \lambda_2^* = \langle\hat{\sigma}_-(0)\rangle \tag{30}$$

From Eqs. (29) it is then easy to derive a system of three equations relating the three expectation values:

$$\langle\hat{\sigma}_x(t)\rangle = 2\Re[\langle\hat{\sigma}_+(t)\rangle], \quad \langle\hat{a}^\dagger\hat{a}(t)\rangle, \quad \langle\hat{a}^\dagger\hat{a}\hat{\sigma}_x(t)\rangle = 2\Re\left[\langle\hat{a}^\dagger\hat{a}\hat{\sigma}_+t)\rangle\right] \tag{31}$$

evaluated at a generic time t, to the variables $\langle\hat{\sigma}_x(0)\rangle, \langle\hat{\sigma}_y(0)\rangle, \langle\hat{\sigma}_z(0)\rangle$, i.e. to the density matrix of the spin-$\frac{1}{2}$ system at time $t=0$. In order to assess if this system of equations has solutions, we have to evaluate the determinant of the matrix $\hat{M}$ made up by the coefficients of $\langle\hat{\sigma}_x(0)\rangle, \langle\hat{\sigma}_y(0)\rangle, \langle\hat{\sigma}_z(0)\rangle$ appearing in the system of equations determined by the evaluation of (31). In order to achieve this goal, it is convenient to define $\hat{M}_{ij} = \sum_{n=0}^{\infty}\frac{e^{-|\alpha|^2}|\alpha|^{2n}}{n!}M_{ij}(n)$ and then determine the coefficients $M_{ij}(n)$. Name $A_{ij}(n)$ the same type of coefficients but relative to the matrix $\hat{A}$ made up by the coefficients multiplying the variables $\lambda_1, \lambda_2, \lambda_2^*$ in the system of equations (29). By replacing the definitions of $f_n^\pm, g_n^\pm, S_n$ and then simplifying, one gets:

$$A_{11}(n) = -i\frac{e^{i\omega t}}{\alpha}(\sqrt{n}\cos\gamma\sqrt{n+1}t\sin\gamma\sqrt{n}t + |\alpha|^2\frac{\cos\gamma\sqrt{n}t\sin\gamma\sqrt{n+1}t}{\sqrt{n+1}}) \tag{32}$$

$$A_{12}(n) = \frac{ie^{i\omega t}\sqrt{n}}{\alpha\sqrt{n+1}}\sin\gamma\sqrt{n+1}t\sin\gamma\sqrt{n}t$$

$$A_{13}(n) = e^{i\omega t}\cos\gamma\sqrt{n+1}t\cos\gamma\sqrt{n}t$$

$$A_{21}(n) = \frac{(1+2n)(1+n-|\alpha|^2)-(1+n+|\alpha|^2)\cos 2\gamma\sqrt{n+1}t}{2(n+1)}$$

$$A_{22}(n) = \frac{i\alpha\cos\gamma\sqrt{n+1}t\sin\gamma\sqrt{n}t}{\sqrt{n+1}}, \qquad A_{23}(n) = A_{22}^*(n)$$

$$A_{31}(n) = nA_{11}(n), \qquad A_{32}(n) = nA_{13}(n), \qquad A_{33}(n) = nA_{12}(n)$$

It is easy to check that these coefficients are related to the coefficients $M_{ij}(n)$ in the following way :

$$M_{i1}(n) = \Re[A_{i1}(n)], \quad M_{i2}(n) = \Re[A_{i2}(n)+A_{i3}^*(n)], \quad M_{i3}(n) = \Im[A_{i2}(n)+A_{i3}^*(n)] \tag{33}$$

If we now calculate the determinant $\Delta(t)$ of the matrix $\hat{M}$, we obtain:

$$\Delta(t) = \sum_{l=0}^{\infty}\sum_{m=0}^{\infty}\sum_{n=0}^{\infty}\frac{e^{-3|\alpha|^2}|\alpha|^{2(l+n+m)}}{l!m!n!}\varepsilon_{ijk}M_{1i}(l)M_{2j}(m)M_{3k}(n) \tag{34}$$

We evaluate this determinant numerically. Convergence within the seventh significant figure is reached by keeping in each of the three sums the first 30 terms when $|\alpha|^2 \leq 9$. For larger values of $|\alpha|^2$ convergence turns out to be much slower. In Figure Fig. 1 the temporal evolution of the determinant is shown. We see that as $|\alpha|^2$ increases, i.e. for larger average number of photons, the determinant has fluctuations of larger amplitude, so that a time with a large enough determinant can be chosen to solve for the initial state of the spin-$\frac{1}{2}$ system. When the matrix elements are determined with some experimental uncertainty such a choice is a sensible one that allows a more accurate determination of the initial state of the spin-$\frac{1}{2}$ system and avoids cases of ill-conditioned matrix inversion.

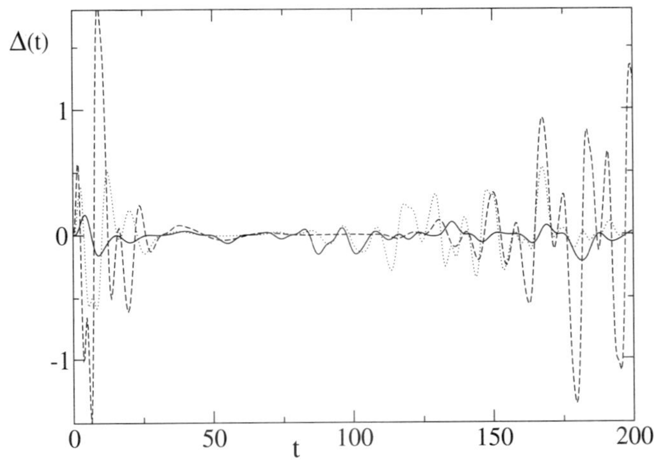

Fig. 1. Evolution of the determinant for $0 < t < 200$, with the following choice of parameters of the model: $\gamma = 0.1, \omega = 0.1$. The continuous line refers to the case $|\alpha|^2 = 1$ the dotted line to $|\alpha|^2 = 4$, the dashed line to $|\alpha|^2 = 9$, with α real and positive.

5. Conclusions

We have illustrated a procedure which allows to reconstruct the state of a spin-$\frac{1}{2}$ system with a simultaneous measurement of the expectation values of three commuting observables, by coupling the system to an assistant. We have also illustrated how the procedure works in the simple case of a spin-$\frac{1}{2}$ system coupled to a coherent laser field. We have shown that in this case, after a proper choice of the commuting observables, it is always possible to reconstruct the initial state of the spin-$\frac{1}{2}$ system. A radiation source with a large average number of photons allows to implement the procedure in an experimentally reliable condition, i.e. with a large absolute value of the determinant of the matrix connecting the expectation values of the commuting observables at time t to the initial state of the spin-$\frac{1}{2}$ system.

Acknowledgments

The authors would like to thank Th. M. Nieuwenhuizen, A. E. Allahverdyan and R. Balian for useful discussions and proofreading. The research of G. Aquino was supported by the EC Network DYGLAGEMEM.

References

1. J. A. Bergou, U Herzog and M. Hillery, *Phys. Rev. A.* **71**, 042314 (2005).
2. C. H. Bennett and D. P. diVincenzo, *Nature* **404**, 247-255 (2000).
3. A. E. Allahverdyan, R. Balian and Th. M. Nieuwenhuizen, in *Foundations of Probability and Physics-4*, A. Khrennikov, ed. (AIP Conference Proceedings, Vol. 750 (2005)).
4. G. M. D'Ariano, *Phys. Lett. A* **300**, 1 (2002).
5. A. E. Allahverdyan, R. Balian, Th. M. Nieuwenhuizen, *Phys. Rev. Lett.* **92**, 120402-1 (2004).
6. D. F. Walls, G. J. Milburn, *Quantum Optics*, (Springer, 1995).
7. U. Leonhardt, *Measuring the Quantum State of Light*, (Cambridge University Press, 1997).
8. C. W. Helstrom, *Quantum Detection and Estimation Theory*, (Academic Press, New York, 1976).

QUANTUM DECOHERENCE AND GRAVITATIONAL WAVES

M.T. JAEKEL*

Laboratoire de Physique Théorique de l'Ecole Normale Supérieure,
24 rue Lhomond, 75231 Paris Cedex 05, France
** E-mail: jaekel@lpt.ens.fr*

B. LAMINE, A. LAMBRECHT and S. REYNAUD**

Laboratoire Kastler Brossel, Université Pierre et Marie Curie
Campus Jussieu, 75252 Paris Cedex 05, France
*** E-mail: reynaud@spectro.jussieu.fr*

P. MAIA NETO

Instituto de Física, UFRJ,
CP 68528, 21945-970 Rio de Janeiro, Brazil

The quite different behaviors exhibited by microscopic and macroscopic systems with respect to quantum interferences suggest the existence of a borderline beyond which quantum systems loose their coherences and can be described classically. Gravitational waves, generated within our galaxy or during the cosmic expansion, constitute a universal environment susceptible to lead to such a quantum decoherence mechanism. We assess this idea by studying the quantum decoherence due to gravitational waves on typical microscopic and macoscopic systems, namely an atom interferometer (HYPER) and the Earth-Moon system. We show that quantum interferences remain unaffected in the former case and that they disappear extremely rapidly in the latter case. We obtain the relevant parameters which, besides the ratio of the system's mass to Planck mass, characterize the loss of quantum coherences.

Keywords: Quantum decoherence; Gravitational waves.

1. Introduction

Quantum decoherence is a universal phenomenon which affects all physical systems as soon as they are coupled to a fluctuating environment. This effect plays an important role in the transition between quantum and classical behaviors, by washing out quantum coherences and thus justifying a purely classical description.[1–5] This implies that quantum decoherence should be very efficient for macroscopic systems, while remaining inefficient for microscopic ones. The quantum/classical transition would then introduce a borderline between microscopic and macroscopic systems.

Existing experimental observations of quantum decoherence confirm these intuitions. Decoherence has only been seen on 'mesoscopic' systems for which the decoherence time is neither too long nor too short, such as microwave photons stored in a high-Q cavity[6] or trapped ions.[7] In such model systems, the environmental fluctuations are particularly well mastered and the quantum/classical transition has been shown to fit the predictions of decoherence theory.[8,9]

It has also been early remarked that Planck mass, that is the mass scale which can be built up on Planck constant $\hbar$, light velocity c and Newton gravitation G, lies at the borderline between microscopic and macroscopic masses

$$m_{\rm P} = \sqrt{\frac{\hbar c}{G}} \sim 22\mu{\rm g} \tag{1}$$

That is to say, one may define microscopic and macroscopic values of a mass m by comparing the associated Compton length $\ell_{\rm C}$ to the Planck length $\ell_{\rm P}$

$$m \lesseqgtr m_{\rm P} \Leftrightarrow \ell_{\rm P} = \sqrt{\frac{\hbar G}{c^3}} \lesseqgtr \ell_{\rm C} = \frac{\hbar}{mc} \tag{2}$$

It is tempting to consider that this property is not just an accidental coincidence but rather reveals a general consequence of fundamental gravitational fluctuations.[10–13] Then, one is led to study the role that the fluctuating gravitational environment might play in the transition from quantum to classical behaviors.

Here, we briefly discuss the quantum decoherence due to our local gravitational environment, namely the stochastic background of gravitational waves surrounding the Earth. Details can be found in previously published work.[14–18] First, taking the example of the atomic interferometer HYPER, we show that gravitational waves do not lead to a significant decoherence at the microscopic level. We then show that, on the contrary, scattering of gravitational waves is the dominant decoherence mechanism, and an extremely efficient one, for macroscopic systems such as the Moon around the Earth. We also go beyond the simple scaling arguments just given above by providing estimates of gravitational quantum decoherence depending not only on the mass of the system, but also on its velocity, on its geometry and on the noise spectrum characterizing the gravitational fluctuations.

2. Gravitational Environment

We first describe the fundamental fluctuations of space-time which originate from our gravitational environment and which are bound to play a crucial role in quantum decoherence. For current quantum systems which are only sensitive to frequencies lying far below Planck frequency, general relativity provides the appropriate description of gravitational phenomena,[19] even if it may ultimately be replaced by a theory of quantum gravity. It follows that the relevant spacetime fluctuations which constitute our gravitational environment are simply the gravitational waves predicted by the linearized theory of gravity[20–22] and which are thoroughly studied in relation with the present development of gravitational wave detectors.[23–26]

Gravitational waves correspond to perturbations of the metric field and can be written in the transverse traceless (TT) gauge

$$g_{\mu\nu} = \eta_{\mu\nu} + h_{\mu\nu}, \qquad \eta_{\mu\nu} = \mathrm{diag}(1, -1, -1, -1)$$
$$h_{00} = h_{\mathrm{i}0} = h_{\mathrm{i}}^{\mathrm{i}} = 0 \tag{3}$$

$\mathrm{i} = 1, 2, 3$ stands for the spatial indices whereas 0 will represent the temporal index; the spatial components $h_{\mu\nu}$ of the metric tensor are directly connected to the Riemann curvature. Gravitational waves are conveniently described through a mode decomposition in space-time (coordinates $(x^\mu), x^0 \equiv ct$)

$$h_{\mu\nu}(x) = \int \frac{\mathrm{d}^4 k}{(2\pi)^4} h_{\mu\nu}[k]\, e^{-ik_\lambda x^\lambda}, \qquad h_{\mu\nu}[k] = \Sigma_\pm \left(\frac{\varepsilon_\mu^\pm \varepsilon_\nu^\pm}{\sqrt{2}}\right)^* h^\pm[k] \tag{4}$$

Each Fourier component is a sum over the two circular polarizations $h^\pm$, which are obtained as products of the polarization vectors $\varepsilon^\pm$ well-known from electromagnetic theory. Gravitational waves correspond to wavevectors k lying on the light cone ($k^2 = k_\mu k^\mu = 0$), they are transverse with respect to this wavevector ($k^\mu \varepsilon_\mu^\pm = 0$) and the metric perturbation has a null trace ($(\varepsilon^\pm)^2 = 0$).

We consider for simplicity the case of stationary, unpolarized and isotropic backgounds. Then, a given metric component, say $h \equiv h_{12}$, is a stochastic variable characterized by a noise spectrum S_h

$$\langle h(t)\, h(0)\rangle = \int \frac{\mathrm{d}\omega}{2\pi} S_h[\omega]\, e^{-i\omega t} \tag{5}$$

S_h is the spectral density of strain fluctuations considered in most papers on gravitational wave detectors (see for example[24]). It can be written in terms of the mean number n_{gw} of gravitons per mode or, equivalently, of a noise temperature T_{gw} with k_{B} the Boltzmann constant and G the Newton constant

$$S_h = \frac{16G}{5c^5} \hbar\omega n_{\mathrm{gw}} = \frac{16G}{5c^5} k_{\mathrm{B}} T_{\mathrm{gw}} \tag{6}$$

Knowledge on gravitational wave backgrounds comes from studies estimating the probability of events which might be observed by interferometric detectors of gravitational waves. An important component is constituted by the 'binary confusion background', that is the estimated level for the background of gravitational waves emitted by unresolved binary systems in the galaxy and its vicinity. This 'binary confusion background' leads to a nearly flat function S_h, that is also to a nearly thermal spectrum, in the μHz to 10mHz frequency range[23]

$$10^{-6}\mathrm{Hz} < \frac{\omega}{2\pi} < 10^{-4}\mathrm{Hz} \qquad S_h \sim 10^{-34}\mathrm{Hz}^{-1} \tag{7}$$

With the conversion factors given above, this corresponds to an extremely large equivalent noise temperature $T_{\mathrm{gw}} \simeq 10^{41}$ K. It is worth stressing that this is only an effective noise temperature. Such a value, larger than Planck temperature ($\sim 10^{32}$ K), does not correspond to an equilibrium temperature and is allowed by the weakness of gravitational coupling.

Previous estimations correspond to the confusion background of gravitational waves emitted by binary systems in our Galaxy or its vicinity. Because of the large number of unresolved and independent sources, and as a consequence of the central limit theorem, they lead to a stochastic noise obeying gaussian statistics. There also exist predictions for gravitational backgrounds associated with a variety of cosmic processes,[24] which are however model dependent and have a more speculative character. Associated temperatures vary rapidly with frequency and are dominated by the confusion binary background in the frequency range considered here.

3. Quantum Decoherence of Atomic Interferometers

Atoms used in interferometry appear as particularly interesting microscopic systems for studying quantum decoherence, as it has recently been suggested that matter-wave interferometers could reveal the existence of intrinsic spacetime fluctuations, through an induced Brownian motion.[27,28] Although it has not been possible to observe such an effect in existing matter-wave interferometers, instruments are now being designed, like the atomic interferometer HYPER for measuring the Lense-Thirring effect in space, which possess a very high sensitivity to gravitation fields.[29] It is thus important, in order to confirm the viability of such instruments, to obtain quantitative estimates of potential decoherence effects, in particular those associated with spacetime or gravitation fluctuations.

We shall consider the atomic field of the matter-wave interferometer HYPER as a typical example of a microscopic system affected by quantum decoherence (see for instance[30–32] for details on atomic interferometry). HYPER is an interferometer with a rhombic geometry which is used as a gyrometer, that is to say, its rotation with respect to inertial frames is measured through the observation of a Sagnac effect. The Sagnac dephasing Φ is proportional to the mass $m_{\rm at}$ of the (non relativistic) atoms, to the area A of the interferometer and to the rotation frequency Ω

$$\Phi = \frac{1}{\hbar} \oint p_{\rm i} dx^{\rm i} = \frac{2m_{\rm at}A}{\hbar}\Omega, \qquad p_\mu = g_{\mu\nu} m_{\rm at} v^\nu_{\rm at}, \qquad A = v^2_{\rm at}\tau^2_{\rm at} \sin\alpha \tag{8}$$

$g_{\mu\nu}$ is the metric field in the frame of the rotating interferometer, $v_{\rm at}$ is the atomic velocity, and the area A is given by the length $v_{\rm at}\tau$ of the rhomb side and the aperture angle α ($\tau_{\rm at}$ is the time of flight on one rhomb side).

According to general relativity, a local inertial frame in the neighborhood of a rotating massive body differs from the celestial frame determined by the 'fixed stars' as a consequence of the dragging of inertial frames. This gravitomagnetic (Lense-Thirring) effect in the Earth neighborhood is measured by HYPER interferometer, by comparing the local inertial measurement performed by the atoms to the indication of a star tracker. A map of the Lense-Thirring effect around the Earth is obtained by recording the dephasings and building the corresponding interferogram for each position of the satellite on its orbit.

Gravitational waves, like other gravitational perturbations such as the Lense-Thirring effect, induce a dephasing of the matter waves within the two arms of the interferometer and thus affect the interference fringes[33]

$$\delta\Phi_{\rm gw} = \frac{m_{\rm at}}{2\hbar}\oint h_{\rm ij} v^{\rm i}_{\rm at} v^{\rm j}_{\rm at} d\tau = \frac{2m_{\rm at}A}{\hbar}\delta\Omega_{\rm gw} \tag{9}$$

Metric components are evaluated in the TT (transverse traceless) gauge. Using the symmetry of the rhomb, this expression may be obtained from the derivative of the metric component h_{12} lying in the spatial plane defined by the interferometer

$$\delta\Omega_{\rm gw}(t) = -\frac{1}{2}\frac{{\rm d}\overline{h_{12}}}{dt}, \qquad \overline{h_{12}}(t) = \int h_{12}(t-\tau)\, g(\tau)\, {\rm d}\tau \tag{10}$$

The linear filtering function g has a triangular shape which reflects the distribution of the time of exposition of atoms to gravitational waves inside the rhombic interferometer. The square of its Fourier transform, which describes linear filtering in frequency space, is an apparatus function characterizing the interferometer[16]

$$|\tilde{g}[\omega]|^2 = \left(\frac{\sin\frac{\omega\tau_{\rm at}}{2}}{\frac{\omega\tau_{\rm at}}{2}}\right)^4 \tag{11}$$

We now consider the degradation of fringe contrast obtained by averaging over stochastic dephasings. This evaluation[16] can be shown to be equivalent to the other approaches to decoherence (see for example[34]). Stochastic gravitational waves with frequencies higher than the inverse of the averaging time identify with the unobserved degrees of freedom which are usually traced over in decoherence theory (see[8] and references therein). When $\delta\Phi_{\rm gw}$ is a gaussian stochastic variable, the degraded fringe contrast is read as

$$\langle\exp(i\delta\Phi_{\rm gw})\rangle = \exp\left(-\frac{\Delta\Phi^2_{\rm gw}}{2}\right), \qquad \Delta\Phi^2_{\rm gw} = \langle\delta\Phi^2_{\rm gw}\rangle \tag{12}$$

Using the expression of $\delta\Phi_{\rm gw}$ in terms of the averaged time derivative of h_{12} we write the variance $\Delta\Phi^2_{\rm gw}$ as an integral over the noise spectrum S_h (5). Particularly interesting is the case of an approximately flat or thermal spectrum S_h (6) which, as discussed in previous section, is approximately realized by the binary confusion background on a significant frequency range. With a white noise assumption, the variance is found to be proportional to the constant value of the noise spectrum S_h

$$\Delta\Phi^2_{\rm gw} = \left(\frac{2m_{\rm at}v^2_{\rm at}}{\hbar}\sin\alpha\right)^2 S_h\, 2\tau_{\rm at} \tag{13}$$

After substitution of the numbers corresponding to HYPER,[29] we deduce that the decoherence due to the scattering of gravitational waves is completely negligible

$$\Delta\Phi^2_{\rm gw} \sim 10^{-20} \quad \ll 1 \tag{14}$$

We have discussed here the decoherence effect on atomic fields. In fact, it appears that the decoherence effect affecting the laser fields, involved in the stimulated

Raman processes used for building up beam splitters and mirrors for matter waves, provides a larger contribution.[16] But this changes neither the mechanism of quantum decoherence which has been discussed here, nor its incidence on the instrument sensitivity. The phase noise induced by the scattering of gravitational waves remains completely negligible with respect to the phase noise induced by mechanical vibrations of the mirrors. In the real instrument, decoherence is expected to be induced by instrumental fluctuations rather than by fondamental fluctuations.

4. Quantum Decoherence of Planetary Systems

After discussing the microscopic case on the example of atomic interferometers, we come to a case which lies at the opposite end, as it can be considered as extremely macroscopic, namely the planetary system built by the Moon orbiting around the Earth. The classicality of such a system may be expected to result from the strong efficiency of decoherence mechanisms acting on it, contrarily to the case of microscopic systems. Indeed, as we show, gravitational waves lead to an extremely rapid decrease of quantum coherences for such macroscopic systems. Moreover, although decoherence may usually be attributed to collisions of residual gaz, to radiation pressure of solar radiation or, even, to the scattering of electromagnetic fluctuations in the cosmic microwave background, we show that, in the case of planetary motions, it is dominated by the scattering of stochastic gravitational waves.

The Earth and Moon constitute a binary system with a large quadrupole momentum, so that its internal motion is highly sensitive to gravitational waves. For the sake of simplicity, we shall describe the Earth-Moon system as a circular planetary orbit in the plane x_1x_2. The reduced mass m, defined from the masses of the two bodies, will be used, such as the radius ρ, that is the constant distance between the two masses, so that the orbital frequency Ω, the normal acceleration a on the circular orbit and the tangential velocity v obey usual relations

$$a = \rho\Omega^2 = \frac{v^2}{\rho} \tag{15}$$

Gravitational waves will be represented as metric perturbations $h_{\mu\nu}$, taken in the TT gauge (3), so that they will be related to Riemann curvature ($R_{0\mathrm{i}0\mathrm{j}} = \partial_t^2 h_{\mathrm{ij}} \equiv \ddot{h}_{\mathrm{ij}}$). The gravitational wave perturbation on the relative position x^{i} in the binary system amounts to a tidal force δF which may also be seen as a geodesic deviation

$$\delta\dot{p}_{\mathrm{i}}(t) = \delta F_{\mathrm{i}}(t) = mc^2 R_{0\mathrm{i}0\mathrm{j}} x^{\mathrm{j}}(t) \tag{16}$$

The stochastic background of gravitational waves then induces a Brownian motion on the relative position of the Moon, which may be characterized by a momentum diffusion with a variance varying linearly with the time of exposition τ

$$< \delta p^2(t) >= 2D_{\mathrm{gw}}\tau \tag{17}$$

The momentum diffusion coefficient D_{gw} is determined by the correlation function of gravitational waves (5,6)[14]

$$D_{\rm gw} = m\Gamma_{\rm gw} k_{\rm B} T_{\rm gw}, \qquad \Gamma_{\rm gw} = \frac{32Gma^2}{5c^5} \tag{18}$$

$T_{\rm gw}$ is the effective noise temperature of the gravitational background, evaluated at twice the orbital frequency, and $\Gamma_{\rm gw}$ is the damping rate associated with the emission of gravitational waves. One recovers with equations (18) the fluctuation-dissipation relation on Brownian motion[35] and the quadrupole formula for gravitational wave emission[36] determined by Einstein. Although gravitational damping can be observed in the case of strongly bound binary systems,[37] it appears to be extremely small for the Moon ($\Gamma_{\rm gw} \approx 10^{-34}$ s^{-1}), with a negligible impact on its mean motion. Moreover, it can be seen to be much smaller than the damping due to other environmental fluctutations, such as electromagnetic radiation pressure or Earth-Moon tides. The latter appear to give the dominant contribution to damping[38]

$$\Gamma_{\rm gw} \ll \Gamma_{\rm em} < \Gamma_{\rm tides} \tag{19}$$

However, as we show now, decoherence processes do not follow the same hierarchy.

Quantum decoherence may be evaluated by considering two neighbouring internal motions of the planetary system which correspond to the same spatial geometry but slightly different values of the epoch, the time of passage at a given space point. For simplicity, we measure this difference by the spatial distance Δx between the two motions, which is constant for uniform motion. The variation of momentum (16) results in a perturbation of the quantum phase one may associate with the relative position in the binary system

$$\delta\Phi_{\rm gw}(t) = \frac{\delta p_{\rm i}(t)}{\hbar} \Delta x^{\rm i} \tag{20}$$

The difference of phase between two neighboring motions then undergoes a Brownian motion,[14] resulting in a random exponential factor $e^{i\delta\Phi_{\rm gw}}$. Averaging this quantity over the stochastic effect of gravitational waves, still supposed to obey gaussian statistics, one obtains a decoherence factor

$$\left\langle e^{i\delta\Phi_{\rm gw}} \right\rangle = \exp\left(-\frac{\Delta\Phi_{\rm gw}^2}{2} \right) \tag{21}$$

The decoherence factor may be expressed in terms of the variables characterizing the Brownian motion (17) and the distance between the two motions Δx

$$\Delta\Phi_{\rm gw}^2 = \frac{2D_{\rm gw}\Delta x^2 \tau}{\hbar^2} \tag{22}$$

Relation (22) agrees with the result expected from general discussions on decoherence:[2] decoherence efficiency increases exponentially fast with τ and Δx^2.

Relation (22) may be rewritten in terms of the gravitational waves spectrum (6) and the geometric parameters of the binary system (15)

$$\Delta\Phi_{\rm gw}^2 = \left(\frac{2mv^2}{\hbar} \sin\alpha \right)^2 S_h \, 2\tau, \qquad \sin\alpha = \frac{\Delta x}{2\rho} \tag{23}$$

$\frac{2mv^2}{\hbar}\sin\alpha$ is a frequency determined by the kinetic energy of the Moon and $\sin\alpha$ is the aperture angle of the equivalent interferometer. In the case of the Earth-Moon system, one finds an extremely short decoherence time up to extremely short distances Δx (in the 10μs range for Δx of the order of the Planck length)

$$\frac{D_{\mathrm{gw}}}{\hbar^2} \approx 10^{75}\ \mathrm{s}^{-1}\mathrm{m}^{-2} \tag{24}$$

The gravitational contribution to decoherence appears to be much larger than the contributions associated with tide interactions and electromagnetic scattering

$$D_{\mathrm{gw}} \gg D_{\mathrm{tides}} > D_{\mathrm{em}} \tag{25}$$

When compared with contributions to damping (19), decoherence contributions obey a modified hierarchy. This results from their further dependence on the level of noise induced by the environment and from the fact that gravitational waves constitute the environment with the largest effective noise temperature (7). To be precise, the ratio $\frac{\Gamma_{\mathrm{gw}}}{\Gamma_{\mathrm{tides}}}$ of the damping constants associated with gravitational waves and tides is of the order of 10^{-16}, while the ratio $\frac{T_{\mathrm{gw}}}{T_{\mathrm{tides}}}$ is of the order of 10^{38}. It follows that the ratio $\frac{D_{\mathrm{gw}}}{D_{\mathrm{tides}}}$ remains very large and that the gravitational contribution to decoherence dominates the other ones.

The dominant mechanism leading to the classical behavior of very macroscopic systems appears to be due to gravitational waves, originating either from the confusion binary background in our galaxy or from extragalactic sources in a larger region of the universe. It is remarkable that the classicality and the ultimate fluctuations of very macroscopic systems appear to be determined by the classical gravitation theory which also explains their mean motion.

5. Gravitational Quantum Decoherence

The results obtained in the previous sections for gravitationally induced decoherence are reminiscent of the qualitative discussions of the Introduction. For microscopic probes, such as the atoms or photons involved in atomic interferometers, decoherence is so inefficient that it can be ignored with the consequence that quantum mechanics remains the appropriate description.. For macroscopic bodies on the contrary, such as the Moon-Earth system, decoherence is extremely efficient with the consequence that potential quantum coherences between different positions can never be observed, leading to an appropriate purely classical description..

The scale arguments sketched in the Introduction may also be associated with precise expressions. In both the microscopic (13) and macroscopic (23) cases, the decoherence factor $e^{-\frac{\Delta\Phi_{\mathrm{gw}}^2}{2}}$ induced by the gravitational environment takes a same form. It involves as an essential factor the gravitational spectral density S_h (6), which may be expressed as an effective noise temperature, putting into evidence its dependence on Planck mass m_P

$$S_h \simeq \Theta_{gw}\, t_P^2, \qquad t_P^2 = \frac{\hbar G}{c^5} = \left(\frac{\hbar}{m_P c^2}\right)^2, \qquad \Theta_{gw} \simeq \frac{k_B T_{gw}}{\hbar} \simeq 10^{52} s^{-1} \quad (26)$$

Θ_{gw} is the temperature of the background measured as a frequency. Relations (13) and (23) may then be rewritten

$$\frac{\Delta\Phi_{gw}^2}{2} \simeq \left(\frac{2mv^2 \sin\alpha}{m_P c^2}\right)^2 \Theta_{gw}\tau \qquad (27)$$

The ratio $\frac{m^2}{m_P^2}$ confirms the preliminary arguments of the Introduction, namely that the Planck mass effectively plays a role in the definition of a borderline between microscopic and macroscopic masses. However, other factors in the formula imply that the scaling argument on masses is not sufficient to obtain correct quantitative estimates. The ratio of the probe velocity over light velocity, the equivalent aperture angle α and the frequency Θ_{gw}, measuring the gravitational noise level, enter the quantum decoherence time on an equal footing. In particular, the very large value of the gravitational noise level implies that the transition between quantum and classical behaviors could in principle be observed for masses smaller than Planck mass. Another interesting feature is that the parameter to be compared with Planck energy $m_P c^2$ is the kinetic energy mv^2 of the probe rather than its mass energy mc^2.

Finally, formula (27) provides a valuable insight into the way to design systems aiming at observing the quantum/classical transition induced by intrinsic gravitational fluctuations. The transition region $\Delta\Phi_{gw}^2 \sim 1$ seems to be best approached by using heavy and fast particles in a matter-wave interferometer. At present, interference patterns have been observed on rather large molecules.[39,40] But one checks that, in these experiments, the kinetic energy of the molecules, the area and aperture angle of the interferometer are such that the gravitational quantum decoherence remains negligible, as in HYPER. Increasing these sensitive parameters so that the transition could be approached appears as a formidable experimental challenge[18] (see[41,42] for using fast molecules). Alternatively, one could consider using quantum condensates,[43,44] an approach however requiring further technological progress.

References

1. H. D. Zeh, *Found. Phys.* **1**, 69 (1970).
2. A. J. Zurek, *Phys. Rev. D* **24**, 1516 (1981); *Phys. Rev. D* **26** 1862 (1982).
3. A. O. Caldeira and A. J. Leggett, *Physica A* **121**, 587 (1983); *Phys. Rev. A* **31**, 1059 (1985).
4. E. Joos and H. D. Zeh, *Z. Phys. B* **59**, 223 (1985).
5. A. E. Allahverdyan, R. Balian and Th. M. Nieuwenhuizen in this issue, page 53.
6. M. Brune, *et al. Phys. Rev. Lett.* **77**, 4887 (1996).
7. C. J. Myatt, *et al. Nature* **403**, 269 (2000).
8. J. M. Raimond, M. Brune and S. Haroche, *Rev. Mod. Phys.* **73**, 565 (2001).
9. F. Laloë in this issue, page 35.

10. R. P. Feynman, F. B. Moringo, W. G. Wagner and B. Hatfield, *Feynman Lectures on Gravitation* (Penguin, 1999).
11. F. Karolyhazy, *Nuovo Cim.* **42 A**, 390 (1966).
12. L. Diosi, *Phys. Rev. A* **40**, 1165 (1989).
13. R. Penrose, *Gen. Rel. Grav.* **28**, 581 (1996).
14. S. Reynaud, P. A. Maia Neto, A. Lambrecht and M. T. Jaekel, *Europhys. Lett.* **54**, 135 (2001).
15. S. Reynaud, *et al. Int. J. Mod. Phys. A* **17**, 1003 (2002).
16. B. Lamine, M. T. Jaekel and S. Reynaud *Eur. Phys. J. D* **20**, 165 (2002).
17. S. Reynaud, *et al. Gen. Rel. Grav.* **36**, 2271 (2004).
18. B. Lamine, R. Hervé, A. Lambrecht and S. Reynaud, *Phys. Rev. Lett.* **96**, 050405 (2006).
19. M. T. Jaekel and S. Reynaud, *Ann. Physik* **4**, 68 (1995).
20. S. Weinberg, *Phys. Rev.* **138** B988 (1965).
21. L. P. Grishchuk, *Usp. Fiz. Nauk* **121**, 629 (1977).
22. Ya. B. Zeldovich and L. P. Grishchuk, *Usp. Fiz. Nauk* **149**, 695 (1986).
23. B. Schutz, *Class. Quant. Grav.* **16**, A131 (1999).
24. M. Maggiore, *Physics Reports* **331**, 283 (2000).
25. C. Ungarelli and A. Vecchio, *Phys. Rev. D* **63**, 064030 (2001).
26. L.P. Grishchuk, et al., *Phys. Uspekhi* **44**, 1 (2001).
27. W. L. Power and I. C. Percival, *Proc. R. Soc. London A* **455**, 991 (2000).
28. G. Amelino-Camelia, *Nature* **408**, 661 (2000).
29. HYPER, *Hyper-precision cold atom interferometry in space*, (Assessment study report **10** ESA-SCI,2000).
30. C. J. Bordé, J. C. Houard and A. Karasiewicz, in *Gyros, Clocks, and Interferometers: Testing Relativistic Gravity in Space*, (eds C. Lämmerzahl, C. W. F. Everitt and F. W. Hehl, Lecture Notes in Physics **562**, 2001).
31. C. J. Bordé, *C. R. Acad. Sci. Paris*, **2-IV**, 509 (2001).
32. A. Peters, K. Y. Chung and S. Chu, *Metrologia* **38**, 25 (2001).
33. B. Linet and P. Tourrenc, *Can. J. Phys.* **54**, 1129 (1976).
34. A. Stern, Y. Aharonov and Y. Imry, *Phys. Rev. A* **41**, 3436 (1990).
35. A. Einstein, *Ann. Physik* **XVII**, 549 (1905).
36. A. Einstein, *Preuss. Akad. Wissen. Sitz.* 154 (1918).
37. J.H. Taylor, A. Wolszczan, T. Damour and J.M. Weisberg, *Nature* **355** 132 (1992).
38. E. Bois, F. Boudin and A. Journet, *Astron. Astrophys.* **314**, 989 (1996).
39. K. Hornberger, *et al.*, *Phys. Rev. Lett.* **90**, 160401 (2003).
40. L. Hackermüller,*et al. Nature* **427**, 711 (2004).
41. O. Houde, D. Kadio and L. Pruvost, *Phys. Rev. Lett.* **85**, 5543 (2000).
42. M. Boustimi, *et al. Eur. Phys. J. D* **17**, 141 (2001).
43. C. H. van der Wal, *et al. Science* **290**, 773 (2000).
44. W. L. Li, A. K. Tuchman, H. C. Chien and M. A. Kasevich, (2006) *arXiv:quant-ph/0609009.*

ROLE OF VARIOUS ENTROPIES IN THE BLACK HOLE INFORMATION LOSS PROBLEM

Th. M. NIEUWENHUIZEN

Institute for Theoretical Physics,
Valckenierstraat 65, 1018 XE Amsterdam, The Netherlands

I. V. VOLOVICH

Steklov Mathematical Institute,
Gubkin St. 8, 119991, Moscow, Russia
E-mail: volovich@mi.ras.ru

We discuss the current status of the black hole information loss paradox and propose a plan for its solution based on analogies with solid state physics and the irreversibility problem. In a recent paper Hawking has argued that there is no information loss in black holes in asymptotically AdS spacetimes. We remind that there are several types of information (entropy) in statistical physics – fine grained (microscopic) and coarse grained (macroscopic) ones which behave differently under unitary evolution. We suggest that the coarse grained information of the rest of the Universe is lost while fine grained information is preserved. A possibility to develop in quantum gravity an analogue of the Bogoliubov derivation of the irreversible Boltzmann and Navier - Stokes equations from the reversible mechanical equations is discussed.

Keywords: Black holes; Information paradox; Black body; Hawking radiation.

1. Introduction

In 1976 Hawking has argued that the black hole creation and evaporation could lead to an evolution of a pure state into a mixed state, which is in contradiction with the rules of quantum mechanics.[1] This has become known as the black hole information loss problem (or black hole information paradox), for a discussion see[2–5] and references therein.

In a recent paper Hawking[6] has suggested that there is no information loss in black holes in asymptotically AdS (anti-de Sitter) spacetimes. The central point in his argument is the assertion that there is no information loss if one has a unitary evolution. Then, since the AdS quantum gravity is dual to a unitary conformal field theory,[7–9] there should be no information loss.

In more detail the proposal looks as follows. Black hole formation and evaporation is considered as a scattering process when all measurements are made at infinity. The evolution operator is defined by means of the path integral for the partition function. In the sum over topologies in the path integral there are trivial topologies, which lead to the unitary evolution, and nontrivial black hole topologies. Information is lost in topologically non-trivial black hole metrics, but their contribution decays to zero at large times, at least for each separate mode,[10] As a result one gets a unitary evolution at large times and the information is preserved in that limit.

We make two remarks concerning this proposal.

(i) Actually it is proposed that we can make observations only at such large time scales where small black holes either do not form yet or have already evaporated. In any case black holes are not present then.

(ii) An important point in the argument is the assertion that, if one considers a theory with a unitary dynamics, there is no information loss problem.

Concerning the point (i) there is a question of how information gets out of a black hole. In fact it is one of the main questions in the whole discussion of the information paradox. Hawking's answer is that there is no sense in asking this question because its answer would require to use a semi-classical metric, which has already lost the information. We discuss this argument below.

About the point (ii) we remind that the problem of how a unitary reversible dynamics can lead to an irreversible behavior (i.e. to information loss) is the subject of numerous studies in statistical physics. One specific application is the problem of relaxation: Even though quantum evolution, believed to describe all condensed matter, is unitary, we observe relaxation in every day life.[12] It is thus clear that much of the irreversible behavior, including probably the expansion of the Universe, should be compatible with unitary quantum mechanics. Point (ii) is the main point which will be discussed in this note.

We consider the black hole information problem as a particular example of the fundamental irreversibility problem in statistical physics. We point out that similar problem occurs when we study ordinary gas or the formation of the ordinary black *body* and its thermal radiation. Actually, one has to give a quantum mechanical explanation for the emergence of the second law of thermodynamics in macroscopic systems.

The irreversibility problem was much studied by Boltzmann and many other authors. There is not yet a complete solution but a deep understanding of the problem has been achieved[11]-.[20]

Information is usually quantified as entropy with a minus sign. There are two classes of entropies in statistical physics – the fine grained (microscopic) one and several coarse grained (macroscopic) ones. They behave differently under unitary evolution. For a given model, unitary dynamics might increase one or more coarse grained entropies while preserving the fine grained one. Whether there is an increase of a certain coarse grained entropy (i.e. loss of *this* information) is a dynamical question and its answer comes for a given model from a thorough investigation of its dynamics, which should show a sort of instability or ergodicity and mixing. Alternatively, coupling to a bath would suffice. This is relevant for the black hole situation, where the "bath" is the rest of the Universe.

We note that the properties of the coarse grained (in particular thermodynamical) entropy are important in statistical physics while the fine grained entropy is not so significant since typically it cannot be determined and, remaining conserved, would not reflect a specification of the dynamical evolution.

2. Different Kinds of Entropies

Let us split up our degrees of freedom in two classes, those of the black hole and those of the rest of the Universe. The latter we call "bath", since for a proper formulation of thermodynamics of black holes, the rest of the Universe indeed plays the role of the thermal bath in condensed matter problems. The density matrix of the total system ρ has several marginals. The reduced density matrix of the black hole is

$$\rho_{\rm BH} = {\rm Tr}_{\rm B}\, \rho$$

while the reduced density matrix of the bath is

$$\rho_{\rm B} = {\rm Tr}_{\rm BH}\, \rho$$

This brings us three von Neumann entropies: the "fine grained" entropy of the total system,

$$S_{\rm total}^{\rm fine} = -{\rm Tr}\, \rho \ln \rho.$$

This quantity is conserved in time for unitary motion. If one starts out from a pure state of incoming matter and bath, it vanishes at all times. Next there is the fine grained von Neumann entropy of the black hole,

$$S_{\rm BH}^{\rm fine} = -{\rm Tr}_{\rm BH}\, \rho_{\rm BH} \ln \rho_{\rm BH}$$

and the fine grained von Neumann entropy of the bath,

$$S_{\rm B}^{\rm fine} = -{\rm Tr}_{\rm B}\, \rho_{\rm B} \ln \rho_{\rm B}$$

When starting from a pure state, entangled or not, the latter two will be equal at all times. Though then vanishing at $t = 0$ they become positive at later times, $S_{\rm BH}^{\rm fine}(t) = S_{\rm B}^{\rm fine}(t) > 0$. At large times one expects them to go to zero again.[25] For $S_{\rm BH}^{\rm fine}$ the reason is simply that matter is radiated, making it smaller and smaller, so that with the matter its entropy evaporates. For $S_{\rm B}^{\rm fine}$ it theoretically expected, because, in the final absence of the hole, it just reflects the purity of the state. But physically it is a surprising and counter-intuitive result, since this vanishing entropy clearly does not reflect the energy radiated into this bath by the black hole evaporation.

Coarse graining can be done at larger and larger scales. Thus there is still another entropy to consider, namely the coarse grained entropy of the bath, where one neglects all correlations between bath modes. One first has to define the reduced density matrix of a given mode, $\rho_{\rm mode} = {\rm Tr}_{\rm all\ other\ modes}\,\rho$ and from it the von Neumann entropy

$$S_{\rm B}^{\rm coarse} = \sum_{\rm mode} S_{\rm mode} = \sum_{\rm mode} -{\rm Tr}_{\rm mode}\,\rho_{\rm mode}\ln\rho_{\rm mode}$$

This is the one entering quasi-classical discussions of the black hole information paradox. Since it probably does not vanish at large times, it is a candidate for our "natural" association of entropy with a measure of disorder. It is then obvious that $S_{\rm B}^{\rm fine}$, which does vanish at large times, takes the correlations between different modes, neglected in $S_{\rm B}^{\rm coarse}$, into account in a very subtle manner.

Boltzmann's famous formula for the entropy reads

$$S = k_B \log W$$

The W in this formula is the number of microstates compatible with the *macroscopic* state. Therefore this formula defines a coarse grained entropy. There is a remarkable computation by Strominger and Vafa of the Bekenstein–Hawking coarse grained entropy by counting of microscopic BPS states in string theory.[21]

Black hole thermodynamics is a problem with two temperatures: the Hawking temperature of the hole and the $3K$ back ground temperature .[22,23] These distinctions automatically show up in condensed matter analogs of the black hole evaporation problem, that we plan to discuss elsewhere.[24]

The thermodynamic entropy of a system in contact with a bath at temperature T is defined in terms of added heat dQ by the Clausius inequality

$$dQ \leq TdS \tag{1}$$

taken as an equality, so $dS = dQ/T$. The "classical intuition" of entropy arose when Bolztmann showed that this thermodynamic entropy agrees with his measure of disorder, more precisely, the logarithm of the number of states. For a closed system, this quantity cannot decrease. For systems with two temperatures a generalized Clausius inequality may hold when there are also two well separated time scales, leading to two different entropies, that enter as: $dQ \leq T_1 dS_1 + T_2 dS_2$. This applies to glasses[26] and black holes.[22]

For nanoscopic and mesoscopic systems, one has led to uncover the field of "quantum thermodynamics".[27]

Page has discussed the microcanonical and canonical entropies of black holes.[28] The first one does not reflect its environment, the second one assumes it to be at the Hawking temperature, which is the case only for a specific black hole size and, moreover, an unstable situation.

We have discussed various entropies which are used to describe the *classical* capacity of a quantum channel. To describe the *quantum* channel capacity the quantum mutual entropy[29] and the quantum coherent information[30] are used. It would be interesting to investigate the role of these entropies in quantum gravity.

3. On the Black Hole Information Paradox

Our proposal for the investigation of the black hole information loss problem is the following. One of the mentioned entropies is the coarse grained bath entropy. It increases during the evolution. In this sense there is information loss. But it does not mean that there is loss of the fine grained information (entropy). The whole picture of the black hole formation and evaporation is similar to the formation and radiation of a black body.

In a specific model the increase of coarse grained entropy has to be demonstrated if the black hole evaporation indeed behaves as a thermodynamic problem.

The next step in the program is to develop in quantum gravity an analogue of the Bogoliubov derivation of the Boltzmann and Navier - Stokes equations from the Liouville equation.[11,13,14] We shall sketch that in the last section. In quantum field theory one derives quantum stochastic differential equations.[15] But we should stress that a much better understanding of the irreversibility problem in various models is required.

It would also be interesting to investigate the role of the quantum mutual entropy[29] and of the quantum coherent information[30] in quantum gravity.

4. Information Loss in Gases

Consider a classical or quantum gas of N particles in a box of the volume V which is described by the Hamiltonian

$$H_N = \sum_{i=1}^{N} \frac{\mathbf{p}_i^2}{2m} + \sum_{i<j} \Phi(\mathbf{x}_i - \mathbf{x}_j).$$

Here $\mathbf{x}_i$ are positions, $\mathbf{p}_i$ are momenta, m is mass and $\Phi(\mathbf{x})$ is the interaction potential between a pair of particles.

One has a reversible classical dynamics and a unitary quantum dynamics. Hence, in neither situation there is a loss of the fine grained information.

However, it is well known that the kinetic theory of gases is based on the Boltzmann equation which reads[13,14]

$$\frac{\partial f}{\partial t} + \frac{\mathbf{p}}{m} \cdot \frac{\partial f}{\partial \mathbf{x}} + \mathbf{F} \cdot \frac{\partial f}{\partial \mathbf{p}} = J(f). \tag{2}$$

Here $f = f(\mathbf{x}, \mathbf{p}, t)$ is the one-particle distribution function, t is time, $\mathbf{F} = \mathbf{F}(\mathbf{x}, t)$ is the force, and $J(f)$ is a bilinear functional in f.

The coarse grained Boltzmann entropy is defined by

$$S_B(t) = -\int f(\mathbf{x}, \mathbf{p}, t) \ln f(\mathbf{x}, \mathbf{p}, t) d\mathbf{x} d\mathbf{p}.$$

Boltzmann has proven (H-theorem) that

$$\frac{dS_B(t)}{dt} \geq 0$$

Moreover the entropy is constant only for an equilibrium distribution function. For a non-equilibrium state the Boltzmann entropy increases and therefore one gets *information loss.*

Now one can ask the same question as for the black hole information loss problem. How is it possible that one gets information loss and irreversibility for a system of N particles which is described by a reversible dynamics?

An important progress in investigation of this question was achieved by Bogoliubov.[11,13,14] He considers a system of equations for s-particle correlation functions

$$f_s(\xi_1, ..., \xi_s, t)$$

where $\xi_i = (\mathbf{x}_i, \mathbf{p}_i), \quad i, s = 1, 2, ..., N$. The system of equations (BBGKI-chain) is equivalent to the Liouville equation and is reversible.

Then for the dilute gases Bogoliubov introduces a kinetic relaxation time scale τ_0 and uses the thermodynamical limit $(N, V \to \infty,\ N/V = \text{const})$ and the factorization of the s- particle correlation functions f_s in terms of the one particle distribution function f:

$$f_s(\xi_1, ..., \xi_s, t) \to \prod_{i=1}^{s} f(\xi_i, t), \quad t > \tau_0, \quad s = 2, 3, ..., N \tag{3}$$

In this way he was able to obtain the Boltzmann equation. Then one can use the Boltzmann equation to derive the hydrodynamical Navier - Stokes equation.

We write the Boltzmann equation Eq. (2) symbolically as

$$\frac{\delta f}{\delta \sigma} = J(f) \tag{4}$$

where $\sigma = (t, \xi)$.

Bogoliubov used a similar approach also for derivation of quantum kinetic equations. In this case one uses the correlation functions

$$f_{sk}(\mathbf{x}_1, ..., \mathbf{x}_s; \mathbf{y}_1, ..., \mathbf{y}_k; t) = Tr[\rho_t \psi(\mathbf{x}_1)...\psi(\mathbf{x}_s)\psi^+(\mathbf{y}_1)...\psi^+(\mathbf{y}_k)]$$

Here ρ_t is the density operator at time t and $\psi(\mathbf{x}),\ \psi^+(\mathbf{y})$ are annihilation and creation operators satisfying the usual commutation relations $[\psi(\mathbf{x}), \psi^+(\mathbf{y})] = \delta(\mathbf{x} - \mathbf{y})$. To derive quantum kinetic equation one uses an approximation similar to the Bogoliubov approximation Eq. (3). It is a kind of the mean field approximation. There are other models in which an irreversible behavior from reversible dynamics was derived, [15,16] Mathematical studying of these questions is also the subject of ergodic theory where for certain dynamical systems the properties of ergodicity and mixing were established and where the notions of classical and quantum Anosov and K-systems and the Kolmogorov - Sinai entropy play an important role.[17–20]

It was demonstrated by Pauli, von Neumann, van Hove, and Prigogine that in quantum mechanics the coarse grained entropy increases as a result of the unitary dynamics.

The transition from the BBGKI chain of equations for the family of correlation functions $\{f_s\}$ to the Boltzmann equation for the one particle distribution function

f is the transition from the fine grained reversible description to the irreversible coarse grained description. We loose information when we describe a gas by means of only the one-particle distribution function $f(\mathbf{x}, \mathbf{p}, t)$, or hydrodynamical variables, which are integrals of f with some weights.

In quantum gravity we interpret the distribution $f[g, \pi]$ of the classical metric $g_{\mu\nu}(x)$ and its conjugate $\pi_{\mu\nu}(x)$ as an analogue of the one particle distribution $f(\mathbf{x}, \mathbf{p}, t)$ or hydrodynamical variables. To get insight into the black hole information loss problem, one has to develop in quantum gravity an analogue of the Bogoliubov derivation of the Boltzmann equation from the Liouville equation. In quantum field theory one derives quantum stochastic differential equations.[15]

5. Information Loss in Quantum Gravity

One can use the Euclidean[6] or the Wheeler - De Witt [31] approach to quantum gravity. Each one has its own advantages and disadvantages. One of problems with the Euclidean approach is that one can define there the Green functions and the partition function but not the scattering matrix. Moreover, path integrals are convenient to write down semiclassical expansion but one can not use this "sane" formalism to solve spectral problems, even to compute the spectrum of the hydrogen atom.

The transition amplitude between configurations of the three-metric h'_{ij} and field Φ' on an initial spacelike surface Σ' and a configuration h''_{ij} and Φ'' on a final surface Σ'' is

$$< h'', \phi'', \Sigma'' | h', \phi', \Sigma' >= \int \; e^{\frac{i}{\hbar} S[g,\Phi]} \; \mathcal{D}\Phi \mathcal{D}g,$$

where the integral is over all four-geometries and field configurations which match given values on two spacelike surfaces, i.e. $\Phi|_{\Sigma'} = \phi'$, $g|_{\Sigma'} = h'$, $\Phi|_{\Sigma''} = \phi''$, $g|_{\Sigma''} = h''$.

The problem of creation of black holes in quantum theory is considered in [32]-.[35] The role of boundary conditions in the path integral describing the creation of black holes is discussed in.[34]

We are interested in the process of black hole creation. Therefore Σ' is a partial Cauchy surface with asymptotically simple past in a strongly asymptotically predictable space-time and Σ'' is a partial Cauchy surface containing black hole(s), i.e. $\Sigma'' - J^-(\mathcal{T}^+)$ is non empty.

Black holes are conventionally defined[36] in asymptotically flat (or AdS) space-times by the existence of an event horizon H. The horizon H is the boundary $\dot{J}^-(\mathcal{I}^+)$ of the causal past $J^-(\mathcal{I}^+)$ of future null infinity $\mathcal{I}^+$. The black hole region B is $B = M - J^-(\mathcal{I}^+)$ and the event horizon $H = \dot{J}^-(\mathcal{T}^+)$. This definition depends on the whole future behavior of the metric. There is a different sort of horizon, trapped horizon, which depends only on the properties of space-time on the surface $\Sigma(\tau)$.[2,36]

We discussed the transition amplitude (propagator) between definite configurations of fields, $< h'', \phi'', \Sigma'' | h', \phi', \Sigma' >$. The transition amplitude from a state described by the wavefunction $\Psi^{in}[h', \phi']$ to a state $\Psi^{out}[h'', \phi'']$ reads

$$< \Psi^{out}|\Psi^{in} >=$$

$$\int \bar{\Psi}^{out}[h'', \phi''] < h'', \phi'', \Sigma''|h', \phi', \Sigma' > \Psi^{in}[h', \phi']\mathcal{D}h'\mathcal{D}\phi'\mathcal{D}h''\mathcal{D}\phi''.$$

Consider a family of asymptotically flat spacetimes. A wave function is a functional of the 3-geometry and the matter fields $\Psi[h_{ij}, \phi]$. It satisfies the Wheeler - De Witt equation

$$\mathcal{H}\Psi = 0$$

where $\mathcal{H}$ is the density of the Hamiltonian constraint.

Let ρ_Σ be the density operator of the Universe at the surface Σ. One defines the fine grained entropy

$$S^{\rm fine}(\Sigma) = -{\rm Tr}\, \rho_\Sigma \ln \rho_\Sigma$$

and correlation functions

$$f^{(s)}_{i_1 j_1 \ldots i_s j_s}(x_1, ..., y_s; \Sigma) = Tr[\rho_\Sigma \hat{g}_{i_1 j_1}(x_1)...\hat{\pi}_{i_s j_s}(y_s)]$$

where $\hat{h}_{ij}, \hat{\pi}_{ij}$ operators of metric and its canonically conjugate.

The correlation functions satisfy a system of equations in superspace. The equations are complicated. We can assume that there is a sort of unitary dynamics which preserves the fine grained entropy but there is no way to determine it. This is similar to the gas dynamics which was discussed in the previous section. But let us try to derive an analogue of the Boltzmann equation. We consider an approximation:

$$f^{(s)}_{i_1 j_1 \ldots i_s j_s}(x_1, ..., y_s; \Sigma) \to \prod_r f_{i_r j_r m_r n_r}(x_r, y_r; \Sigma), \quad \Sigma > \Sigma_0,$$

where

$$f_{i_r j_r m_r n_r}(x_r, y_r; \Sigma) = Tr[\rho_\Sigma \hat{g}_{i_r j_r}(x_r)\hat{\pi}_{m_r n_r}(y_r)]$$

This is similar to the Bogoliubov approximation Eq. (3). Quantum Boltzmann equation in quantum gravity will have the form

$$\frac{\delta f}{\delta \sigma} = J(f)$$

where $f = f_{i_r j_r m_r n_r}(x_r, y_r; \Sigma)$ and $\sigma = (x, y, g, \pi, \Sigma)$. The problem is to determine an explicit form of the functional $J(f)$.

The coarse grained entropy is

$$S^{\rm coarse}(\Sigma) = -\int tr f(\cdot; \Sigma) \ln f(\cdot; \Sigma)$$

where an appropriate normalization for f is assumed.

Further, if we make also the semiclassical approximation, or assume the coherent pure states, then, in principle, we could get the classical Einstein equations for the metric $g_{\mu\nu}(x)$. However the form of classical equations depend on the chosen state.

For classical gravity the known laws of black hole thermodynamics are valid. In this case the entropy, which is proportional to the area of horizon, is a coarse grained entropy. This entropy increases during the classical evolution, so one gets a loss of the coarse grained information.

We speculate that the coarse grained entropy which is obtained in the Bogoliubov approximation increases in the classical regime. Note that the Hamiltonian in mini–superspace[31] is an N- particle hamiltonian but it has a form more complicated then the Hamiltonian for gases. One can try to study this problem for the Matrix model.[37] Note however that the irreversibility problem is a rather difficult problem even for such a simple and well studied system as quantum baker's map.[38]

On later times the evaporation is the only relevant but slow process (quantum regime). Then the black hole looses matter and its coarse grained entropy decreases. This matter will go to the bath, so its coarse grained entropy is expected to increase such that the total coarse grained entropy also increases.

6. Conclusions

The information paradox of the black hole problem has bothered scientists since the discovery of the evaporation process. It has often not been realized, however, that "the" entropy of a system does not exist. The thermodynamic entropy of a system in contact with a bath at temperature T is defined in terms of added heat dQ as $dS = dQ/T$. The "classical intuition" of entropy arose when Boltzmann showed that this thermodynamic entropy agrees with a measure of disorder, namely the logarithm of the number of relevant states. For a closed system, this quantity cannot decrease. On the other hand, it is known in quantum mechanics that the von Neumann entropy *of a closed system* is a constant due to unitary motion. This is the quantum analog of the classical fine grained entropy, which is also conserved in time.

A second complicating factor it that the setup for black hole thermodynamics is one of systems far from equilibrium.[22] Thus the evaporation process is a problem of thermodynamics far from equilibrium, to which Gibbsian thermodynamics does not apply and for which, in general, few tools are available.

The problem of black hole evaporation is one of the field of *quantum thermodynamics*: the target system is small (the hole), but the bath is large (the rest of the Universe) and also the work source is large (here it would stand for the incoming matter, or work done externally on the hole), see e.g.[27] In this field one can imagine a condensed matter analog where, starting from a ground state, work is from the outside put into a certain degree of freedom (" growing of a toy black hole"), that is later taken out (its disappearance). In this process, not all work can be recovered due to Thomson's formulation of the second law: cyclic processes done on an equilibrium system (here: in its ground state) cannot yield work, and typically will cost work. This work will end up as phonons running away in the infinite condensed matter bath, in the very same way as matter and photons evaporated from the black hole will run in the otherwise empty, infinite Universe. We plan to discuss this setup in future, considering the separate entropies in detail.[24] The recent work on a realistic quantum measurement[39] shows that measurement problems are probably disconnected from black holes issues.

There is a little hope that quantum gravity or string theory could help to get an insight into the fundamental irreversibility problem until the further progress in the considerations of simple models will be achieved. However we should remind that Boltzmann has predicted the cosmological Big Bang just from the consideration of the irreversibility problem.[16] Not only the black hole problem but also the recent discovery of the cosmological acceleration and the mystery of dark energy indicates, it seems, to the necessity of the unified treatment of the basic problems in cosmology, quantum gravity/string theory, high energy physics, statistical physics and quantum information theory.

As to the black hole problem itself, we have outlined how to derive an equivalent of the Boltzmann entropy for gravitation. Also this subject deserves further attention. One outstanding question is to show that, taken together with the coarse grained bath entropy, it is non-decreasing.

Acknowledgments

I. V. Volovich acknowledges hospitality at the University of Amsterdam, on a grant by Stichting voor Fundamenteel Onderzoek der Materie (FOM), financially supported by the Nederlandse Organisatie voor Wetenschappelijk Onderzoek (NWO). He is grateful to A. Arvilsky and B. Dragovich for fruitful discussions and to grants RFFI 05-01-00884 and NS-6705.2006 for support.

References

1. S. W. Hawking, *Phys. Rev.* **43**, 199 (1976)
2. V. P. Frolov and I. D. Novikov, *Black Hole Physics: Basic Concepts and New Developments*, (Kluwer, Utrecht, 1998).
3. B. L. Hu, *Correlation Dynamics of Quantum Fields and Black Hole Information Paradox*, Invited talk at the International School of Astro-fundamental Physics, Sept. 1995. Proceedings edited by N. Sanchez and Zichichi, Physics, Sept. 1995. (Kluwer Publishers, Dordrecht, 1996); gr-qc/9511075.
4. M. Srednicki, *On the Observability of Quantum Information Radiated from a Black Hole*, hep-th/0207090.
5. H. D. Zeh, *Where has all the information gone?*, gr-qc/0507051.
6. S. W. Hawking, *Information Loss in Black Holes*, hep-th/0507171.
7. J. M. Maldacena, *Ad. Theor. Math. Phys.* **2**, 231 (1998).
8. S. S. Gubser, I. R. Klebanov, and A. M. Polyakov, *Phys. Lett. B* **428**, 105 (1998).
9. E. Witten, *Adv. Theor. Math. Phys.* **2**, 253 (1998); hep-th/9802150.
10. J. M. Maldacena, *Eternal Black Holes in Anti-de Sitter*, JHEP 0304, 21 (2003); hep-th/0106112.
11. N. N. Bogoliubov, *Problems of dynamical theory in statistical physics*, (Gostekhizdat, Moscow, 1946).
12. R. Balian, *From Microphysics to Macrophysics*, in 2 volumes, (Springer, Berlin, 1992).

13. A. Isihara, *Statistical Physics*, (Academic Press, 1971).
14. D. Zubarev, V. Morozov, and G. Ropke, *Statistical Mechanics of Nonequilibrium Processes*, (Academic Verlag, 1996).
15. L. Accardi, Yu. G. Lu, I. Volovich, *Quantum Theory and Its Stochastic Limit*, (Springer, 2002).
16. J. L. Lebowitz, *Rev. Mod. Phys.* **71**, S346 (1999).
17. D. V. Anosov, *Geodesic flows on closed Riemannian manifolds of negative curvature*, (Proceedings of Steklov Mathematical Institute, vol. **90**, 1967).
18. I. P. Cornfeld, S. V. Fomin, and Ya. G. Sinai, *Ergodic Theory*, (Springer-Verlag, 1982).
19. V. I. Arnold, V. V. Kozlov, and A. N. Neischtadt, *Mathematical Aspects of Classical and Celestial Mechanics*, (Springer-Verlag, 1988).
20. G. Gallavotti, F. Bonetto, G. Gentile, *Aspects of the ergodic, qualitative and statistical properties of motion*, (Springer-Verlag, 2004).
21. A. Strominger and C. Vafa, *Phys. Lett. B* **379** (1996) 99; hep-th/9601029.
22. Th. M. Nieuwenhuizen, *Phys. Rev. Lett.* **81**, (1998) 2201.
23. In Ref. 22 a difference is pointed out between the *third law of black hole dynamics*, which is about extremal black holes, and the *third law of black hole thermodynamics*, which is about the fate of a black hole in the expanding Universe.
24. Th. M. Nieuwenhuizen, I. V. Volovich, and A. E. Allahverdyan (in preparation).
25. There are also scenario's in Ref. 24 where the toy-black-hole entropy remains constant, or even diverge logarithmically with time.
26. Th. M. Nieuwenhuizen, *Phys. Rev. Lett.* **80**, 5580 (1998).
27. A. E. Allahverdyan, R. Balian and Th. M. Nieuwenhuizen, *J. Mod. Optics* (2004); cond-mat/0402387.
28. Don N. Page, *Hawking Radiation and Black Hole Thermodynamics*, in "Spacetime 100 Years Later," to be published in New Journal of Physics; hep-th/0409024.
29. M. Ohya, *IEEE Trans. Inform. Theory* **29**, 770 (1983).
30. M. A. Nielsen and I. L. Chuang, *Quantum Computation and Quantum Information*, (Cambridge, 2000).
31. D. R. Brill and R. H. Gowdy, *Rep. Progr. Phys.* **33**, 413 (1970).
32. G. 't Hooft, *Phys. Lett. B* **198** (1987) 61; *Nucl. Phys. B* **304** (1988) 867.
33. D. Amati, M. Ciafaloni and G. Veneziano, *Phys. Lett. B* **197** (1987) 81; *Nucl. Phys. B* **403** (1993)703.
34. I. Ya. Aref'eva, K. S. Viswanathan and I. V. Volovich, *Nucl. Phys. B* **452** (1995) 346-368; *Erratum-ibid. B* **462** (1996) 613; hep-th/9412157.
35. M. Kowalski, A. Ringwald, H. Tu, *Phys. Lett. B* **529**, 1 (2002); hep-ph/0201139.
36. S. W. Hawking and G. F. R. Ellis, *The Large Scale Structure of Space-time* (Cambridge University Press, Cambridge, 1973).
37. T. Banks, W. Fischler, S. H. Shenker and L. Susskind, *Phys. Rev. D* **55** (1997) 5112; hep-th/9610043.
38. K. Inoue, M. Ohya, I. V. Volovich, *J. Math. Phys.* **43**, 734 (2002); quant-ph/0110135.
39. A. E. Allahverdyan, R. Balian and Th. M. Nieuwenhuizen, *Europhys. Lett.* **61**, 452 (2003).

QUANTUM AND SUPER–QUANTUM CORRELATIONS

G. S. JAEGER

College of General Studies,
Boston University, 871 Commonwealth Ave.,
Boston, MA 02215, U.S.A.
E-mail: jaeger@bu.edu

and

Quantum Imaging Laboratory, Dept. Electrical and Computer Engineering,
Boston University, 8 St. Mary's St.,
Boston, MA 02215, U.S.A.

Recent investigations of correlations within compound physical systems are considered in the broad context, which includes "superquantum" correlations, namely, those that are stronger than predicted by standard quantum mechanics. Although the significance of these results in the search for deeper principles underlying quantum physics remains uncertain, the results do improve our understanding of quantum correlations.

Keywords: Correlation; Non-locality; Entanglement; Bell's theorem.

1. Introduction

In the history of mechanics, it has been useful to consider correlations of physical systems in successively broader contexts. The strength of correlations of properties between microscopic subsystems allowed by quantum mechanics was found by Bell, Einstein, Podolsky, and Rosen, Schrödinger, and others, somewhat surprisingly, to surpass that allowed by local causal correlations.[1] Recently, it was noted that the range of causally allowed correlations includes not only the realm of quantum correlations, that is, those correlations stronger than can be produced within the constraints of local realism, but also to include "superquantum" correlations, which are *stronger* than those predicted by the standard quantum formalism.[2,3] Taking a perspective in this sense "beyond the quantum" allows one to better understand the principles underlying quantum mechanics and to better understand the relationship between physical systems and the communication and information processing tasks they can perform.[2] This understanding may also provide information based on which possible successor theories of microphysics may be constructed.

2. Quantum Correlations

John Bell helped distinguish quantum behavior from classical physical behavior by deriving an inequality that must be obeyed by local realistic physical systems, which are characteristic of classical physics.[4] Bell considered local physical models as local hidden-variables theories, for situations similar to those earlier considered by Einstein, Podolsky, and Rosen,[5] who doubted the completeness of the quantum-mechanical formalism, as having every complete physical state assign a definite probability to a positive measurement outcome for a bivalent property of one subsystem, when the hidden parameter describing it takes a given value independently of measurements performed on the other.

John Clauser, Michael Horne, Abner Shimony, and Richard Holt (CHSH) modified Bell's original treatment so as to be applicable to any practical experimental arrangement that could be described as performing coincidence measurements of bivalent properties, for example particle polarization, in particle pairs such as Bell considered, to obtain the CHSH inequality,

$$|S| \leq 2 \ , \tag{1}$$

$$\text{for} \quad S \equiv E(\theta_1, \theta_2) + E(\theta_1', \theta_2) + E(\theta_1, \theta_2') - E(\theta_1', \theta_2') \ , \tag{2}$$

where the Es are expectation values of the products of measurement outcomes given parameter values θ_i and θ_i' of the two different directions $\hat{\mathbf{n}}_i$ for the same one side i of the two sides jointly constituting the joint-detection apparatus[6] The correlation coefficients contributing to S in terms of experimental detection rates from which S is obtained are

$$E(\theta_i, \theta_j) = \frac{C(\theta_i, \theta_j) + C(\theta_i^\perp, \theta_j^\perp) - C(\theta_i, \theta_j^\perp) - C(\theta_i^\perp, \theta_j)}{rC(\theta_i, \theta_j) + C(\theta_i^\perp, \theta_j^\perp) + C(\theta_i, \theta_j^\perp) + C(\theta_i^\perp, \theta_j)} \ , \tag{3}$$

where the $C(\cdot, \cdot)$ are coincidence detection count rates, i is the index for particle 1, j the index for particle 2, and the parameter $\theta^\perp$ is the direction perpendicular to θ in the plane normal to particle propagation in this scheme.

The value of S characterizes just how "quantum mechanical" the system observed is. Quantum mechanics provides a maximum violation of this inequality by a factor of $\sqrt{2}$, which is achievable when, for example, one prepares the quantum state $|\Phi^+\rangle = \frac{1}{\sqrt{2}}\big(|00\rangle + |11\rangle\big)$, and measures with $\theta_1 = \frac{\pi}{4}$, $\theta_1' = 0$, $\theta_2 = \frac{\pi}{8}$, and $\theta_2' = \frac{3\pi}{8}$. For this set of parameters, one finds $|S| = 2\sqrt{2}$. Once $|S|$ exceeds 2, the behavior of a system is no longer considered classical in nature but is instead considered quantum mechanical. This quantity is an example of the sort of quantity useful for placing quantum mechanics in a broader context. Indeed, it has been shown that such states allow communications tasks to be performed with improved efficiency.[7,8]

3. Super–Quantum Correlations

Popescu and Rohrlich first explicitly considered the possibility of broadening the context even further, by considering theories exceeding the maximum quantum-mechanical value of S by introducing the "non-local box" schema.[2] These "boxes" were taken up specifically for the purpose of considering the question of whether quantum mechanics is uniquely distinguished by its reconciliation of nonlocality and causality. As it turns out, they answered the question in the negative. Much as Bell himself had constructed an explicit hidden-variables model violating the assumptions of von Neumann's putative proof of the nonexistence of hidden-variables theories underlying quantum mechanics in order to sharpen the analysis the assumptions underlying quantum theory,[9] so Popescu and Rohrlich constructed the nonlocal box to find that the reasonable, that is, causally acceptable violation of the Bell inequality is not sufficient to characterize *all* nonlocal causal physical descriptions. Although Bell had not asserted that causal violation of his inequality distinguishes quantum mechanics uniquely, the question is certainly an important one. Notably, Popescu and Rohrlich referred to correlations stronger than those of quantum mechanics as "superquantum."

As Shimony put the matter, "*nonlocality* plus *no signalling* plus *something else simple and fundamental*" might then help one get at what are the basic conceptual underpinnings of quantum mechanics.[10] This initial foray into the realm of super-quantum correlations took place one decade ago. More recently, after the ascension of quantum information science as a field of its own,[11,12] these correlations have once again been studied, this time virtually as an end in themselves and with the tools made available by this new area of study.[13–16] It turns out that particularly pertinent in this endeavor is the structure of communication complexity theory under the assumption of the availability of both quantum[7] and super-quantum correlations as shared "resources" for communication.[17]

The results of Bell and CHSH can be reformulated in the following form, which is more convenient than the original versions. Any local realistic hidden-variables theory describing a joint system AB must obey an inequality of the form

$$s = \sum_{x,y\in\mathbb{Z}_2} p(m_x^A + m_y^B \equiv x \cdot y) \leq 3 \,, \tag{4}$$

where $p(m_x^A + m_y^B \equiv x \cdot y)$ is the probability of obtaining measurement outcomes m_x^A, $m_y^B \in \mathbb{Z}_2$ summing (mod 2) to the product of "measurement setting" parameters x and y in $\mathbb{Z}_2$, that is, $\{0, 1\}$. Quantum mechanics violates such an inequality by reaching the value $s = 3.41 = 2 + \sqrt{2}$. As it turns out, by sharing a Bell state, two agents Alice and Bob are able to approximate nonlocal boxes with probability $\cos^2 \frac{\pi}{8} \approx 0.854$.[18]

4. The Collapse of Super–Quantum Complexity Theory

There are many ways to probe theoretical alternatives to quantum mechanics. Our concern here is what may be understood in the realm of correlations "beyond the quantum." The non-local box of Popescu and Rohrlich causally providing nonlocal correlations is defined by the characteristic that if the two-bit string xy is a member of $\{00, 01, 10\}$ then $p(m_x^A = 0, m_y^B = 0) = \frac{1}{2}$ and $p(m_x^A = 1, m_y^B = 1) = \frac{1}{2}$, and when xy takes the remaining possible value 11 then $p(m_x^A = 0, m_y^B = 1) = \frac{1}{2}$ and $p(m_x^A = 1, m_y^B = 0) = \frac{1}{2}$, all other pertinent probabilities being zero. For this schema, on the left-hand side of the above inequality, one sees that the value 4 can be achieved. This "box" still obeys causality because the outcomes on one side of the coincidence-counting apparatus still occur locally at random, as they do in the case of Bell states such as the singlet state of spin, for example, but with stronger correlations between joint measurement outcomes than in any quantum state.

If one chooses, as Popescu and Rohrlich did, the most natural route of super-quantum correlations, one finds that nonlocal causal correlations are, in the sense of the distributed computations this enables, *too* powerful. The area in which such ideas are explicated, communication complexity theory, is based on the amount of information that must necessarily be communicated in order to obtain the value of a function f distributed between at least two parties.[19] Directly pertinent here are situations where f is a Boolean function from $\mathbb{Z}_2^{\times 2}$ to $\mathbb{Z}_2$. To be specific, allowing *maximally* nonlocal such correlations to two spacelike separated agents, Alice and Bob, allows them to perform *all* distributed computations with perfect accuracy given a *trivial* amount of communication, namely, one bit; that one bit is necessary for preserving causality.[20] This result is due to van Dam, who reduced all possible distributed functions in the standard inner-product form and proceeded to show that the correlations provided by nonlocal boxes allowed the inner-product functions to be solved with just one bit of communication, with a maximum of an exponential amount of prior-shared nonlocal boxes.

Given the fundamental role of communication complexity theory, such a collapse of the classes of communication problems seems remarkable. Indeed, van Dam points out that this classification of communications problems, along with that of computational complexity theory, is fundamental to theoretical computer science, and that "their absence...goes squarely against the worldview and experience of probably all researchers in the field of complexity theory".[17] It is natural then to ask whether what distinguishes quantum mechanics is not *causality*—a physical constraint—but rather nontrivial quantum complexity theory—an information-theoretic constraint—or something physical associated with *it*, as Shimony's "something else." This result certainly does suggest that nonlocal boxes themselves described correlations that don't exist on the grounds of *computer science* rather than physics. On the other hand, presumably most researchers in theoretical computer science have little experience even with previously shared *quantum* correlations much less nonlocal correlations constrained only by causality. Furthermore, the associated "collapse" of communication complexity might be prevented by its being based not simply on the number of bits that need to be communicated but also the quantity of shared resources required. The connection between information theory and physics in this situation bears further scrutiny.

5. Beyond the Quantum

The primary concern for physicists at this juncture is the original question posed by Shimony and investigated by Popescu and Rohrlich: what *physical* principle can, within the constraint of causality, provide nonlocal correlations of the strength predicted by quantum mechanics *but only that strength*. Recently, a bound has been demonstrated as following from the requirement of nontrivial computational complexity, but one still *beyond* the strength of quantum correlations, namely, the theorem that "In any world in which it is possible, without communication, to implement an approximation to the [nonlocal box] that works correctly with probability greater than $\frac{3+\sqrt{6}}{6} \approx .908$, every Boolean function has trivial probabilistic communication complexity".[18] This result is to be considered in light of the fact that a quantum Bell state approximates a nonlocal box with probability 0.854.

This result leaves a gap of about 5 percent that may yet be closed. Nonetheless, even should this gap be closed, there still remains the question of the *physical concept* associated with a trivial communication complexity hierarchy.

6. The Simulation of Quantum Correlations

To develop a better understanding of the nature of the remarkable communication power associated with nonlocal boxes, it is valuable to consider the relationship between prior-shared nonlocal boxes and better known information-theoretic resources. Because entanglement (as in the form of prior-shared Bell states of two quantum bits), the quantum property associated with violation of Bell-type theorems, has been proven to constitute an information-theoretic resource surpassing local classical resources, such as shared randomness (as in the form of prior-shared random bit strings), one may hope to also to gain inside into the constitution of quantum mechanics "from the outside" by considering the relationship between prior-shared nonlocal boxes, prior-shared Bell states of qubit pairs, and prior-shared random bits.

In 2003, the question of the resources necessary for simulating the correlations arising from prior-shared Bell states, that is "Bell correlations" by means not quantum mechanical in nature, was taken up.[13] Bell's theorem establishes that local theories cannot cannot simulate the behavior of outcomes of measurements on a singlet without communication. It was demonstrated, however, that one bit of (superluminal) communication and prior-shared classical randomness *are* sufficient to produce such correlations. In 2005, it was shown that a single prior-shared nonlocal box is capable to simulating all outcomes of projective measurements on a Bell state *without communication*.[15] Quantitative relationships between these resources are clear because the resource of one shared Bell singlet state has been formally identified, namely, the "e-bit," which was done by Schumacher.[21] The above result establishes that the nonlocal box information unit, the "nl-bit," is a stronger resource than the e-bit.[14,22] Of all the information-theoretic resources under consideration, one bit of superluminal communication is the strongest resource, since that bit, unlike a nonlocal box, is not constrained by causality. However, this unit is *nonphysical.* Whether physicists or theorists in computer science should be particularly concerned about the implications of shared super-correlations is, therefore, suspect.

What has been established by the investigation of correlations in the broad context that has emerged from the study of nonlocal boxes is that both groups of scientists do need to find out *what in physics* corresponds to nontrivial communication complexity, and whether this is quantum mechanical or "beyond" quantum mechanical. Causality is one example of a physical concept pertinent to both physics and information theory; this may be another. The physical correspondent of the information-theoretic concept of non-trivial communication complexity may help provide the physical principle that is the missing element of the set of principles underlying quantum mechanics, or at least to aid in the identification of the physical "something else" the study of the foundations of quantum theory yet requires. It therefore requires our attention.

References

1. A. Aspect, P. Grangier, and G. Roger, *Phys. Rev. Lett.* **47**, 460 (1981).
2. S. Popescu, and D. Rohrlich, *Potentiality, entanglement and passion-at-a-distance* (Kluwer; Dordrecht, 1997); also http://xxx.lanl.gov quant-ph/9605004 (3 May, 1996).
3. K. Svozil,*Phys. Rev. A* **72**, 050302(R) (2005).
4. J. S. Bell,*Physics* **1**, 195 (1964).
5. A. Einstein, B. Podolsky, and N. Rosen *Phys. Rev.* **47**, 777 (1935).
6. J. F. Clauser, M. Horne, A. Shimony, and R. A. Holt, *Phys. Rev. Lett.* **23**, 880 (1973).
7. H. Buhrman, R. Cleve, and W. van Dam, *SIAM J. Comput.* **30**, 1829 (2001).
8. C. Brukner, M. Zukowski, and A. Zeilinger, *Phys. Rev. Lett.* **89**, 19701 (2002).
9. J. S. Bell, *Rev. Mod. Phys.* **38**, 447 (1966).
10. A. Shimony, *Foundations of quantum mechanics in light of the new technology*, (Physical Society of Japan; Tokyo, 1983).
11. G. Jaeger, *Quantum information: An overview*, (Springer; Berlin, 2006).
12. M. Nielsen and I. Chuang, *Quantum computation and quantum information*, (Cambridge University Press; Cambridge, 2000).
13. B. F. Toner and D. Bacon *Phys. Rev. Lett.* **91**, 187904 (2003).
14. J. Barrett and S. Pironio *Phys. Rev. Lett.* **95**, 140401 (2005).
15. N. J. Cerf, N. Gisin, S. Massar, and S. Popescu *Phys. Rev. Lett.* **94**, 220403 (2005).
16. J. Barrett, N. Linden, S. Massar, S. Pironio, S. Popescu, and D. Roberts *Phys. Rev. A* **71**, 022101 (2005).
17. W. Van Dam, *http://xxx.lanl.gov quant-ph/0501159* (27 Jan., 2005)
18. G. Brassard, H. Buhrmann, N. Linden, A. A. Méthot, A. Tapp, and F. Unger, *http://xxx.lanl.gov quant-ph/0508042* (4 Aug., 2005).
19. H. Buhrman, W. van Dam, P. Høyer, and A. Tapp, *Phys. Rev. A* **60**, 2737 (1999).
20. W. Van Dam, *Nonlocality and Communication complexity*, D. Phil. thesis, University of Oxford, Dept. of Physics (2000), Ch. 9.
21. B. W. Schumacher, *Phys. Rev. A* **51**, 2738 (1995).
22. N. Jones and L. Masanes, *Phys. Rev. A* **72**, 052312 (2005).

PART C

Long Distance Correlations and Bell Inequalities

UNDERSTANDING LONG-DISTANCE QUANTUM CORRELATIONS

L. MARCHILDON

Département de physique, Université du Québec, Trois-Rivières, QC, G9A 5H7, Canada
E-mail: marchild@uqtr.ca

The interpretation of quantum mechanics (or, for that matter, of any physical theory) consists in answering the question: How can the world be for the theory to be true? That question is especially pressing in the case of the long-distance correlations predicted by Einstein, Podolsky and Rosen, and rather convincingly established during the past decades in various laboratories. I will review four different approaches to the understanding of long-distance quantum correlations: (i) the Copenhagen interpretation and some of its modern variants; (ii) Bohmian mechanics of spin-carrying particles; (iii) Cramer's transactional interpretation; and (iv) the Hess–Philipp analysis of extended parameter spaces.

Keywords: Long-distance correlations; Quantum mechanics; Interpretation.

1. Introduction

In one of his thought-provoking discussions of the two-slit experiment, Feynman[1] expressed the view that "it is safe to say that no one understands quantum mechanics. [...] Nobody knows how it can be like that." Yet 80 years of research in the foundations of the theory have led a growing number of investigators not to share Feynman's fatalism. They, in fact, have turned his assessment into a challenge, by asking "How can the world be for quantum mechanics to be true?" I have argued elsewhere,[2] following others,[3] that interpreting the theory consists in providing a precise answer to this question. Moreover, I believe that providing more than one possible and consistent answer, far from introducing confusion, brings instead additional understanding, and may even stimulate the imagination.

Long-distance quantum correlations, first pointed out by Einstein, Podolsky and Rosen (EPR),[4] and sharply investigated by Bell,[5] have long been considered paradoxical and in need of explanation. It is the purpose of this contribution to briefly review and analyze four different approaches through which one can make sense of them.

2. Long-Distance Correlations

Following Bohm,[6] I consider two spin 1/2 particles prepared in the singlet state $|\chi\rangle$ and leaving in opposite directions (Fig. 1). On each side, an apparatus can measure the component of the spin of the associated particle either along axis $\hat{n}$ or along axis $\hat{n}'$.

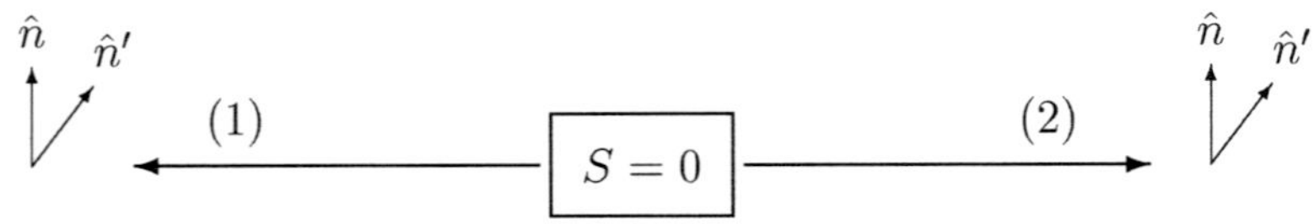

Fig. 1. Two particles prepared in the singlet state and leaving in opposite directions.

In a nutshell, the paradoxical features of the arrangement can be expressed as follows:

(1) Quantum mechanics predicts, and experiments confirm, that there is perfect anticorrelation when the spins of both particles are measured along the same axis.
(2) This seems to suggest, against conventional quantum mechanics, that all spin components have values even before they are measured. This assertion is an instance of *local realism.*
(3) Quantum mechanics predicts, and experiments confirm, that the spin correlations are in general given by

$$\langle\chi|\vec{\sigma}_1 \cdot \hat{n} \otimes \vec{\sigma}_2 \cdot \hat{n}'|\chi\rangle = -\hat{n} \cdot \hat{n}'. \qquad (1)$$

(4) Conventional wisdom holds that local realistic theories imply, against quantum mechanics, that the spin correlations satisfy the Bell inequalities. These are inconsistent with Eq. 1 and are experimentally violated.

3. Copenhagen and Related Views

In his reply to the EPR paper, Bohr[7] emphasized the holistic aspect of measurement. For him, the whole experimental setup is inseparable. The measurement of a physical quantity on one side fundamentally alters the conditions of the measurement of a conjugate variable on the other side. It prevents the definition of meaningful "elements of reality" pertaining to one part of a system only. Since no value can be ascribed to an observable outside its measurement context, the inference from statement (1) to statement (2) in the last section is, for Bohr, unwarranted.

Among the leading exponents of the Copenhagen interpretation, Heisenberg[8] has expressed the view that the state vector represents knowledge rather than the state of an independent object. The development of quantum information theory has led to renewed interest in this *epistemic view* of quantum states. To quote a recent column,[9] "the time dependence of the wave function does not represent the evolution of a physical system. It only gives the evolution of our probabilities for the outcomes of potential experiments on that system." In the epistemic view, or at least in the more radical variants of it, there are no microscopic carriers of elements of reality. The state vector is simply a device to predict correlations between distant local measurements. The correlations don't stand in need of further explanation.

That quantum mechanics is about information is also stressed in the relational view advocated by Rovelli.[10,11] In relational quantum mechanics, all systems (including apparatus) are quantum mechanical. An observable can have a value with respect to an observer and not with respect to another.

Specifically, spin $\vec{\sigma}_2 \cdot \hat{n}'$ has a value for observer O_1 only if O_1 measures it, or measures the result obtained by O_2. The meaningful correlations are not those between $\vec{\sigma}_1 \cdot \hat{n}$ measured by O_1 and $\vec{\sigma}_2 \cdot \hat{n}'$ measured by O_2, but (say) those between $\vec{\sigma}_1 \cdot \hat{n}$ and $\vec{\sigma}_2 \cdot \hat{n}'$ both measured by O_1. Hence there is no problem with locality. The price one has to pay for this resolution of the paradox is, however, a rather significant weakening of realism. Specifically, statements like "the measurement of this observable has yielded that value" no longer hold in an absolute way.

4. Bohmian Mechanics

The Schrödinger wave function for two spinless particles can be written as

$$\Psi(\vec{r}_1, \vec{r}_2, t) = \rho(\vec{r}_1, \vec{r}_2, t) \exp\left\{\frac{i}{\hbar} S(\vec{r}_1, \vec{r}_2, t)\right\}. \tag{2}$$

In Bohmian mechanics, the particles follow deterministic trajectories governed by[12,13]

$$\vec{v}_1 = \frac{1}{m_1}\vec{\nabla}_1 S, \qquad \vec{v}_2 = \frac{1}{m_2}\vec{\nabla}_2 S. \tag{3}$$

The statistical predictions of quantum mechanics are recovered by postulating that the particles are drawn from an ensemble with probability density $|\Psi|^2 = \rho^2$.

For a factorizable wave function like

$$\Psi(\vec{r}_1, \vec{r}_2, t) = \psi_1(\vec{r}_1, t)\psi_2(\vec{r}_2, t), \tag{4}$$

we get $S(\vec{r}_1, \vec{r}_2, t) = S_1(\vec{r}_1, t) + S_2(\vec{r}_2, t)$, and the motion of particle 1 is independent of what happens to particle 2. But for an entangled wave function like

$$\Psi(\vec{r}_1, \vec{r}_2, t) = \sum_k \psi_1^{(k)}(\vec{r}_1, t)\psi_2^{(k)}(\vec{r}_2, t), \tag{5}$$

the function $S(\vec{r}_1, \vec{r}_2, t)$ does not break up into the sum of a function of $\vec{r}_1$ and a function of $\vec{r}_2$. What happens to particle 2 instantaneously affects the motion of particle 1. From this one may be tempted to conclude that Bohmian mechanics will allow for superluminal transfer of information. This is indeed the case if state preparation is not suitably restricted.[14] But if particles are prepared with a probability density $|\Psi(\vec{r}_1, \vec{r}_2, t_0)|^2$ at time t_0, they evolve into a density $|\Psi(\vec{r}_1, \vec{r}_2, t)|^2$ at any time t, and one can show that no superluminal transfer of information is possible.

To incorporate spin in Bohmian mechanics, one adds spinor indices to the wave function, in such a way that $\Psi \to \Psi_{i_1 i_2}$. There can be several ways to associate particle spin vectors with the wave function,[12] but one way or other they involve the expressions

$$\vec{s}_1 = \frac{\hbar}{2\Psi^\dagger\Psi}\Psi^\dagger\vec{\sigma}_1\Psi, \qquad \vec{s}_2 = \frac{\hbar}{2\Psi^\dagger\Psi}\Psi^\dagger\vec{\sigma}_2\Psi. \tag{6}$$

In the singlet state, the initial wave function typically has the form

$$\Psi = \psi_1(\vec{r}_1)\psi_2(\vec{r}_2)\frac{1}{\sqrt{2}}(u_{1+}u_{2-} - u_{1-}u_{2+}), \tag{7}$$

in obvious notation. With such wave function, it is easy to show that $\vec{s}_1 = 0$ and $\vec{s}_2 = 0$. That is, both particles initially have spin zero. This underscores the fact that in Bohmian mechanics, values of observables outside a measurement context do not in general coincide with operator eigenvalues.

Spin measurement was analyzed in detail in Refs. 15 and 16. In the EPR context, in particular, Dewdney, Holland and Kyprianidis first wrote down the two-particle Pauli equation adapted to the situation shown in Fig. 1. With Gaussian initial wave packets ψ_1 and ψ_2, the equation can be solved under suitable approximations. Bohmian trajectories can then be obtained by solving Eq. 3. These involve the various components of the two-particle wave function in a rather complicated way, and must be treated numerically.

Suppose that the magnetic field in the Stern–Gerlach apparatus on the left of Fig. 1 is oriented in the $\hat{n}$ direction. Consider the case where particle 1 enters that apparatus much before particle 2 enters the one on the right-hand side. What was shown was the following. When particle 1 enters the apparatus along a specific Bohmian trajectory, the various forces implicit in Eq. 3 affect both the trajectory and the spin vector, the latter building up through interaction with the magnetic field. The beam in which particle 1 eventually ends up depends on its initial position. If particle 1 ends up in the upper beam of the Stern–Gerlach apparatus, its spin becomes aligned with $\hat{n}$. Meanwhile there is an instantaneous action on particle 2, simultaneously aligning its spin in the $-\hat{n}$ direction. Similarly, if particle 1's initial position is such that it ends up in the lower beam, its spin becomes aligned with $-\hat{n}$, and the spin of particle 2 simultaneously aligns in the $\hat{n}$ direction.

Thus the nonlocal forces inherent in Bohmian mechanics have, once the measurement of the spin of particle 1 has been completed, resulted in particle 2 having a spin exactly opposed. It is then easy to see that if particle 2 later enters a Stern–Gerlach apparatus with magnetic field oriented in the $\hat{n}'$ direction, its deflection in the upper or lower beam will precisely reproduce the correlations of Eq. 1.

5. The Transactional Interpretation

Cramer's transactional interpretation[17,18] is inspired by the Wheeler–Feynman electromagnetic theory, in which advanced electromagnetic waves are as important as retarded waves.

In this interpretation, a quantum process (e.g. the emission of an α particle, followed by its absorption by one of several detectors) is held to involve the exchange of offer waves (solutions of the Schrödinger equation) and confirmation waves (complex conjugates of the former). The confirmation waves propagate backward in time.

Suppose that D, at point $\vec{r}$, is one of a number of detectors that can absorb the particle. The offer wave, emitted at t_0 from the α particle source, will arrive at D with an amplitude proportional to $\psi(\vec{r}, t)$, the Schrödinger wave function. The confirmation wave produced by D is stimulated by the offer wave, and Cramer argues that it arrives back at the source with an amplitude proportional to $\psi(\vec{r}, t)\psi^*(\vec{r}, t)$ $= |\psi(\vec{r}, t)|^2$. Similar offer and confirmation waves are exchanged between the source and all potential detectors, and all confirmation waves reach the source exactly at t_0, the time of emission. Eventually, what Cramer calls a *transaction* is established between the source and one of the detectors, with a probability proportional to the amplitude of the associated confirmation wave at the source. The quantum process is then completed.

Fig. 2 is a space-time representation of an EPR setup, in the transactional interpretation. Arrows pointing in the positive time direction label offer waves, and those pointing in the negative direction label confirmation waves. Two particles are emitted by the source, and in Cramer's sense each particle can be absorbed by two detectors, corresponding to the two beams in which each particle can emerge upon leaving its Stern–Gerlach apparatus.

Let us focus on what happens on the left-hand side. An offer wave is emitted by the source, and in going through the Stern–Gerlach apparatus it splits into two parts. One part goes into the detector labelled +, and the other goes into detector −. Each detector sends back a confirmation wave, propagating backward in time through the apparatus and reaching the source at the time of emission. A transaction is eventually established, resulting in one of the detectors registering the particle. A similar process occurs on the right-hand side, with one of the two detectors on that side eventually registering the associated particle.

If offer and confirmation waves represent causal influences of some sort, one can see that these influences can be transmitted between the spacelike-separated detectors on different sides along paths that are entirely timelike or lightlike. In this way, the EPR correlations are explained without introducing any kind of superluminal motion.

6. Extended Parameter Space

Bohmian mechanics and Cramer's transactional interpretation explain the long-distance quantum correlations by means of channels which, although not allowing

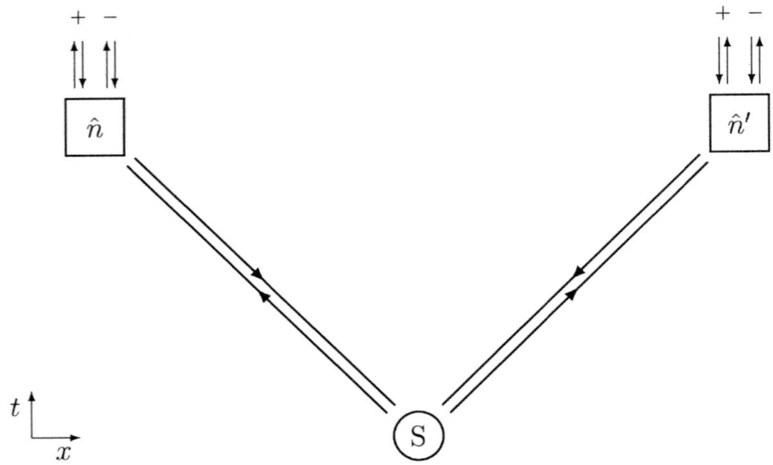

Fig. 2. Offer waves (upward arrows) and confirmation waves (downward arrows) in the EPR setup.

for the superluminal transfer of classical bits of information, involve causal links of some sort between spacelike-separated instruments. In recent work, Hess and Philipp[19] have argued that the correlations might be understood without appealing to such links.

In the original proof of his inequality, Bell[5] assumed that the state of the particle pair is characterized by a hidden variable λ, which represents one of the values of a random variable Λ. He further assumed that the result of measuring (say) the $\hat{n}$ component of the spin of particle 1 is fully determined by λ and $\hat{n}$. He supposed, however, that the result of measuring the $\hat{n}$ component of the spin of particle 1 does not depend on which component of the spin of particle 2 is measured. The latter assumption embodies the prohibition of superluminal causal influences, and is fully endorsed by Hess and Philipp.

Hess and Philipp point out that Bell's proof, as well as all subsequent proofs of similar inequalities, make use of parameter spaces that are severely restricted. They introduce much more general spaces. Like Bell, they assume that pairs of particles emitted in the singlet state are characterized by a random variable Λ, which is stochastically independent on the settings on both sides. But then they associate with each measuring instrument random variables $\Lambda^{(1)}_{\hat{n}}(t)$ and $\Lambda^{(2)}_{\hat{n}'}(t)$, which depend both on the setting of the instrument and on the time. The result of measuring, say, the $\hat{n}$ component of the spin of particle 1 at time t, is taken to be a deterministic function of Λ and $\Lambda^{(1)}_{\hat{n}}(t)$.

Several important remarks should be made at this stage. Firstly, and in the spirit of standard quantum mechanics, neither particle has a precise value of any of its spin components before measurement. Rather, the particles and the instruments jointly possess information that is sufficient for deterministic values to obtain upon measurement. Secondly, the dependence of the instruments' random variables on some universal time allows for a stochastic dependence of measurement results on one another, conditional on Λ, if the measurements are performed at correlated times in the two wings. And yet thirdly, the measurement result on one side can be stochastically independent on the setting on the other side.

With such extended parameter spaces, Hess and Philipp have shown that the standard proofs of the Bell inequalities come to a halt. Such proofs typically assume that the two particles, once they have left the source, simultaneously have well-defined values of more than one spin component. But in the extended parameter space approach, spin components get values only upon measurement. Counterfactual reasoning is allowed only in the sense that had a different spin component been measured, it would have yielded a definite and deterministic value. But that value does not exist before measurement. And since the measurement of different spin components requires incompatible apparatus, different spin components of the same particle cannot have values at the same time. But spin components of both particles measured at correlated times in the two wings can be stochastically dependent, through the dependence of the instrument random variables on time.

In experimental tests of the Bell inequalities, spin measurements on a given pair were performed in a time frame many orders of magnitude smaller than the time interval between successive measurements on two different pairs. It is therefore conceivable that a time dependence of the instrument random variables, having no effect on such properties as perfect anticorrelation for particles in the same run, could reproduce the quantum-mechanical long-distance correlations observed on runs performed at different times. Such runs would not sample the quantities that appear in the standard forms of the Bell inequalities.

Hess and Philipp also proposed an explicit model of Einstein-local random variables that lead to violations of the Greenberger–Horne–Zeilinger equations,[20] violations that experiments claimed to have observed.[21]

7. Summary and Conclusion

The long-distance quantum correlations and the violation of Bell inequalities can be understood in a number a different ways, four of which were reviewed here.

In the Copenhagen and epistemic views, correlations are basically dealt with by relaxing the requirements of explanation. In Bohmian mechanics, instantaneous interactions orient the spin of the second particle while the spin of the first one is measured, but restrictions on state preparation prevent the superluminal transfer of information. In the transactional interpretation, advanced waves provide for a communication channel between spacelike–separated detectors. In the Hess–Philipp approach, finally, correlations are explained through instrument random variables that depend both on setting and on time.

I have attempted to illustrate the idea that a theory is made clearer through the display of various models that make it true. This is the process of interpretation, and in connection with it one should be wary of identifying consequences of the formalism of quantum mechanics with consequences of specific interpretations of it. This, unfortunately, has not always been done, as the example of Ref. 22 still shows.

Acknowledgments

This work was supported by the Natural Sciences and Engineering Research Council of Canada. I thank Karl Hess for comments on the manuscript. I am grateful to many participants in the workshop for discussions, and especially to Walter Philipp, who has sadly left us since and will be keenly remembered.

References

1. R. P. Feynman, *The Character of Physical Law* (MIT, Cambridge, MA, 1967), p. 129.
2. L. Marchildon, *Found. Phys.* **34**, 1453 (2004).
3. B. C. van Fraassen, *Quantum Mechanics: An Empiricist View* (Clarendon Press, Oxford, 1991).
4. A. Einstein, B. Podolsky and N. Rosen, *Phys. Rev.* **47**, 777 (1935).
5. J. S. Bell, *Physics* **1**, 195 (1964).
6. D. Bohm, *Quantum Theory* (Prentice-Hall, Englewood Cliffs, 1951), Chap. 22.
7. N. Bohr, *Phys. Rev.* **48**, 696 (1935).
8. W. Heisenberg, *Physics and Philosophy. The Revolution in Modern Science* (Harper, New York, 1958), Chap. 3.
9. C. A. Fuchs and A. Peres, *Phys. Today* **53**, No. 3, 70 (2000).
10. C. Rovelli, *Int. J. Theor. Phys.* **35**, 1637 (1996).
11. M. Smerlak and C. Rovelli, *quant-ph/0604064*.
12. D. Bohm and B. J. Hiley, *The Undivided Universe* (Routledge, London, 1993).
13. P. R. Holland, *The Quantum Theory of Motion* (Cambridge University Press, Cambridge, 1993).
14. A. Valentini, *Phys. Lett. A* **158**, 1 (1991).
15. C. Dewdney, P. R. Holland and A. Kyprianidis, *Phys. Lett. A* **119**, 259 (1986).
16. C. Dewdney, P. R. Holland and A. Kyprianidis, *J. Phys. A: Math. Gen.* **20**, 4717 (1987).
17. J. G. Cramer, *Phys. Rev. D* **22**, 362 (1980).
18. J. G. Cramer, *Rev. Mod. Phys.* **58**, 647 (1986).
19. K. Hess and W. Philipp, *Proc. Nat. Acad. Sci. USA* **101**, 1799 (2004).
20. K. Hess and W. Philipp, "Bells theorem: Critique of proofs with and without inequalities," *quant-ph/0410015*.
21. J. W. Pan, D. Bouwmeester, M. Daniell, H. Weinfurter and A. Zeilinger, *Nature* **403**, 515 (2000).
22. A. Zeilinger, *Nature* **438**, 743 (2005).

CONNECTION OF PROBABILITY MODELS TO EPR EXPERIMENTS: PROBABILITY SPACES AND BELL'S THEOREM

K. HESS

Beckman Institute for Advanced Science and Technology, University of Illinois, 405 N. Mathews Avenue, Urbana, Illinois 61801, USA
E-mail: k-hess@uiuc.edu

W. PHILIPP*

University of Illinois, Department of Mathematics, USA

We compare quantum mechanics as a theory involving probabilities to the framework of Kolmogorov's probability theory with emphasis on the connections of these theories to actual experiments. We find crucial differences in the way incompatible experiments are defined and treated in these two approaches and show that these differences are the origin for difficulties and apparent contradictions that are encountered when considering so called no-go proofs particularly that of John Bell. For example, Bell was convinced that in a theory in which parameters are added to quantum mechanics to determine the results of individual measurements, violations of Einstein-locality must occur. Based on our comparative study, we show that rather the opposite is true and that a precise space-time treatment based on relativity uncovers contradictions in the assumptions for Bell's no-go proof and resolves the difficulties.

Keywords: Hidden variables; Bell's Theorem; Einstein locality.

1. Introduction

John Bell[1] developed a mathematical model for EPR-types of experiments[2] that led to his well known inequalities. In this model, Bell attempted to complete quantum mechanics by the introduction of the Einstein-local hidden variable Λ. The failure of this model to agree with the predictions of quantum mechanics and experimental results[2] was taken as proof (no-go proof) that no hidden parameters of this type exist. The inclusion of Λ served the purpose to effect the complete specification of the outcomes of individual measurements. Bell[1] states that "It is a matter of indifference in the following whether Λ denotes a single variable or a set, or even a set of functions, and whether the variables are discrete or continuous." This statement led Bell's followers to believe that Λ can be "anything". In the next step Bell introduced possible outcomes for spin measurements $A_{\mathbf{a}}(\Lambda) = \pm 1$ and $B_{\mathbf{b}}(\Lambda) = \pm 1$.

---

*Deceased.

Here **a** and **b** denote the various orientations of the Stern-Gerlach magnets and other orientations are also used. Bell then proceeded to treat all variables thus introduced as random variables. Because all of these variables are functions of Λ and Λ is also a random variable, all random variables are then defined on one common probability space and Bell uses this fact in his algebraic manipulations and integrations.

One of the purposes of this paper is to show that Bell's assumption of a completely general Λ, that includes for example general time dependencies, and the second assumption that all random variables A, B involved are functions of Λ and therefore defined on one common probability space are the real reason behind the mathematical and physical contradictions. The mathematical contradictions have their basis in the well known consistency problem that was worked out in full generality by Bass[3] and Vorob'ev[4] many years before Bell's work but has never been cited or included in the considerations of Bell and his followers. With this statement we do not wish to indicate any diminishment of the importance of Bell's inequalities. We do wish to indicate, however, that Bell was not aware of the precise mathematical conditions for his inequalities to be valid. A careful examination of the consistency problem shows that Bell's assumptions are not general and can, without serious contradictions, cover only a sub-set of the physically possible parameters. In fact, Λ can not even be related to the times of the measurement in the laboratory frame as we will show in detail below. We also show that Einstein locality cannot justify Bell's assumptions but rather the opposite. Therefore Einstein locality is not a basic pillar in the derivation of Bell's inequalities. On the contrary, a careful inclusion of Einstein locality uncovers logical contradictions in Bell's assumptions. Key to the understanding of all of this is the concept of the probability space in Kolmogorov's theory and how this concept serves to connect the mathematical model to the actual experiments.

Mathematical models that involve probability require special care for the treatment of their relation to actual experiments. A major reason arises from the fact that by their very definition probability models are based on sequences of experiments and not just on a single experiment. The connection of these sequences to mathematical theory, e.g. set theory and the theory of measurable functions, necessitates then provisions that go beyond those of the single experiment. In Kolmogorov's framework, these additional provisions are taken care of by the introduction of a sample space. The sample space set contains indecomposable elements that are uniquely related to each experiment of a sequence.[5] Random variables are functions on the sample space. Sample space, sigma field, probability measures and random variables complete the relation to the actual experiments and form the basis for a set theoretic approach. Quantum mechanics has proceeded along a different path. The quantum mechanical state forms the basis for the theory and repeated state preparation as well as repeated measurement of identically prepared states form the basis for the connection to the experiments.[6] The mathematical treatment involves now state-vectors and operators (characteristic for the actual measurement) and computes expectation values without using random variables. The comparison of the quantum mechanical approach to Kolmogorov's is a main subject of this paper. We show that there exists a clear correspondence between these two approaches. However, this correspondence requires special care when incompatible (to be defined in the bulk of the paper) experiments are involved. We show that relativity and the physics of light-cones is important for a careful comparison and, when taken correctly into account, resolves many if not all of the puzzling difficulties surrounding no-go proofs such as the theorem of Bell. We certainly are confident that the mathematical and physical difficulties of the Bell theorem that are discussed below give ample of reasons not to take it as an obstacle standing in the way of visionary approaches such as the mathematical basis for deterministic quantum mechanics as given by Gerard 't Hooft in these proceedings and in.[7]

2. Random Variables, EPR Experiments and the Theorem of Bell

Consider a typical EPR type experiment with Fermion spins as discussed in Bell's paper[1] i.e. measurement of an entangled pair of spin 1/2 particles prepared in the Bell state $|\psi_B\rangle$. The measurement is performed in two wings with the magnet in one wing set in direction **a** and, in the other wing, in direction **b** where **a** and **b** are unit vectors of three dimensional Euclidean space. The results of such experiments are in agreement with quantum theory and the expectation value M for the Bell wave function is:[6]

$$M(|\psi_B\rangle) = \langle\psi_B|\sigma_{\mathbf{a}} \otimes \sigma_{\mathbf{b}}|\psi_B\rangle = -\mathbf{a} \cdot \mathbf{b}. \tag{1}$$

Here $\sigma_{\mathbf{a}}, \sigma_{\mathbf{b}}$ are the well known spin operators e.g. $\sigma_{\mathbf{a}} = a_x\sigma_x + a_y\sigma_y + a_z\sigma_z$ and $\sigma_x, \sigma_y, \sigma_z$ are the Pauli spin matrices given below.

Bell investigated the question whether the quantum result could be obtained for at least two different settings on each side by introducing the random variable Λ that assumes outcomes λ for any given experiment involving an entangled pair. Thus Bell made a transition from the probability treatment of quantum mechanics and used random variables and the rules of Kolmogorov's probability theory. To represent the possible outcomes of the spin measurements for a given pair of measurements in the two stations S_1, S_2, Bell introduced random variables $A_{\mathbf{a}} = \pm 1$ and $B_{\mathbf{b}} = \pm 1$ respectively. These random variables were in turn assumed to be functions of Λ the random variable characterizing the entangled pair i.e. $A_{\mathbf{a}} = A_{\mathbf{a}}(\Lambda)$ for S_1 and $B_{\mathbf{b}} = B_{\mathbf{b}}(\Lambda)$ for S_2. From quantum mechanics, and experimentally confirmed, we have for the expectation values on each side $M(A_{\mathbf{a}}) = M(B_{\mathbf{b}}) = 0$. In this notation the expectation value $M(|\psi_B\rangle)$ of Eq. (1) represents the expectation value of the product $M(A_{\mathbf{a}}(\Lambda)B_{\mathbf{b}}(\Lambda))$. As mentioned, the parameter random variable Λ is usually considered to be quite general and it is often remarked that it can be "anything".

It is now important to note that Bell proceeded in his analysis beyond a single pair of settings in spite of the fact of having only one entangled pair per measurement and considered, for example, integrals of the form

$$\int [A_{\mathbf{a}}(\lambda)B_{\mathbf{b}}(\lambda) - A_{\mathbf{a}}(\lambda)B_{\mathbf{c}}(\lambda)]\rho(\lambda)d\lambda \tag{2}$$

and algebraically manipulated this expression to give

$$\int A_{\mathbf{a}}(\lambda)B_{\mathbf{b}}(\lambda)[1 + A_{\mathbf{b}}(\lambda)B_{\mathbf{c}}(\lambda)]\rho(\lambda)d\lambda. \tag{3}$$

Here ρ is a "normalized probability distribution".[1] Clearly such algebraic steps and integrations can not be taken if Λ, A and B can be anything. They must be a set of functions *on one domain*; they must be random variables. Thus, there is a hidden assumption in Bell's treatment: random variables are defined on a probability space and all random variables ($A_{\mathbf{a}}$, $A_{\mathbf{b}}$, $B_{\mathbf{b}}$ and $B_{\mathbf{c}}$ with $A_{\mathbf{b}} = -B_{\mathbf{b}}$[1]) of Bell's treatment must therefore be defined and definable on one common probability space. Then and only then can one perform the mathematical operations that Bell has performed in order to derive his inequalities. As we will see in more detail below, a common probability space requires a common domain Ω (the sample space) with elements ω of which the random variables are functions. In addition we need a probability measure $P(\Omega) = 1$. The above integrals have then a set theoretic basis and are commonly written in the following notation:[8]

$$\int_{\omega\in\Omega} [A_{\mathbf{a}}(\Lambda(\omega))B_{\mathbf{b}}(\Lambda(\omega)) - A_{\mathbf{a}}(\Lambda(\omega))B_{\mathbf{c}}(\Lambda(\omega))]P(d\omega) \tag{4}$$

and

$$\int_{\omega\in\Omega} A_{\mathbf{a}}(\Lambda(\omega))B_{\mathbf{b}}(\Lambda(\omega))[1 + A_{\mathbf{b}}(\Lambda(\omega))B_{\mathbf{c}}(\Lambda(\omega))]P(d\omega) \tag{5}$$

Note that we have indicated the common domain as subscript to emphasize that the assumption of a common domain was made by Bell. There is, however, no physical reason to assume a common domain for distinctly different experimental arrangements. After all we deal with hidden parameters and therefore do not know their precise nature. A common domain and probability measure may also not exist for purely mathematical reasons. Kolmogorov's existence proof says nothing about the consistency of a system of joint probability distributions (e.g. joint probability distributions that lead to the expectation values of Eq. (1) for a variety of settings) but *assumes* consistency. If a system of probability distributions is consistent then one can find a probability space and random variables that reproduce the probability distributions. Questions surrounding this fact are known as "consistency problem" in probability theory and have not been recognized or cited in the physics literature dealing with the Bell theorem. We have shown recently,[9] based on the work of Bass[3] and Vorob'ev[4] that the four random variables $A_{\mathbf{a}}$, $A_{\mathbf{b}}$, $B_{\mathbf{b}}$, $B\mathbf{c}$ with pair expectations as given by Eq. (1) can not always be defined on one common probability space. In fact, one can use the results of Bass and Vorob'ev and express them in terms of inequalities that are now known as Bell inequalities[9] to prove the following theorem which gives the precise conditions for these inequalities to be valid:

Theorem 2.1. *The random variables of the EPR experiment considered by Bell can be defined on one common probability space if and only if they fulfill all possible Bell inequalities.*

Thus the Bell inequalities are already guaranteed to be valid as soon as definition on one common probability space is assumed i.e. as soon as it is assumed that $A_{\mathbf{a}}$, $A_{\mathbf{b}}$, $B_{\mathbf{b}}$ and $B_{\mathbf{c}}$ are random variables. This is a consequence of the simple range of the involved functions (± 1) and agrees with findings of Fine,[10,11] and Pitovsky.[12] Khrennikov[13] has also emphasized the role of a single Kolmogorov space. For details on this theorem in its full generality we refer the reader again to our paper.[9] As we will show in detail below, the physical problem that Bell has considered does not justify the mathematical framework that he used. The physics does justify that Λ is taken to be independent of the settings because of the delayed choice of the settings. However, the physics of EPR experiments does not justify the assumption that all functions involved are a function of a single random variable Λ on one common probability space. We will show in section 7 by formulating a summary statement that this latter assumption and the assumption that Λ is a very general physical parameter lead to a logical contradiction. There is also a very special aspect to all of this. The Bell type inequality does not just give a numerical rule for values that can be assumed by the parameters involved but really states that in case of a violation the whole system of random variables is not and can not be consistently defined on a common domain, a common probability space; and the mathematical operations that are used to derive the inequalities can then not be performed. As mentioned Vorob'ev has studied this consistency problem.

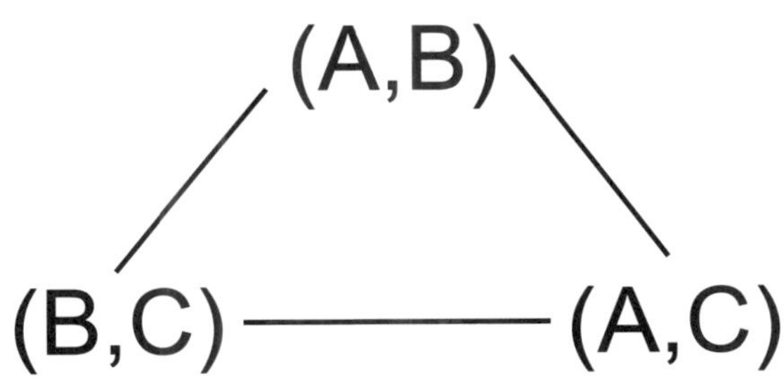

Fig. 1. A Vorob'ev-type closed loop of random variables

Vorob'ev showed in complete generality that one can not prescribe arbitrary pair correlations to a set of random variables that form a closed loop. For example take the random variables A, B, and C and prescribe arbitrary pair correlations to (A, B), (A, C) and (B, C). The formation of the closed loop is shown in Fig. 1. Vorob'ev shows that the arbitrary prescription of pair correlations can lead to a contradiction because once the joint distributions of (A, B), (A, C) are chosen, there is no complete freedom to choose the joint distribution of (B, C). This applies, of course to the above EPR experiment by putting $A = A_{\mathbf{a}}, B = B_{\mathbf{b}}$, $C = B_{\mathbf{c}}$ and $A_{\mathbf{b}} = -B$. Similar considerations apply to the Clauser-Horne-Shimony-Holt inequalities.[14] For details on this see.[9]

Thus from a purely mathematical point of view, the Bell inequalities just tell us that it is not always possible to model a number of different EPR experiments on one common probability space. Questions of Einstein locality or the existence of influences or even action at a distance can then be answered as soon as we have a clear physical understanding of what it means that certain experimental results can not be explained by a random variable model on one common probability space. It is the purpose of this paper to show that the use of different probability spaces for certain incompatible experiments is suggested not only by quantum mechanics but also by classical relativistic physics and as just discussed also for abstract mathematical reasons, the closed loops of Vorob'ev. Violations of Einstein locality may, of course, also suggest the involvement of different probability spaces but the converse is not true as can be seen from the summary statement in section 7 which indicates the opposite: that obeying Einstein locality in presence of time dependencies negates the existence of one common probability space for certain EPR experiments. We take now several steps of explanation by starting with the connection of the various probability approaches to experiments.

3. Connection of Quantum Mechanics to Experiments

Quantum mechanical theory can be connected to experiments in the following way.[6] For each distinct well defined experiment of a series of experiments a state $|\psi\rangle$ (corresponding uniquely to a vector of Hilbert space) is prepared repeatedly and identically. That state is subjected to a measurement which is fully described by a quantum mechanical operator of Hilbert space that we denote, for example, by $\sigma_{\mathbf{a}}$. This situation defines the expectation value $M(|\psi\rangle, \sigma_{\mathbf{a}})$ for the outcomes of the repeated experiments by:

$$M(|\psi\rangle, \sigma_{\mathbf{a}}) = \langle\psi|\sigma_{\mathbf{a}}|\psi\rangle \tag{6}$$

In the following we restrict ourselves to the case where the operators have only discrete eigenvalues and will deal mostly with spin operators, the Pauli spin matrices, and with spinors (vectors in two dimensional Hilbert space) for the states. Generalizations are straightforward but may be mathematically cumbersome. We further assume that the operators have a complete ortho-normal system of eigenvectors $|n\rangle, n = 1, 2, ..., i, ...$ such that any state can be written as:

$$|\psi\rangle = \sum_{n=1}^{\infty} c_n|n\rangle \tag{7}$$

where c_n are complex coefficients and we have

$$\sum_{n=1}^{\infty} |c_n|^2 = 1 \tag{8}$$

The scalar product of Eq. (6) therefore becomes

$$\langle \psi | \sigma_{\mathbf{a}} | \psi \rangle = \sum_{n=1}^{\infty} E_n |c_n|^2 \tag{9}$$

where E_n is, for a given n, the eigenvalue of the operator (now $\sigma_{\mathbf{a}}$) and the eigenvector $|n\rangle$, all as usual. $|c_n|^2$ can be interpreted as the frequency (or probability in that sense) that a measurement of $|\psi\rangle$ will return the eigenvalue E_n.

A well defined experiment requires therefore a well defined state preparation and measurement procedure. That procedure is idealized in the quantum mechanical model and represented by an operator; e.g. $\sigma_{\mathbf{a}}$ for a spin measurement. The significance of the index $\mathbf{a}$ of our notation is that $\mathbf{a}$ describes the macroscopic experimental arrangement as assessed by the observer (experimenter). A different experimental arrangement (with a different spatial arrangement of the macroscopic equipment) would then be described, for example, by a subscript $\mathbf{b}$. For the Stern-Gerlach type spin measurement $\mathbf{a}$ and $\mathbf{b}$ would just denote unit vectors, e.g. $\mathbf{a} = (a_x, a_y, a_z)$, describing different directions of the magnet. The operator $\sigma_{\mathbf{a}}$ is given, as noted already above, by

$$\sigma_{\mathbf{a}} = a_x\sigma_x + a_y\sigma_y + a_z\sigma_z \tag{10}$$

where σ_x, σ_y and σ_z are the Pauli spin matrices:

$$\sigma_x = \begin{pmatrix} 0 & 1 \\ 1 & 0 \end{pmatrix} \tag{11}$$

$$\sigma_y = \begin{pmatrix} 0 & -i \\ i & 0 \end{pmatrix} \tag{12}$$

$$\sigma_z = \begin{pmatrix} 1 & 0 \\ 0 & -1 \end{pmatrix} . \tag{13}$$

Quantum mechanics separates experiments into compatible and incompatible experiments. Two experimental sequences are compatible if their corresponding operators commute. If they do not commute the experiments are incompatible. The precise meaning of this compatibility (incompatibility) in the approach of Kolmogorov and also from a relativistic point of view will be discussed below. Elementary quantum texts usually state that incompatible measurements are those which can not be performed simultaneously. However, a spin measurement can certainly not be performed simultaneously in both $\mathbf{a}$ and $-\mathbf{a}$ direction (at least not in a direct fashion) whereas $\sigma_{\mathbf{a}}$ and $\sigma_{-\mathbf{a}}$ do commute. This indicates some of the complexities that are involved if a general definition of "incompatible" is attempted.

4. Connection of Kolmogorov Type Models to Experiments

The connection of Kolmogorov's probability theory to experiments proceeds along lines that are quite different from those of quantum mechanics. Key to the connection is the definition of a probability space and of random variables.

Mathematically a probability space is a triple (Ω, F, P) consisting of a set Ω, a σ-algebra F, and a measure P on F with $P(\Omega) = 1$. The set Ω is called the sample space and the elements of F are called the events. The measure P is called the probability measure, and $P(E_v)$ is the probability of the event E_v. In the case of a single particle spin measurement, the sample space Ω can be assumed to be countable. Then we can choose F to be the collection of all subsets of Ω and thus avoid mathematical complexities. We can write the probability space in this simple case as a pair (Ω, P). The elements ω of the sample space are indecomposable entities[5] and represent, for example, the propagation of a given entangled pair with certain physical properties.

The random variables are functions on the probability space. If we denote them e.g. by $A_{\mathbf{a}}$ or $A_{\mathbf{b}}$ then these are functions returning a single result for a given ω^{act} chosen by the goddess of chance Tyche.[8] The superscript *act* just indicates Tyche's actual choice for a given experiment. For example $A_{\mathbf{a}}(\omega^{act}) = +1$ could stand for a spin measurement performed on an electron that results in a spin of $+\frac{\hbar}{2}$ where $\hbar$ is Planck's constant divided by 2π.

5. Connecting Quantum Mechanics and Kolmogorov's Probability Theory

It is now easy to construct a Kolmogorov model for a given well defined quantum experiment. Consider the quantum experiment from above where the state is prepared repeatedly and denoted by $|\psi\rangle$ and the measurement is performed corresponding to certain equipment settings denoted by **a**. The quantum mechanical operator $\sigma_{\mathbf{a}}$ corresponds to these measurements. We can then choose a probability space consisting of elements ω and corresponding probabilities P related to ω as follows:

$$P(\omega : A^j_{\mathbf{a}}(\omega) = E_n) = |c_n|^2 \qquad (14)$$

Here the index j numbers the act of state preparation and subsequent measurement. Obviously this model will return exactly the expectation values of quantum mechanics.

The correspondence of elements of the quantum model to elements of the Kolmogorov model and the meaning of the constituent elements of one framework in the other is not as straightforward. It is clear that the values that the random variables can assume correspond to the possible experimental outcomes of eigenvalues E_n. The probability of occurrence of these eigenvalues is also uniquely determined by the state preparation and the mode of measurement. However, quantum mechanics has no obvious mathematical entity that corresponds to ω. If it had, this would give us the outcome of a single experiment. Bell, of course, attempted to complete quantum mechanics just to include the outcomes of single experiments. As we discussed above and in[9] and will see in more detail below this is just not always possible on one common probability space if closed loops are involved.

Kolmogorov type models, on the other hand, do not intrinsically contain an entity that corresponds to the quantum state. However one could construct such an entity. If the measurements correspond to commuting operators in the quantum model, then these operators have a set of orthonormal eigenvectors in common and $|\psi\rangle$ can be described by the same linear combination of these common eigenvectors. This means that these experiments may all be described by using the same probability space. Therefore we can say that compatible experiments in the sense of quantum mechanics can be described by a Kolmogorov model on one common probability space. If the measurements correspond to non-commuting operators, then these operators do not have a common set of eigenvectors. Therefore, in order to determine the probabilities of Eq. (14) we need to use expansions of $|\psi\rangle$ in terms of different sets of eigenvectors corresponding to the non-commuting operators. Then we obtain expansion coefficients c_n for one type of measurement and c'_n for the incompatible other type of measurement with $c_n \neq c'_n$ for some n. Therefore, the use of two different probability spaces is suggested by quantum mechanics in order to describe incompatible experiments within a Kolmogorov model. In other words, we can define incompatible experiments in Kolmogorov type of approaches but we may have to use different probability spaces for the different experiments or experimental sequences. The question of whether or not different probability spaces must in fact be used is more difficult to answer and is discussed below. Even without answering this question, we are now in a position to describe the quantum state $|\psi\rangle$ in the language of Kolmogorov. The quantum state $|\psi\rangle$ can be seen as representing the set of all possible probability spaces $(\Omega_{\mathbf{x}}, P_{\mathbf{x}})$ for all experimental settings $\mathbf{x} = \mathbf{a}, \mathbf{b}, ...$ so that the choice of a given experiment results in the choice of precisely one probability space corresponding to that experiment. In quantum language, $|\psi\rangle$ represents the catalogue of all expectations for all possible measurements. Of course it needs to return the expectations with the correct probability and therefore, together with the measurement equipment, define the probability space uniquely.

Can one describe the results of any number of experiments and of corresponding quantum theory on one common probability space? Contrary to frequent belief the answer to this question is no. Probability spaces are constructed to describe certain well defined experiments not all experiments in the universe at all times. We elucidate this fact now by three examples, (i) the change of probability spaces with time evolution (ii) the non-existence of joint distributions (a common probability space) for certain experiments corresponding to non-commuting operators and (iii) the non-existence of joint distributions for certain EPR experiments and the relation to Vorob'ev's closed loops and Bell's inequalities.

6. Probability Spaces, Time Evolution, Non-Commuting Operators, Closed Loops

The physics literature frequently assumes, without justification, the existence of a common probability space. We show by an example below that quantum mechanics suggests the use of different probability spaces even for certain experiments involving states representing solutions of one and the same Schrödinger equation. Subsequently we show that if quantum mechanically incompatible experiments are involved (non-commuting operators) then a common probability space may not and for certain experiments can not exist and that a key for the understanding is given by the closed loops of Vorob'ev.

6.1. *Different probability spaces and unitary time evolution*

Consider the following experiment. The state of an electron is prepared at time $t_0 = 0$ to be the ground state $|t_0\rangle$ of a quantum dot of nanometer size volume and a defining potential that permits tunnelling of the electron out of the dot. A measurement of the kind to be able to detect the electron in a given volume V surrounding the dot is made and described by a random variable A with $A = 1$ if the electron is detected and $A = 0$ if it is not. This volume V of measurement is contained in a bigger volume V_b that is confined by an infinite repulsing potential so that we know the electron will stay within this bigger volume for all times. The state of the electron for times $t > t_0$ is given by solutions of the Schrödinger equation that are denoted by $\psi(\mathbf{r}, t)$ where $\mathbf{r}$ denotes the space coordinate. To describe the time evolution we use the unitary time evolution operator $U(t, t_0)$. We define a density $\rho(\mathbf{r}, t) = |\psi(\mathbf{r}, t)|^2$ and normalize such that

$$\int_{V_b} \rho(\mathbf{r}, t) = 1 \tag{15}$$

which is possible because of the unitary time evolution and indicates the conservation of the number of particles. One often says that *probability* is conserved. This, however, does not necessarily mean the conservation of the *probability space* for any measurement connected to $\psi(\mathbf{r}, t)$ as one can see from the following. The probability $P_V(t)$ to find an electron in any smaller volume V is time dependent and given by:

$$P_V(t) = \int_V \rho(\mathbf{r}, t) d\mathbf{r} \tag{16}$$

To see the time dependence explicitly we recall the equation of continuity and integrate over the volume V that is assumed to be bounded by a surface S:[6]

$$\frac{\partial}{\partial t} \int_V \rho(\mathbf{r}, t) d\mathbf{r} = - \int_S \mathbf{j}(\mathbf{r}, t) \cdot d\mathbf{S} \tag{17}$$

where $\mathbf{j}(\mathbf{r}, t)$ is the probability current given by $\mathbf{j} = \psi^* \nabla \psi - \psi \nabla \psi^*$ and the integral $\int_S$ is taken over the surface S of the volume V. One can see that a time dependence exists for the probability to find the electron in the smaller volume V and therefore for the corresponding random variable whenever a current channel out of this volume exists. For the bigger volume V_b we have no current channel and therefore no time dependence.

We can therefore describe the repeated measurement in volume V at time t_1 of the repeatedly prepared electron by a random variable $A^j = 0, 1$ and a sample space with elements ω and corresponding probabilities

$$P(\omega : A^j(\omega) = 1) = P_V(t_1) \tag{18}$$

and

$$P(\omega : A^j(\omega) = 0) = 1 - P_V(t_1) \tag{19}$$

where, as before, the index j represents just the number of the act of preparation of a state. It is now important to notice that for a measurement sequence at a different time, say t_2 we obtain a different probability space because $P_V(t_1)$ in the above equations changes now to $P_V(t_2)$. This shows that the"conservation of probability" as dictated by Eq. (15) implies, in general, nothing for probability spaces that describe certain experiments in a smaller volume. Even the same type of experiment performed at different times can not necessarily be described by random variables (for the different times) on one given probability space. We note also that the introduction of different probability spaces has nothing to do with non-locality. The experiments above are performed for the same volume V.

6.2. *Probability spaces, joint distributions and commutativity*

Consider the measurement of the observable $[\sigma_{\mathbf{a}}, \sigma_{\mathbf{b}}]/(2i)$ where $[\cdot]$ denotes the commutator. If we now assume $A_{\mathbf{a}}$ and $A_{\mathbf{b}}$ to be functions that describe the measurement outcome corresponding to the non-commuting operators $\sigma_{\mathbf{a}}$ and $\sigma_{\mathbf{b}}$ respectively we have the following problem. Let $A_{\mathbf{a}}$ and $A_{\mathbf{b}}$ be random variables defined on a common probability space. Then they have a joint distribution and the expectation value of the product fulfills

$$M(A_{\mathbf{a}}A_{\mathbf{b}}) = M(A_{\mathbf{b}}A_{\mathbf{a}}) \tag{20}$$

However, there exist $|\psi\rangle$ so that the expectation value of $[\sigma_{\mathbf{a}}, \sigma_{\mathbf{b}}]/(2i)$ is equal to a positive number δ:[15]

$$\langle\psi|[\sigma_{\mathbf{a}}, \sigma_{\mathbf{b}}]/(2i)|\psi\rangle = \delta \tag{21}$$

where again $[\sigma_{\mathbf{a}}, \sigma_{\mathbf{b}}]$ denotes the commutator and i is the imaginary unit. This is "difficult to square"[15] with Eq. (20) and we therefore can, in general, not form a one to one correspondence of the operators to random variables that are defined on a common probability space. A joint distribution does therefore, in general, not exist. It still may exist depending on the details of the experiments that are considered (see next section). More extensive discussions involving quantum mechanical incompatibility can be found in the work of Willem de Muynck (these proceedings and references therein). Of course, the Kolmogorov framework also admits and even postulates the necessity of different probability spaces for incompatible experiments.

6.3. *Probability spaces and EPR experiments*

We have already pointed out in section 2 that the incompatible experiments involved in EPR experiments and particularly in the derivation of Bell's inequalities do not always permit definition of all involved random variables on a common probability space. For the experiments considered by Bell it was shown that a common probability space exists if and only if all Bell inequalities are valid. We have also traced these facts to the consistency problem studied in great generality by Vorob'ev.[4] Vorob'ev pointed to the problem with closed loops of random variables as discussed in section 2. We would like to highlight the importance of closed loops by the following example. One could think, particularly after what was discussed in the previous section that the involvement of non-commuting operators determines that the experiments are incompatible not only by the definition of quantum mechanics but also in the sense of Kolmogorov. Then, the involvement of a non-commuting operator would simply mean that definition on one probability space is impossible. As discussed above, this is indeed the case for certain measurements that involve non-commuting operators. EPR experiments, however, are more complex. The correlated pair is measured in two stations and the measurements corresponding to non-commuting operators in the same station are performed for different correlated pairs. In this situation one can fulfill the Bell inequalities in spite of the involvement of non-commuting operators. In fact it is well known that violations occur only for certain sets of direction of the Stern-Gerlach magnets and not for others. This shows that for the EPR experiments it is not the non-commutativity by itself that leads to the impossibility to define the random variables on a common space. It is the existence of closed loops and the involvement of certain pair expectation values that cause the problems as shown in great detail in.[9] Incompatibility of experiments in quantum mechanics is therefore not the same as incompatibility defined by the consistency problem in Kolmogorov's framework. This fact has, in our opinion, led to some of the difficulties in the explanations surrounding EPR experiments. We show next how these difficulties can possibly be resolved.

7. Probability Spaces and Relativity: Physics and the Sample Space Ω

The outcome of the idealized experiment i.e. the value that a random variable like $A_{\mathbf{a}}$ assumes is fully determined by Tyche's choice of an element of the sample space because $A_{\mathbf{a}}$ is a function on that space. We take now certain basic elements of the special theory of relativity (light-cones, limiting role of the speed of light in vacuo) as a physical basis for the idealized mathematical model. These elements of the special theory are sufficient for all purposes of our explanations and simple enough to illustrate the interplay of physics and mathematics when constructing probability models. The outcome of a given single localized observation is then determined by all causal factors that can be attributed to the corresponding backward light-cone. These causal factors may be a spatial setting **a** like the direction of the magnets, they may be any "element" of the light-cone in the sense of Mach or they may represent any other scientifically constructed entity such as a particle prepared in a certain way. We can therefore label the outcome by the space-time coordinates (x_j, y_j, z_j, t_j) of the tip of the backward light-cone that characterizes the given observation (for example the detection of a particle). As before the index $j = 1, 2, 3...$ just labels the experimental procedure involving a given entangled pair. The quadruple (x_j, y_j, z_j, t_j) is then an indecomposable "entity" that could possibly be taken as a physical interpretation of what ω^{act} represents because the physical outcomes of experiments can be described in terms of the space-time coordinates in the light-cone according to longstanding physical tradition. However, such an interpretation of Tyche's ω^{act} is not necessary. We can continue with the more abstract definition of ω^{act} and introduce the connection to physics by using special random variables ST. In the inertial frame of e.g. station S_1, we define space-time related random variables $ST = ST(X, Y, Z, T)$ by:

$$X(\omega^{act}) = x_j \ , \ Y(\omega^{act}) = y_j \ , \ Z(\omega^{act}) = z_j \ , \ T(\omega^{act}) = t_j \tag{22}$$

In this way we are flexible and have defined a random variable ST that relates to the physics of the backward light-cone of a given observation in S_1. We also have established a physical connection to the sample space.

Note, however, that the parameters used by Bell to model EPR experiments require further physical and mathematical assumptions. Consider again for the moment only station S_1. Let us introduce a random variable $\mathbf{V}$ that characterizes the magnet settings and can assume the outcomes $\mathbf{a}, \mathbf{b}$. We then can describe the possible experimental outcomes of the spin measurements in S_1 by the random variable $A_\mathbf{V}$. This was in essence done by Bell who also assumed that his random variables would be functions of an arbitrary general additional random variable Λ. If Λ is indeed general, then we certainly should be permitted to relate it to the tip of the light-cone of the observation by letting e.g. $\Lambda = ST$. Because we consider now only station S_1 we can disregard the constant space coordinates and deal with time dependencies only. Thus we let

$$A_\mathbf{V} = A_\mathbf{V}(T). \tag{23}$$

and have arrived at a time and setting dependent random variable. (For a more detailed discussion of time and setting dependent random variables see[9] and the paper by Marchildon in these proceedings.) While this innocuous expression appears to be valid and free of contradictions under very general conditions, it definitely is not. From the viewpoint of relativity we have now the symbol $\mathbf{V}$ that signifies the possible macroscopic space-like orientations of a massive body (the magnet in S_1) as one random variable, and the measurement time of the tip of the backward light-cone as the other (additional parameters related to the given entangled pair can easily be included). Note that the random variables $\mathbf{V}, T$ are not independent from a physical point of view. Certain light-cones will simply not contain certain magnet orientations. The turning of the massive magnet requires time because of the finite speed of light which has the consequence that certain magnet settings can not be achieved for certain light-cones. Consider now the consequences of these physical facts for a Kolmogorov model. We are, of course free to choose $\mathbf{V}$ on a given sample space and to choose the probabilities for an appropriate probability space. We can do the same for T using a different probability space. However, we may not be able to unite these two probability spaces to form one common probability space for all random variables because certain magnet orientations are impossible for certain space-time values and we certainly can not have two different orientations of the same magnet for the same measurement time. These facts can be seen already from the elementary definitions of any function F. As an example let F be a function of two variables e.g. the magnet settings and the measurement times that we denote by t_j for the j's experiment. Assume that the function is defined on a domain D that contains the two settings $\mathbf{V} = \mathbf{a}, \mathbf{b}$ and the possible measurement times (or equivalently short time periods) that will be contained in an interval Δ_t. Therefore we have for the function F the domain $D = [\mathbf{V}, \Delta_t]$. We must then be able to assign a value to both $F(\mathbf{a}, t_j)$ and $F(\mathbf{b}, t_j)$ and we must be able to use these values in our mathematical model for the experiments. However, the physics of a particular experiment may not permit this because $\mathbf{a}$ and $\mathbf{b}$ may be the settings of one and the same magnet that can not assume two positions for the same measurement time. This renders the mathematical model invalid for the given experiment, the reason being that such functions can not be defined on such a given domain without physical contradictions. A fortiori the same is true if we replace the domain of the function F by a probability space, which is the domain for the random variables. For the case of our EPR experiment we can replace the function F by random variables $F_1 = A_{\mathbf{a}}(T)$ and $F_2 = A_{\mathbf{b}}(T)$ for the measurement outcomes in S_1 and similar functions (not presently needed for the argument) for S_2. Clearly a contradiction arises if we assume that these functions (random variables) are defined on one common probability space because we can have then the same backward light-cone tip and corresponding measurement time for different magnet settings. These facts can also be collected and formulated as the following summary statement:

Summary Statement:

Theorem 7.1. *Consider a mathematical model for a sequence of EPR experiments and consider a single experimental station S_1. Assume that the mathematical model is based on a set of random variables $A_{\mathbf{a}}$ and $A_{\mathbf{b}}$ that are labelled or characterized by the different orientations* $\mathbf{a}$ *and* $\mathbf{b}$ *of massive objects such as Stern-Gerlach magnets. Let these random variables in turn be functions of a random variable T that represents the measurement time for a given experimental observation (of one part of a correlated pair) and that assumes values t_j for the j's experiment. It is then impossible to find a single probability space on which $A_{\mathbf{a}}(T)$ and $A_{\mathbf{b}}(T)$ are defined without creating physical contradictions that render the mathematical model inapplicable.*

Proof. If T is a random variable on a probability space then there must exist a sample space element ω^{act} chosen by Tyche[8] that corresponds to the propagation of a given entangled pair with certain physical properties such that both $A_{\mathbf{a}}(T(\omega^{act}))$ and $A_{\mathbf{b}}(T(\omega^{act}))$ represent outcomes of the experiment. Because $T(\omega^{act})$ represents exactly one measurement time t_j this means that two magnet orientations of the same magnet need to exist for the same measurement time t_j and therefore for the same light-cone tip and the given correlated pair. This, however, is not possible because the time period necessary to turn the massive magnets is bounded away from zero due to the finite velocity c of light and because any measurement using one magnet direction may alter the correlated pair. Therefore the mathematical model of the summary statement does not correspond to the physical data but contains outcomes that are physically impossible. □

Thus, a logical contradiction is obtained from Bell's assumptions without any use of quantum mechanics. The assumptions (i) that Λ is a random variable as defined by Kolmogorov, (ii) that Λ can be "anything" and (iii) that the spin measurement outcomes are functions of Λ are inconsistent as can be seen from a few elementary facts of relativity. Note that a similar summary statement and proof can be formulated also for a number of different EPR arrangements and correlated pairs from different sources and even for moving magnets (polarizers in optical experiments). However, to obtain contradictions in these cases one needs to directly involve Bell's algebraic steps of Eqs.(2) and (3). We will present such a general proof in a future paper. Here we just want to emphasize the principle that the laws of physics may prevent the definability of certain random variables on one probability space. This brings home Einstein's famous line "Gott wuerfelt nicht (God does not play dice)".

Note also that even if massive magnets are not involved but Kerr cells or the like are switched instead in optical experiments, the time periods necessary for the switching are always bounded away from zero.

Thus Bell's innocuous assumption that Λ is entirely general leads necessarily to the inclusion of un-physical domains for the random variables, in particular to the inclusion of different magnet settings for the same time of measurement. Stephen Adler[16] has described necessary cautions for the Conway-Kochen argument that are also based on time periods bounded away from zero by relativity considerations. A careful study of these facts led us to the conjecture that probably all or at least most paradoxa in this area of no-go proofs can be removed by a careful relativistic treatment of the involved functions and their domains. In this connection, see also the discussion of continuous random fields and of Bell inequalities by Peter Morgan.[17]

8. Discussion and Conclusions

Thus for a probability model of physical reality, the probabilities and the elements of the sample space need to be constructed according to the rules of physics and according to corresponding set-theoretic considerations that link the physics to the mathematical model. The important question is then, what does mathematics and what does physics suggest and even dictate for probability spaces. In particular when can we describe experiments on the same probability space, when should we use and when do we have to use different probability spaces?

Mathematics and, in particular, Kolmogorov's probability theory is unambiguous in its determination of when it may be and when it is impossible to use one common probability space for a number of random variables: The use of different probability spaces for different experiments is suggested by the necessity of using different indecomposable elements of the sample space for the random variables that describe the possible outcomes of the experiments in consideration. If, for example in an EPR experiment the indecomposable event is the propagation of a given entangled pair with certain physical properties, then different magnet positions (characterizing the random variables on a given side) can not be used to measure the same indecomposable event, the same pair. Such experiments are therefore loosely speaking incompatible and the use of different probability spaces may be necessary to describe such experiments depending on circumstance. At this point, it may still be possible, even for very different experimental sequences, to consistently construct one common space e.g. a product space. However, it is never guaranteed from the start that one can achieve consistency. The work of Vorob'ev has shown that a particular danger exists if pair expectation values are given and the random variables involved form a closed loop. Vorob'ev's examples make it clear when it is impossible to define certain sets of random variables on one probability space. Then the experiments linked to the random variables are guaranteed incompatible. Thus Kolmogorovian approaches partition experiments into compatible and incompatible ones, with incompatibility guaranteed if a Vorob'ev type contradiction can be derived or, for the special case of the discussed EPR experiment if the Bass criteria or equivalently any of the Bell-type inequalities are violated. Note that Vorob'ev type contradictions are not limited to quantum mechanics but occur, as discussed by Vorob'ev, also in classical situations.

Quantum mechanics has also a clear division into compatible and incompatible experiments. Two experiments are incompatible if their corresponding operators do not commute. It is then also, in general, not possible to construct a common probability space even for a single pair of corresponding variables. The closed loops of Vorob'ev are particularly relevant for quantum measurements as outlined above and always imply the possibility of incompatible experiments. The involvement of non-commuting operators, however, does not necessarily by itself mean that it is impossible to describe certain experiments by a model that uses only one common probability space.

The use of physical space-time pictures and relativistic light-cones also does suggest when different probability spaces should be used: they should in principle be used whenever a measurement or a sequence of measurements involves relevant differences in the light-cones. In particular we have shown that it is impossible to construct time dependent experiments that involve different macroscopic space-like configurations of massive objects on one probability space because a common probability space would then involve different configurations of the massive objects at the same time. Experiments are thus incompatible if they involve time dependencies and a space-like change of the configuration of macroscopic massive bodies (or macroscopic voltages as in some of the optics experiments).

All three distinctions of incompatible experiments, the Kolmogorovian distinction through indecomposable events, the quantum mechanical one through non-commuting operators and the relativistic one through configuration changes of massive bodies and the presence of time dependencies stigmatize the same type of experiments as somehow incompatible. Whether or not incompatible experiments can or can not be defined on one common probability space without contradiction, and therefore are mathematically truly incompatible, can be determined using Vorob'ev's criteria. Closed loops should always be regarded as a warning sign to be careful with assumptions of the existence of common probability spaces.

Note added by K. Hess: This manuscript was in its essence written by Walter Philipp and myself. Walter still corrected the manuscript up to section 6. Further corrections of his were made impossible by the caesura that Walter faced on July 19 2006; a finality that we all will be facing one day. No one in our community will feel the loss as much as I do.

Acknowledgement: Heartfelt thanks go to Steve Adler, Gerard t' Hooft, Louis Marchildon, Andrei Khrennikov and Arkady Plotnitsky for their encouragement, comments and constructive criticism that helped me with the task to finish the manuscript without the help of Walter. The work was supported by the Office of Naval Research (N00014-98-1-0604).

References

1. J. S. Bell, *Physics* **1**, 195 (1964).
2. A. Einstein, B. Podolsky and N. Rosen, *Phys. Rev.* **47**, 777 (1935)
3. J. Bass, *C. R. Acad. Sci.(Paris)* **240**, 839 (1955).
4. N. N. Vorob'ev, *Theory of probability and its Applications* **VII**, pp 147-162 (1962).
5. W. Feller, *"An Introduction to Probability Theory and its Applications"*, (Wiley Series in Probability and Mathematical Statistics **1**), pp 1-9 (1968).
6. L. Marchildon, *Quantum Mechanics*, (Springer, Berlin 2002).
7. G. t' Hooft, *quant-ph/0604008* (2006).
8. D. Williams, *Weighing the Odds*, (Cambridge University Press , Cambridge, 2001).
9. K. Hess and W. Philipp, *Found. Phys.* **35**, 1749 (2005).
10. A. Fine, *J. Math. Phys.* **23**, 1306 (1982).
11. A. Fine, *Phys. Rev. Lett.* **48**, 291 (1982).
12. I. Pitovsky, "Quantum Probability - Quantum logic", *Lecture Notes in Physics 321*, (Springer Verlag, Berlin-Heidelberg-New York 1989).
13. A. Yu. Khrennikov, *Interpretations of Probability* VSP Int. Sc. Publishers, Utrecht and Tokyo, pp 91 (1999).
14. J. F. Clauser, M. A. Horne, A. Shimony, *Phys. Rev. Lett.* **23**, 880 (1969).
15. D. Williams, *Weighing the Odds*, (Cambridge University Press , Cambridge, 2001).

16. S. Adler, *quant-ph/0604122* (2006).
17. P. Morgan, *quant-ph/0607165* (2006).

FAIR SAMPLING VS NO-SIGNALLING PRINCIPLE IN EPR EXPERIMENTS

G. ADENIER and A. YU. KHRENNIKOV

International Center for Mathematical Modelling in Physics and Cognitive Sciences
Växjö University, 351 95 Växjö, Sweden

We present some recent results of a new statistical analysis of the optical EPR experiment performed by Weihs *et al* in Innsbruck 1997-1998. Under the commonly used assumption of fair sampling, we show that the coincidences counts exhibit a small and anomalous non-signalling component, which seems impossible to explain by using conventional quantum mechanics, and we discuss some possible interpretations of this phenomenon.

Keywords: EPR experiments; Fair sampling; No-signalling Principle.

1. Introduction

The experimental violation of Bell Inequalities[1–4] in optical EPR experiments[5,6] can only be validated under the additional assumption of *fair sampling*.[4,8,9] This type of test being crucial for modern quantum communication research,[10,11] it is unfortunate that the result of a test meant to disprove Local Realism would depend on such an additional assumption, since Local Realism is a priori no less plausible an assumption than Fair Sampling is.

We have recently shown that the validity of the fair sampling assumption can be questioned on experimental ground,[12] after analyzing some anomalies in the data extracted from the optical EPR performed by Weihs *et al.*[6,7] Indeed, even when relaxing this assumption to only require that the efficiency for a pair of photons to be detected would factor into two local terms on each side (independent errors), we have observed a slight but anomalous non-signalling component in the marginal coincidence probabilities extracted from experimental data.[7,12]

We recall here the theoretical ground behind the normalization procedure that led us to these results, and discuss some possible interpretations.

2. Channel efficiencies

In a real experiment, not all photons are detected and one should take account of the efficiencies of each of the four channels involved in the coincidence detections. This problem is at the heart of the detection efficiency loophole, and can be used to design local realistic models that reproduce the main experimental features of optical EPR experiments.[13]

We label $\eta_A^+(\alpha)$ and $\eta_A^-(\alpha)$ the single-count channel efficiencies of Alice's plus and minus channels respectively, and similarly $\eta_B^+(\beta)$ and $\eta_B^-(\beta)$ for Bob's plus and minus channels; the parameter dependence in α and β reflects the fact that these efficiencies may, in principle, depend on the local settings. It is important to stress that we can put in the concept of *channel efficiencies* all possible variations in the experimental conditions that are not directly related to the quantum state associated to the source (such as signal intensity variations, detector efficiencies, optical misalignments, collection efficiencies, etc.). This idea is important since it is precisely the aim of our paper to analyze experimental quantities that can directly be compared with Quantum predictions — under the assumption of Fair Sampling — independently of all these experimental contingencies encapsulated in the channel efficiencies.

We can expect the number of single counts to be proportional to the respective channel efficiencies:

$$N_{\exp}^{A,\varepsilon_1}(\alpha) \approx \eta_A^{\varepsilon_1}(\alpha) N_A/2$$
$$N_{\exp}^{B,\varepsilon_2}(\beta) \approx \eta_B^{\varepsilon_2}(\beta) N_B/2,$$

where N_A and N_B are the (unknown) number of photons actually sent respectively to Alice and Bob, and where ε_1 and ε_2 can each be either $+$ or $-$, to shorten the forthcoming equations[a].

We now consider the number of coincidence counts experimentally registered. If we consider a pair of photons as a whole, the probability that it is detected in each specific combination of two channels should depend as well on a combined channel efficiency. We label these combined channel efficiencies as $\eta^{++}(\alpha,\beta)$, $\eta^{+-}(\alpha,\beta)$, $\eta^{-+}(\alpha,\beta)$, and $\eta^{--}(\alpha,\beta)$. The number of coincidences in a pair of channels $(\varepsilon_1,\varepsilon_2)$ should thus be proportional to the relevant combined efficiency $\eta^{\varepsilon_1,\varepsilon_2}(\alpha,\beta)$, the (unknown) number of pairs sent N_{AB}, and the relevant joint probability predicted by Quantum Mechanics $P_{QT}^{\varepsilon_1,\varepsilon_2}(\alpha,\beta)$, that is,

$$N_{\exp}^{\varepsilon_1,\varepsilon_2}(\alpha,\beta) \approx \eta^{\varepsilon_1,\varepsilon_2}(\alpha,\beta)\; N_{AB}\; P_{QT}^{\varepsilon_1,\varepsilon_2}(\alpha,\beta). \tag{1}$$

Here the angular dependence of the efficiencies is clearly unwanted since we are interested in an experimental test of the predictions of Quantum Mechanics, independently of the inaccuracies involved in any particular experimental setup. Our immediate purpose is therefore to get rid of these angular dependencies due to the combined efficiencies.

In order to do so, we assume that the ensemble of detected pairs of photon provides a fair statistical sample of the ensemble of emitted pairs (Fair Sampling Assumption). A consequence of this assumption is that the probabilities of non detection for Alice and Bob should be independent of one another. Indeed, for fixed

[a]The use of the symbol $\approx$ is here to remind of the statistical variability in any experimental results that is naturally expected to induce small deviation from the predictions. The amplitude of these deviations are expected to decrease with the number of trials obtained experimentally.

settings (α, β), the probability of a non detection should be independent of the polarization state of the photon (otherwise the sampling would clearly be unfair), and should thus be independent of the fate of the distant correlated photon.

Hence, the channel efficiencies should be the same for all photons going into a specific channel, independently of whether a photon happens to be single or to be paired with a distant detected photon (independent errors). That is, the above combined efficiencies for pairs of particles should be equal to the product of the relevant channel efficiencies for the single counts:

$$\eta^{\varepsilon_1,\varepsilon_2}(\alpha, \beta) = \eta_A^{\varepsilon_1}(\alpha)\eta_B^{\varepsilon_2}(\beta). \tag{2}$$

With this independent errors condition, we can rewrite the predicted number of coincidence counts as:

$$N_{\text{exp}}^{\varepsilon_1,\varepsilon_2}(\alpha, \beta) \approx N_{AB}\, \eta_A^{\varepsilon_1}(\alpha)\eta_B^{\varepsilon_2}(\beta)\, P_{QT}^{\varepsilon_1,\varepsilon_2}(\alpha, \beta) \tag{3}$$

3. Normalizing using the single counts

The probabilities derived from the standard normalization procedure with coincidence counts still depend explicitly on the channel efficiencies, so that some properties of the photon pairs remain out of reach of the experimenter.

This problem makes it necessary to circumvents these dependencies in all cases by means of the new normalization procedure. Our idea is that most of the counts registered in the channels are non-coincidence events, because the channel efficiencies are low. Since there are many times more non-coincident events than coincident ones, they provide a useful and accurate additional statistical information about the relative efficiency of the channels.

In order to get rid of channel efficiencies in the above equations, we thus define the following experimental quantities, which are proportional to the ratio of the number of coincidence counts in a combined channel over the product of the two corresponding single counts:

$$f_{\text{exp}}^{\varepsilon_1,\varepsilon_2}(\alpha, \beta) \equiv \frac{1}{4}\frac{N_{\text{exp}}^{\varepsilon_1,\varepsilon_2}(\alpha, \beta)}{N_{\text{exp}}^{A,\varepsilon_1}(\alpha)N_{\text{exp}}^{B,\varepsilon_2}(\beta)} \tag{4}$$

Replacing the single counts and coincidence counts by their expressions given respectively in (1) and (3), we obtain:

$$f_{\text{exp}}^{\varepsilon_1,\varepsilon_2}(\alpha, \beta) \approx \frac{N_{AB}}{N_A N_B}\, P_{QT}^{\varepsilon_1,\varepsilon_2}(\alpha, \beta) \tag{5}$$

for which the only angular dependence is the one due to the Quantum Mechanical term.

Since the four joint probabilities $P_{QT}^{++}(\alpha, \beta)$, $P_{QT}^{+-}(\alpha, \beta)$, $P_{QT}^{-+}(\alpha, \beta)$ and $P_{QT}^{--}(\alpha, \beta)$ add up to unity, summing these four equation together yields

$$\sum_{\varepsilon_1,\varepsilon_2} f_{\text{exp}}^{\varepsilon_1,\varepsilon_2}(\alpha, \beta) \approx \frac{N_{AB}}{N_A N_B}. \tag{6}$$

and finally,

$$P_{QT}^{\varepsilon_1,\varepsilon_2}(\alpha,\beta) \approx \frac{f_{\exp}^{\varepsilon_1,\varepsilon_2}(\alpha,\beta)}{\sum_{\varepsilon_1,\varepsilon_2} f_{\exp}^{\varepsilon_1,\varepsilon_2}(\alpha,\beta)} \tag{7}$$

Hence, by using the Fair Sampling Assumption together with experimental statistics for the single counts, we are able to obtain experimental quantities that should coincide with the prediction given by Quantum Mechanics for the joint probabilities, independently of any channel efficiency imbalance[b]. In particular this new normalization procedure allows us to obtain experimental quantities that should directly coincide with the marginal probabilities:

$$\begin{aligned}
\frac{f_{\exp}^{++}(\alpha,\beta)+f_{\exp}^{+-}(\alpha,\beta)}{\sum_{\varepsilon_1,\varepsilon_2} f_{\exp}^{\varepsilon_1,\varepsilon_2}(\alpha,\beta)} &\approx P_{QT}^{++}(\alpha,\beta)+P_{QT}^{+-}(\alpha,\beta)\\
\frac{f_{\exp}^{-+}(\alpha,\beta)+f_{\exp}^{--}(\alpha,\beta)}{\sum_{\varepsilon_1,\varepsilon_2} f_{\exp}^{\varepsilon_1,\varepsilon_2}(\alpha,\beta)} &\approx P_{QT}^{-+}(\alpha,\beta)+P_{QT}^{--}(\alpha,\beta)\\
\frac{f_{\exp}^{++}(\alpha,\beta)+f_{\exp}^{-+}(\alpha,\beta)}{\sum_{\varepsilon_1,\varepsilon_2} f_{\exp}^{\varepsilon_1,\varepsilon_2}(\alpha,\beta)} &\approx P_{QT}^{++}(\alpha,\beta)+P_{QT}^{-+}(\alpha,\beta)\\
\frac{f_{\exp}^{+-}(\alpha,\beta)+f_{\exp}^{--}(\alpha,\beta)}{\sum_{\varepsilon_1,\varepsilon_2} f_{\exp}^{\varepsilon_1,\varepsilon_2}(\alpha,\beta)} &\approx P_{QT}^{+-}(\alpha,\beta)+P_{QT}^{--}(\alpha,\beta).
\end{aligned} \tag{8}$$

An important prediction of Quantum Mechanics for these marginal probabilities is the no-signalling principle. A correlation $P(\varepsilon_1,\varepsilon_2|\alpha,\beta)$ is non-signaling if and only if its marginal probabilities are independent of the other side input: $\sum_{\varepsilon_2} P(\varepsilon_1,\varepsilon_2|\alpha,\beta)$ is independent of β and $\sum_{\varepsilon_1} P(\varepsilon_1,\varepsilon_2|\alpha,\beta)$ is independent of α.[14]

This non-signalling property does not depend on the state of the source of photons. Whatever is the source sent to Alice and Bob, they cannot use it to communicate. Alice's marginal probabilities, represented by the two first equations above, cannot depend on Bob's measurement setting β. Similarly, Bob's marginal probabilities, represented by the two last equations above, cannot depend on Alice's measurement setting α.

Hence, if only one local parameter varies (say, α), then only two out of these four quantities can vary accordingly, the other two remaining constant. It is precisely this important prediction that we have checked with the experimental results at hand.

4. Experimental Results

We only briefly recall here the main results of our investigation. The reader is invited to refer to our previous publication for more details.[12]

[b]Note that the normalization procedure that we propose here is only necessary in the case of imbalanced channel efficiencies. To some extent, this procedure is equivalent to experimentally balancing the four channels so that they exhibit the same constant efficiency, before actually starting to register the coincidences.

In all the runs, we observed that the coincidence rate exhibit minima close to zero, and cosine-squared shape, as expected from the predictions of Quantum Mechanics. However, the maxima of the four coincidence curves differed significantly (see figure 1a). In spite of this anomalous behavior, the correlation function computed with the standard normalization (i.e., with the sum of all coincidence counts) coincided very well with the quantum mechanical predictions. It is interesting to note that possibly similar anomalies were observed in the two-channel EPR experiments performed in Orsay in the early 1980's by Alain Aspect.[12,15]

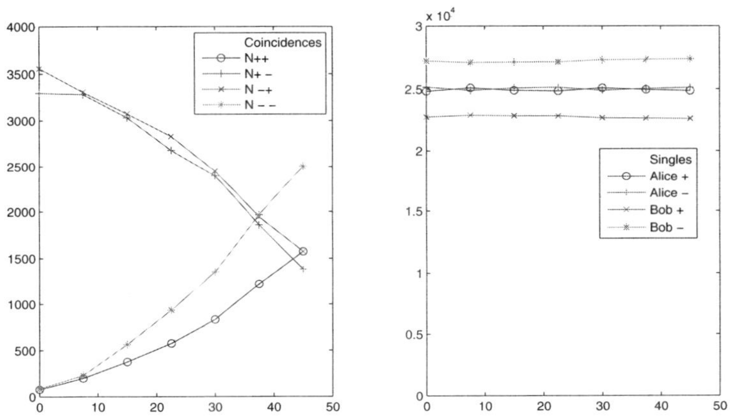

Fig. 1. **(a)** Coincidence counts as Alice varies her measurement angle (in degrees). Note that N_{++} and N_{--} differ significantly while N_{+-} and N_{-+} roughly coincide. (Sourcefile: bluesine), **(b)** Single Counts for Alice and Bob as Alice varies her measurement angle α.

Such differences in the maxima of the four coincidence curves can in principle be explained by a non-rotationally invariant source state,[7,12] and by some imbalance in the channel efficiencies.

In order to remove the latter possibility, we extracted the single counts: they were typically balanced on one side, while imbalanced on the other (see figure 1b). With our normalization procedure described above, we were able to remove this channel efficiency dependencies from the coincidences.

The remaining anomalies were problematic to explain using a non-rotationally invariant source state. Indeed, the obtained experimental quantities that should have coincided with the marginal probabilities for both Alice and Bob exhibited a dependence on the one local setting that was varied (see figure 2), in apparent violation of the no-signalling principle.

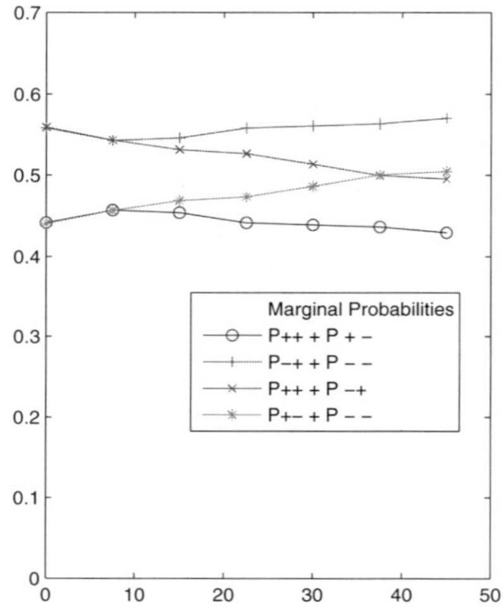

Fig. 2. Experimental result that should coincide with the marginal probabilities, assuming Fair Sampling. Bob's marginals clearly depend on Alice's measurement setting.

5. Interpretation

The no-signalling principle being a fundamental feature of Quantum Mechanics, as well as of Local Realism, the most reasonable interpretation of these anomalies is that the Fair Sampling assumption should be rejected, as it is the only extra assumption that we used to observe them. Once the Fair Sampling assumption is rejected, there is no evidence of violation of the no-signalling principle, and therefore no evidence whatsoever of a possible superluminal communication. It should nevertheless be stressed that if one prefers to maintain the Fair Sampling assumption at all cost, the most striking result that can be derived under this assumption is not so much the violation of Bell Inequalities, but rather the violation of the no-signalling principle.

It is not clear yet how the violation of the Fair Sampling assumption that is hinted by our results can be explained. From an experimental point of view, it has been pointed out that the possibility of a nonlinearity in the high-voltage amplifier used to drive the electro-optic modulators should be investigated, as well as the influence of accidental coincidences.[7]

From a theoretical local realistic point of view, the violation of the fair sampling assumption is a simple way to reproduce EPR correlations, but requiring that the model reproduces as well the anomalies described here represents an additional challenge. We are currently working on providing such a model. Some preliminary simulations shows that a simple symmetrical model (that is, with the same detection pattern for Alice and Bob) subjected to our normalization procedure can indeed reproduce the anomalies, but at the cost of having as well some slight variations in the total rates of coincidences.

6. Conclusion

Our results show that more EPR experiments are definitely required, with as clean as possible coincidences rates and balanced channels, but without necessary involving fast time-varying analyzer settings, in order to understand whether these anomalies are unique or not in EPR experiments.

Acknowledgments

We are most grateful to Gregor Weihs from Waterloo University (Canada) as he had the courtesy to send us two CDs containing the raw data gathered in Innsbruck between 1997 and 1998 in various version of the EPR setup. This work would not have been possible otherwise. We are grateful to Alain Aspect and Gregor Weihs for stimulating discussions on the structure of of statistical data in EPR experiments. This work was supported by the Profile Mathematical Modelling of Växjö University and the EU-network on QP and Applications.

References

1. A. Einstein, B. Podolsky, and N. Rosen, *Phys. Rev.* **47**, 777, (1935)
2. D. Bohm, *Quantum Theory*, Englewood Cliffs: Prentice-Hal, (1951)
3. J. S. Bell, *Physics*, **1**, 195, (1964)
4. J. F. Clauser, M. A. Horne, A. Shimony, and R. A. Holt, *Phys. Rev. Lett.*, **23**, 880, (1969)
5. A. Aspect, J. Dalibard, and G. Roger, *Phys. Rev. Lett.*, **49**, 1804-1807, (1982)
6. G. Weihs, T. Jennewein, C. Simon, H. Weinfurter, and A. Zeilinger, *Phys. Rev. Lett.*, **81**, 5039-5043, (1998)
7. G. Weihs, *in Foundations of Probability and Physics-4* (AIP Conference Proceedings), 2006
8. J. F. Clauser and A. Shimony, *Rep. Prog. Phys.*, **41**, 1881-1927, (1978)
9. G. Adenier and A. Yu. Khrennikov *Proc. Conf. Quantum Theory: Reconsideration of Foundations-2* (Växjö Univ. Press)(2003)
10. W. Tittel and G. Weihs, *Photonic Entanglement for Fundamental Tests and Quantum Communication* (QIC), 3-56, (2001)
11. A. V. Sergienko and G. Jaeger, *Contemporary Physics*, **44**, 341, (2003)
12. G. Adenier and A. Yu. Khrennikov, *J. Phys.* B, **40**, 131, (2007)
13. J. -Å. Larsson, *Phys. Rev.* A., **57**, 33043308, (1998)
14. N. Gisin, *Preprint* quant-ph/0512168, (2005)
15. A. Aspect, *PhD. Thesis*, **2674**, (Orsay), 1983.

PART D

Mathematical Foundations

WHERE THE MATHEMATICAL STRUCTURE OF QUANTUM MECHANICS COMES FROM

G. M. D'ARIANO

Dipartimento di Fisica "A. Volta" dell'Università di Pavia, via Bassi 6, 27100 Pavia, Italy
and
Dept. of Electrical and Computer Engineering, Northwestern University, Evanston, IL 60208
E-mail:dariano@unipv.it

The mathematical formulation of Quantum Mechanics is derived from purely operational axioms based on a general definition of *experiment* as a set of transformations. The main ingredient of the mathematical construction is the postulated existence of *faithful states* that allows one to calibrate the experimental apparatus. Such notion is at the basis of the operational definitions of the scalar product and of the *adjoint* of a transformation.

Keywords: Quantum Mechanics; Axiomatics; Hilbert spaces; Banach spaces; C*-algebras.

1. Introduction

In spite of its unprecedented predicting power in the whole physical domain, the starting point of Quantum Mechanics is purely mathematical, with no direct physical interpretation of the formalism. Undeniably Quantum Mechanics is not based on a set of physical laws or principles from which the mathematical framework is derived—as we would expect from a theory. Considering the universality of Quantum Mechanics, its "physical" axioms should be of very general nature, transcending Physics itself, at the higher epistemological level, and should be related to *observability principles* that must be satisfied independently on the specific physical laws object of the experiment. In previous works[1–3] I showed how it is possible to derive the Hilbert space formulation of Quantum Mechanics from five operational Postulates concerning *experimental accessibility and simplicity.* In the present paper I will give a synthetical presentation of this axiomatization: additional details and mathematical proofs can be found in Ref. 3. The mathematical formulation of Quantum Mechanics in terms of complex Hilbert space for finite dimensions is derived starting from the five Postulates. For the infinite dimensional case a C*-algebra representation of physical transformations is derived from only four of the five Postulates, via a Gelfand-Naimark-Segal (GNS) construction.[4] The starting point for the axiomatization is a seminal definition of *physical experiment*, which, as first shown in Ref. 2, entails a thorough series of notions that lie at the basis of the axiomatization. The postulated existence of a *faithful state*, which allows one to calibrate the experimental apparatus, provides operational definitions for the scalar product and for the *adjoint* of a transformation at the core of the C*-algebra representation of transformations via the Gelfand-Naimark-Segal (GNS) construction. This crucial ingredient from the present axiomatization comes from modern Quantum Tomography,[6] and concerns the possibility of performing a complete quantum calibration of measuring apparatuses[7] and transformations[8] by using a single pure bipartite state.[9]

2. The Postulates

The general background is that in any experimental science we make *experiments* to get *information* on the *state* of an *object physical system*. Knowledge of such a state will allow us to predict the results of forthcoming experiments on the same object system. Since we necessarily work with only partial *a priori* knowledge of both system and experimental apparatus, the rules for the experiment must be given in a probabilistic setting.

General Axiom: On what is an experiment. An experiment on an object system consists in making it interact with an apparatus. The interaction between object and apparatus produces one of a set of possible transformations of the object, each one occurring with some probability. Information on the "state" of the object system at the beginning of the experiment is gained from the knowledge of which transformation occurred, which is the "outcome" of the experiment signaled by the apparatus.

- **Postulate1: Independent system**
  There exist independent physical systems.
- **Postulate2: Informationally complete observable**
  For each physical system there exists an informationally complete observable.
- **Postulate3: Local observability principle**
  For every composite system there exist informationally complete observables made only of local informationally complete observables.
- **Postulate4: Informationally complete discriminating observable**
  For every system there exists a minimal informationally complete observable that can be achieved using a joint discriminating observable on the system + an ancilla (i.e. an identical independent system).
- **Postulate5: Symmetric faithful state**
  For every composite system made of two identical physical systems there exist a symmetric joint state that is both dynamically and preparationally faithful.

3. The Statistical and Dynamical Structure

According to our definition of **experiment**—the starting point of our axiomatization—the experiment is identified with the set $\mathbb{A} \equiv \{\mathscr{A}_j\}$ of possible transformations $\mathscr{A}_j$ that can occur on the object system. The apparatus will signal the **outcome** j labeling which transformation actually occurred. The experimenter cannot control which transformation occurs, but he can decide which experiment to perform, namely he can choose the set of possible transformations $\mathbb{A} = \{\mathscr{A}_j\}$. For example, in an Alice&Bob communication scenario Alice will encode the different bit values by choosing between two experiments $\mathbb{A} = \{\mathscr{A}_j\}$ and $\mathbb{B} = \{\mathscr{A}_j\}$ corresponding to two different sets of transformations $\{\mathscr{A}_j\}$ and $= \{\mathscr{B}_j\}$. The experimenter has control on the transformation itself only in the special case when the transformation $\mathscr{A}$ is deterministic. In the following, wherever we consider a nondeterministic transformation $\mathscr{A}$ by itself, we always regard it in the context of an experiment, namely assuming that there always exists at least a complementary transformation $\mathscr{B}$ such that the overall probability of $\mathscr{A}$ and $\mathscr{B}$ is unit.

Now, since the knowledge of the state of a physical system allows us to predict the results of forthcoming possible experiments on the system (more generally, on another system in the same physical situation), namely it would allow us to evaluate the probabilities of any possible transformation for any possible experiment, then, by definition, a **state** ω for a physical system is a rule that provides the probability for any possible transformation, namely ω is a state means that $\omega(\mathscr{A})$ is the probability that the transformation $\mathscr{A}$ occurs. We clearly have the completeness condition $\sum_{\mathscr{A}_j \in \mathbb{A}} \omega(\mathscr{A}_j) = 1$, and we will assume that the identical transformation $\mathscr{I}$ occurs with probability one, i. e. $\omega(\mathscr{I}) = 1$, corresponding to a special choice of the lab reference frame as in the *Dirac picture*. In the following for a given physical system we will denote by $\mathfrak{S}$ the set of all possible states and by $\mathfrak{T}$ the set of all possible transformations. In order to include also non-disturbing experiments, we must conceive situations in which all states are left invariant by each transformation. It is convenient to extend the notion of state to that of **weight**, i. e. a nonnegative bounded functionals $\tilde{\omega}$ over the set of transformations with $0 \leq \tilde{\omega}(\mathscr{A}) \leq \tilde{\omega}(\mathscr{I}) < +\infty$ for all transformations $\mathscr{A}$. To each weight $\tilde{\omega}$ it corresponds the properly normalized state $\omega = \tilde{\omega}/\omega(\mathscr{I})$. Weights make the convex cone $\mathfrak{W}$ generated by the convex set of states $\mathfrak{S}$.

When composing two transformations $\mathscr{A}$ and $\mathscr{B}$, the probability $p(\mathscr{B}|\mathscr{A})$ that $\mathscr{B}$ occurs conditional on the previous occurrence of $\mathscr{A}$ is given by the Bayes rule for conditional probabilities $p(\mathscr{B}|\mathscr{A}) = \omega(\mathscr{B} \circ \mathscr{A})/\omega(\mathscr{A})$. This sets a new probability rule corresponding to the notion of **conditional state** $\omega_{\mathscr{A}}$ which gives the probability that a transformation $\mathscr{B}$ occurs knowing that the transformation $\mathscr{A}$ has occurred on the physical system in the state ω, namely $\omega_{\mathscr{A}} \doteq \omega(\cdot \circ \mathscr{A})/\omega(\mathscr{A})$ (in the following we will make extensive use of the functional notation with the central dot corresponding to a variable transformation). One can see that the present definition of "state", which logically follows from the definition of experiment, leads to the identification *state-evolution≡state-conditioning*, entailing a *linear action of transformations on states* (apart from normalization) $\mathscr{A}\omega := \omega(\cdot \circ \mathscr{A})$: this is the same concept of **operation** that we have in Quantum Mechanics, giving the conditional state as $\omega_{\mathscr{A}} = \mathscr{A}\omega/\mathscr{A}\omega(\mathscr{I})$. In other words, this is the analogous of the Schrödinger picture evolution of states in Quantum Mechanics. One can see that in the present context linearity of evolution is just a consequence of the fact that the evolution of states is pure state-conditioning: this will include also the deterministic case $\mathscr{U}\omega = \omega(\cdot \circ \mathscr{U})$ of transformations $\mathscr{U}$ with $\omega(\mathscr{U}) = 1$ for all states ω—the analogous of quantum unitary evolutions and channels.

From the Bayes conditioning it follows that we can define two complementary types of equivalences for transformations: the *dynamical* and *informational* equivalences. The transformations $\mathscr{A}_1$ and $\mathscr{A}_2$ are *dynamically equivalent* when $\omega_{\mathscr{A}_1} = \omega_{\mathscr{A}_2}$ $\forall\omega \in \mathfrak{S}$, whereas they are *informationally equivalent* when $\omega(\mathscr{A}_1) = \omega(\mathscr{A}_2)$ $\forall\omega \in \mathfrak{S}$. The two transformations are then completely equivalent when they are both dynamically and informationally equivalent, corresponding to the identity $\omega(\mathscr{B} \circ \mathscr{A}_1) = \omega(\mathscr{B} \circ \mathscr{A}_2)$, $\forall\omega \in \mathfrak{S}$, $\forall\mathscr{B} \in \mathfrak{T}$. We call **effect** an informational equivalence class of transformations (this is the same notion introduced by Ludwig[5]). In the following we will denote effects with the underlined symbols $\underline{\mathscr{A}}$, $\underline{\mathscr{B}}$, etc., or as $[\mathscr{A}]_{\text{eff}}$, and we will write $\mathscr{A}_0 \in \underline{\mathscr{A}}$ meaning that "the transformation $\mathscr{A}$ belongs to the equivalence class $\underline{\mathscr{A}}$", or "$\mathscr{A}_0$ corresponds to the effect $\underline{\mathscr{A}}$", or "$\mathscr{A}_0$ is informationally equivalent to $\mathscr{A}$". Since, by definition one has $\omega(\mathscr{A}) \equiv \omega(\underline{\mathscr{A}})$, we will legitimately write $\omega(\underline{\mathscr{A}})$ instead of $\omega(\mathscr{A})$. Similarly, one has $\omega_{\mathscr{A}}(\mathscr{B}) \equiv \omega_{\mathscr{A}}(\underline{\mathscr{B}})$, which implies that $\omega(\mathscr{B} \circ \mathscr{A}) = \omega(\underline{\mathscr{B}} \circ \mathscr{A})$, which gives the chaining rule $\underline{\mathscr{B}} \circ \mathscr{A} \in \underline{\mathscr{B} \circ \mathscr{A}}$ corresponding to the "Heisenberg picture" evolution of transformations acting on effects (notice that in this way transformations act from the right on effects). Now, by definitions effects are linear functionals over states with range $[0, 1]$, and, by duality, we have a convex structure over effects. We will denote the convex set of effects by $\mathfrak{P}$.

The fact that we necessarily work in the presence of partial knowledge about both object and apparatus corresponds to the possibility of incomplete specification of both states and transformations, entailing the convex structure on states and the addition rule for *coexistent transformations*, namely for transformations $\mathscr{A}_1$ and $\mathscr{A}_2$ for which $\omega(\mathscr{A}_1) + \omega(\mathscr{A}_2) \leq 1$, $\forall\omega \in \mathfrak{S}$ (i. e. transformations that can in principle occur in the same experiment). The addition of the two coexistent transformations is the transformation $\mathscr{S} = \mathscr{A}_1 + \mathscr{A}_2$ corresponding to the event $e = \{1, 2\}$ in which the apparatus signals that either $\mathscr{A}_1$ or $\mathscr{A}_2$ occurred, but does not specify which one. Such transformation is specified by the informational and dynamical equivalence classes $\forall\omega \in \mathfrak{S}$: $\omega(\mathscr{A}_1 + \mathscr{A}_2) = \omega(\mathscr{A}_1) + \omega(\mathscr{A}_2)$ and $(\mathscr{A}_1 + \mathscr{A}_2)\omega = \mathscr{A}_1\omega + \mathscr{A}_2\omega$. Clearly the composition "$\circ$" of transformations is distributive with respect to the addition "+". We will also denote by $\mathscr{S}(\mathbb{A}) := \sum_{\mathscr{A}_j \in \mathbb{A}} \mathscr{A}_j$ the deterministic transformation $\mathscr{S}(\mathbb{A})$ corresponding to the sum of all possible transformations $\mathscr{A}_j$ in $\mathbb{A}$. We can also define the multiplication $\lambda\mathscr{A}$ of a transformation $\mathscr{A}$ by a scalar $0 \leq \lambda \leq 1$ as the transformation which is dynamically equivalent to $\mathscr{A}$, but occurs with rescaled probability $\omega(\lambda\mathscr{A}) = \lambda\omega(\mathscr{A})$. Now, since for every couple of transformation $\mathscr{A}$ and $\mathscr{B}$ the transformations $\lambda\mathscr{A}$ and $(1 - \lambda)\mathscr{B}$ are coexistent for $0 \leq \lambda \leq 1$, the set of transformations also becomes a convex set. Moreover, since the composition $\mathscr{A} \circ \mathscr{B}$ of two transformations $\mathscr{A}$ and $\mathscr{B}$ is itself a transformation and there exists the identical transformation $\mathscr{I}$ satisfying $\mathscr{I} \circ \mathscr{A} = \mathscr{A} \circ \mathscr{I} = \mathscr{A}$ for every transformation

$\mathscr{A}$, the transformations make a semigroup with identity, i. e. a *monoid*. Therefore, the set of physical transformations is a convex monoid.

It is obvious that we can extend the notions of coexistence, sum and multiplication by a scalar from transformations to effects via equivalence classes.

A purely dynamical notion of **independent systems** coincides with the possibility of performing local experiments. More precisely, we say that two physical systems are *independent* if on the two systems 1 and 2 we can perform *local experiments* $\mathbb{A}^{(1)}$ and $\mathbb{A}^{(2)}$ whose transformations commute each other (i. e. $\mathscr{A}^{(1)} \circ \mathscr{B}^{(2)} = \mathscr{B}^{(2)} \circ \mathscr{A}^{(1)}$, $\forall \mathscr{A}^{(1)} \in \mathbb{A}^{(1)}, \forall \mathscr{B}^{(2)} \in \mathbb{B}^{(2)}$). Notice that the above definition of independent systems is purely dynamical, in the sense that it does not contain any statistical requirement, such as the existence of factorized states. Indeed, the present notion of dynamical independence is so minimal that it can be satisfied not only by the quantum tensor product, but also by the quantum direct sum. As we will see in the following, it is the local observability principle of Postulate 3 which will select the tensor product. In the following, when dealing with more than one independent system, we will denote local transformations as ordered strings of transformations as follows $\mathscr{A}, \mathscr{B}, \mathscr{C}, \ldots := \mathscr{A}^{(1)} \circ \mathscr{B}^{(2)} \circ \mathscr{C}^{(3)} \circ \ldots$. For effects one has the locality rule $([\mathscr{A}]_{\text{eff}}, [\mathscr{B}_{\text{eff}}) \in [(\mathscr{A}, \mathscr{B})]_{\text{eff}}$. The notion of independent systems now entails the notion of *local state*—the equivalent of partial trace in Quantum Mechanics. In the presence of many independent systems in a joint state Ω, we define the **local state** $\Omega|_n$ of the n-th system as the probability rule $\Omega|_n(\mathscr{A}) \doteq \Omega(\mathscr{I}, \ldots, \mathscr{I}, \underbrace{\mathscr{A}}_{n\text{th}}, \mathscr{I}, \ldots)$ of the joint state Ω with a local transformation $\mathscr{A}$ only on the n-th system and with all other systems untouched. For example, for two systems we write $\Omega|_1 = \Omega(\cdot, \mathscr{I})$.

We conclude this section by noticing that our definition of dynamical independence implies the acausality of correlations between independent systems—the so-called *no-signaling*— i. e. : Any local "action" (i. e. experiment) on a system does not affect another independent system. In equations: $\forall \Omega \in \mathfrak{S}^{\times 2}, \forall \mathbb{A}$, $\Omega_{\mathscr{S}(\mathbb{A}),\mathscr{I}}|_2 = \Omega|_2$. Notice that even though the no-signaling holds, the occurrence of the transformation $\mathscr{B}$ on system 1 generally affects the local state on system 2, i. e. $\Omega_{\mathscr{B},\mathscr{I}}|_2 \neq \Omega_2$, and such correlations can be checked *a posteriori*. We emphasize that the no-signaling is a mere consequence of our minimal notion of dynamical independence.

4. Banach Structure

We can extend the convex cone of weights to its embedding linear space by taking differences of weights, and forming *generalized weights*. We will denote the linear space of generalized weights as $\mathfrak{W}_{\mathbb{R}}$. Likewise we can extend effects and transformations to generalized effects and transformations, whose linear spaces will be denoted by $\mathfrak{P}_{\mathbb{R}}$ and $\mathfrak{T}_{\mathbb{R}}$, respectively. The linear space $\mathfrak{T}_{\mathbb{R}}$ of generalized transformations inherits a real algebra structure from the convex monoid of physical transformations $\mathfrak{T}$. On the linear spaces $\mathfrak{W}_{\mathbb{R}}$, $\mathfrak{P}_{\mathbb{R}}$, and $\mathfrak{T}_{\mathbb{R}}$ we can now superimpose a Banach space structure, by introducing norms in form of supremum. We start from physical effects for which we define the norm as the supremum of the respective probability over all possible physical states. We then extend the norm to generalized effects $\underline{\mathscr{A}} \in \mathfrak{P}_{\mathbb{R}}$ by taking the absolute value, i. e. $\|\underline{\mathscr{A}}\| := \sup_{\omega\in\mathfrak{S}} |\omega(\underline{\mathscr{A}})|$. It is easy to check that this is indeed a norm. We can now introduce the unit ball $\mathfrak{B}_1 := \{\underline{\mathscr{A}} \in \mathfrak{P}_{\mathbb{R}},\ \|\underline{\mathscr{A}}\| \leq 1\}$ and define the norm for weights as $\|\tilde{\omega}\| := \sup_{\underline{\mathscr{A}}\in\mathfrak{B}_1} |\tilde{\omega}(\underline{\mathscr{A}})|$. For transformations we then introduce the norm in the standard way used for linear operators over Banach spaces, namely $\|\mathscr{A}\| := \sup_{\underline{\mathscr{B}}\in\mathfrak{B}_1} \|\underline{\mathscr{B}}\| \circ \mathscr{A}$, which is equivalent to the double supremum $\|\mathscr{A}\| = \sup_{\underline{\mathscr{B}}\in\mathfrak{B}_1} \sup_{\omega\in\mathfrak{S}} |\omega(\underline{\mathscr{B}}\circ\mathscr{A})|$. It is then easy to check that $\mathfrak{T}_{\mathbb{R}}$ becomes a real Banach algebra (i. e. it satisfies the norm inequality $\|\mathscr{B} \circ \mathscr{A}\| \leq \|\mathscr{B}\|\|\mathscr{A}\|$). It is crucial to perform the supremum over the unit ball, instead of just physical effects: this guarantees the Banach algebra structure for generalized transformations. It is also clear that physical transformation correspond to contractions, i. e. they have bounded norm $\|\mathscr{A}\| \leq 1$, whence the convex monoid of physical transformations $\mathfrak{T}$ has the form of a truncated convex cone. As a corollary, we have that two physical transformations $\mathscr{A}$ and $\mathscr{B}$ are coexistent iff $\mathscr{A} + \mathscr{B}$ is a contraction. We also have the bound between transformation and effect norms $\|\underline{\mathscr{A}}\| \leq \|\mathscr{A}\|$, with the identity for $\mathscr{A}$ in the double cone. Operationally all norm closures correspond to assume preparability (of effects, states, and transformations) by an approximation criterion in-probability. The norm closure may not be required operationally, however, as any other kind of extension, it is mathematically very convenient. The convex set of states $\mathfrak{S}$ and the convex sets of effects $\mathfrak{P}$ are dual each other under the pairing $\omega(\underline{\mathscr{A}})$ giving the probability of effect $\underline{\mathscr{A}}$ in the state ω. Therefore, the convex set of effects is a truncated convex cone of positive linear contractions over the convex set of states, namely the set of bounded positive functionals $0 \leq l \leq 1$ on $\mathfrak{S}$, with $l_{\underline{\mathscr{A}}}(\omega) := \omega(\underline{\mathscr{A}})$. Such duality can be trivially extended to generalized effects and generalized weights via the pairing $|\omega(\underline{\mathscr{A}})|$, and $\mathfrak{W}_{\mathbb{R}}$ and $\mathfrak{P}_{\mathbb{R}}$ become a dual Banach pair. This Banach space duality is the analogous of the duality between bounded operators and trace-class operators in Quantum Mechanics. It is worth noticing that this dual Banach pair is just a consequence of the probabilistic structure that is inherent in our starting definition of experiment.

5. Observables

The *observable* is just a complete set of effects $\mathbb{L} = \{l_i\}$ of an experiment $\mathbb{A} = \{\mathscr{A}_j\}$, namely one has $l_i = \underline{\mathscr{A}_j}\ \forall j$. Clearly, one has the completeness relation $\sum_i l_i = 1$. The observable $\mathbb{L} = \{l_i\}$ is *informationally complete* when each effect l can be written as a linear combination $l = \sum_i c_i(l) l_i$. of elements of $\mathbb{L}$, or, in other words, $\mathfrak{P}_{\mathbb{R}} \equiv \mathsf{Span}(\mathbb{L})$. We will call the informationally complete observable *minimal* when its effects are linearly independent. Clearly, using an informationally complete observable we can reconstruct any state ω from just the probabilities $l_i(\omega)$ as $\omega(\underline{\mathscr{A}}) = \sum_i c_i(l_{\underline{\mathscr{A}}}) l_i(\omega)$: this is just the Bloch representation of states. In such representation the Banach structure manifests itself in a vector representation for states and effects, and in a matrix representation for transformations, the physical transformations corresponding to affine linear maps.

We will call an effect (and likewise a transformation) $\underline{\mathscr{A}}$ *predictable* if there exists a state for which $\underline{\mathscr{A}}$ occurs with certainty and another state for which it never occurs, and *resolved* if there is only a single pure state for which it occurs with certainty. Similarly an experiment will be called *predictable* when it is made only of predictable effects, and *resolved* when all its effects are resolved. For a predictable effect $\underline{\mathscr{A}}$ one has $\|\underline{\mathscr{A}}\| = 1$, and for a predictable transformation $\mathscr{A}$ one has $\|\mathscr{A}\| = 1$. Notice that a predictable transformation is not necessarily deterministic. Predictable effects $\underline{\mathscr{A}}$ correspond to affine functions $f_{\mathscr{A}}$ on the state space $\mathfrak{S}$ with $0 \leq f_{\mathscr{A}} \leq 1$ achieving both bounds. We call a set of states $\{\omega_n\}_{n=1,N}$ *perfectly discriminable* if there exists a predictable and resolved experiment $\mathbb{L} = \{l_j\}_{j=1,N}$ which discriminates the states, i. e. $\omega_m(l_n) = \delta_{nm}$. We call *informational dimension* of the convex set of states $\mathfrak{S}$, denoted by $\dim_{\#}(\mathfrak{S})$, the maximal cardinality of perfectly discriminable set of states in $\mathfrak{S}$. Clearly, an observable $\mathbb{L} = \{l_j\}$ is *discriminating* and *resolved* for $\mathfrak{S}$ when $|\mathbb{L}| \equiv \dim_{\#}(\mathfrak{S})$, i. e. $\mathbb{L}$ discriminates a maximal set of discriminable states.

We now come to the notions of faithful state. We say that a state Φ of a composite system is *dynamically faithful* for the nth component system when for every transformation $\mathscr{A}$ the map $\mathscr{A} \leftrightarrow (\mathscr{I}, \ldots, \mathscr{I}, \underbrace{\mathscr{A}}_{n\text{th}}, \mathscr{I}, \ldots)\Phi$ is one-to-one, with the transformation $\mathscr{A}$ acting locally only on the nth component system. Physically, the definition corresponds to say that the output conditioned weight (i. e. the conditioned state multiplied by the probability of occurrence) is in one-to-one correspondence with the transformation. Restricting attention to bipartite systems, a state is dynamically faithful (for system 1) when $(\mathscr{A}, \mathscr{I})\Phi = 0 \iff \mathscr{A} = 0$, which means that for every bipartite effect $\underline{\mathscr{B}}$ one has $\Phi(\underline{\mathscr{B}} \circ (\mathscr{A}, \mathscr{I})) = 0 \quad \iff \quad \mathscr{A} = 0$. Clearly the correspondence remains one-to-one when extended to $\mathfrak{T}_{\mathbb{R}}$. On the other hand, we will call a state Φ of a bipartite system *preparationally faithful* for system 1 if every joint bipartite state Ω can be achieved by a suitable local transformation $\mathscr{T}_\Omega$ on system 1 occurring with nonzero probability. Clearly a bipartite state Φ that is preparationally faithful for system 1 is also locally preparationally faithful for system 1, namely every local state ω of system 2 can be achieved by a suitable local transformation $\mathscr{T}_\omega$ on system 1.

In Postulate 5 we also use the notion of *symmetric* joint state. This is simply defined as a joint state of two identical systems such that for any couple of transformations $\mathscr{A}$ and $\mathscr{B}$ one has $\Phi(\mathscr{A}, \mathscr{B}) = \Phi(\mathscr{B}, \mathscr{A})$.

6. Dimensionality Theorems

We now consider the consequences of Postulates 3 and 4. The *local observability principle* (Postulate 3) is operationally crucial, since it reduces enormously the experimental complexity, by guaranteeing that only local (although jointly executed) experiments are sufficient to retrieve a complete information of a composite system, including all correlations between the components. The principle reconciles

Fig. 1. **Left:** Illustration of the notion of *dynamically faithful* state for a bipartite system. The state Φ is dynamically faithful when the output weight (conditioned state multiplied by the probability of occurrence) is in one-to-one correspondence with the transformation. **Right:** Illustration of the notion of preparationally faithful state for a bipartite system. The state Φ is *preparationally faithful* for system 1 if every joint bipartite state Ω can be achieved by a suitable local transformation $\mathscr{T}_\Omega$ on system 1 occurring with nonzero probability.

holism with reductionism, in the sense that we can observe an holistic nature in a reductionistic way— i. e. locally. This principle implies identity (D_3) in Table 1 for the affine dimension of the convex set of a bipartite systems as a function of the dimensions of the components. This identity is the same that one obtains in Quantum Mechanics due to the tensor product structure. We conclude that the tensor product is not a consequence of dynamical independence in Def. 1, but follows from the local observability principle.

Table 1. Dimensionality identities implied by Postulates.

$\Longrightarrow$		
Postulate 2	$\dim(\mathfrak{P}_\mathbb{R}) = \dim(\mathfrak{S}) + 1$	(D_2)
Postulate 3	$\dim(\mathfrak{S}_{12}) = \dim(\mathfrak{S}_1)\dim(\mathfrak{S}_2) + \dim(\mathfrak{S}_1) + \dim(\mathfrak{S}_2)$	(D_3)
Postulate 4	$\dim(\mathfrak{S}) = \dim_\#(\mathfrak{S}^{\times 2}) - 1$	(D_4)
(D_3)+(D_4)	$\dim(\mathfrak{S}^{\times 2}) = \dim_\#(\mathfrak{S}^{\times 2})^2 - 1$	(D_{34})
(D_{34})	$\dim(\mathfrak{S}) = \dim_\#(\mathfrak{S})^2 - 1$	(D'_{34})[a]
$(D_4+D'_{34})$	$\dim_\#(\mathfrak{S}^{\times 2}) = \dim_\#(\mathfrak{S})^2$	$(\otimes)$
Postulate 5	$\dim(\mathfrak{T}) = \dim(\mathfrak{S}^{\times 2}) + 1$	$(\mathfrak{T})$
(D_2)+(D'_{34})	$\dim(\mathfrak{P}_\mathbb{R}) = \dim_\#(\mathfrak{S})^2$	$(\mathfrak{P})$

Note: [a] Generalizing from convex sets of states of bipartite systems to any convex set of states.

Postulate 4 now gives identity (D_4) in Table 1. By comparing this with the affine dimension of the bipartite system, we get identity (D_{34}), and generalizing to any convex set we get identity (D'_{34}) corresponding to the dimension of the quantum convex sets $\mathfrak{S}$ due to the underlying Hilbert space. Moreover, upon substituting identity (D_4) one obtains identity $(\otimes)$ which is the quantum product rule for informational dimensionalities corresponding to the quantum *tensor product.* To summarize, it is worth noticing that the *quantum dimensionality rules* (D_3) and $(\otimes)$ follow from Postulates 3 and 4. Postulate 5, on the other hand, implies identity $(\mathfrak{T})$.

7. The Complex Hilbert Space Structure for Finite Dimensions

The faithful state Φ provides a symmetric bilinear form $\Phi(\underline{\mathscr{A}},\underline{\mathscr{B}})$ over $\mathfrak{P}_{\mathbb{R}}$, from which one can extract a positive scalar product over $\mathfrak{P}_{\mathbb{R}}$ as $|\Phi|(\underline{\mathscr{A}},\underline{\mathscr{B}})$, where $|\Phi| := \Phi_+ - \Phi_-$ is the absolute value of Φ (the absolute value can be defined thanks to the fact that Φ is real symmetric, whence it can be diagonalized over $\mathfrak{P}_{Reals}$). Upon denoting by $\mathscr{P}_\pm$ the orthogonal projectors over the linear space corresponding to positive and negative eigenvalues, respectively, one has $|\Phi|(\underline{\mathscr{A}},\underline{\mathscr{B}}) = \Phi(\underline{\mathscr{A}},\varsigma(\underline{\mathscr{B}}))$, where $\varsigma(\underline{\mathscr{A}}) := (\mathscr{P}_+ - \mathscr{P}_-)(\underline{\mathscr{A}})$. The map ς is an involution, namely $\varsigma^2 = \mathscr{I}$. The fact that the state is also preparationally faithful implies that the scalar product is strictly positive, namely $|\Phi|(\underline{\mathscr{C}},\underline{\mathscr{C}}) = 0$ implies that $\underline{\mathscr{C}} = 0$ (see Ref. 3). Now, being $|\Phi|(\underline{\mathscr{A}},\underline{\mathscr{B}})$ a strictly positive real symmetric scalar product, the linear space $\mathfrak{P}_{\mathbb{R}}$ of generalized effects becomes a <u>real</u> pre-Hilbert space, which can be completed to a Hilbert space in the norm topology. For finite dimensional convex set $\mathfrak{S}$ one has Eq. (D_2) in Table 1, which follows from the fact that since $\mathfrak{P}_{\mathbb{R}}$ is just the space of the linear functionals over $\mathfrak{S}$, it has an additional dimension corresponding to normalization. But from Eq. (D_2) and (D'_{34}) one has identity $(\mathfrak{P})$, which implies that $\mathfrak{P}_{\mathbb{R}}$ as a real Hilbert space is isomorphic to the real Hilbert space of Hermitian complex matrices representing selfadjoint operators over a complex Hilbert space H of dimensions $\dim(\mathsf{H}) = \dim_{\#}(\mathfrak{S})$. This last assertion is indeed the Hilbert space formulation of Quantum Mechanics, from which one can recover the full mathematical structure. In fact, once the generalized effects are represented by Hermitian matrices, the physical effects will be represented as elements of the truncated convex cone of positive matrices, the physical transformations will be represented as CP identity-decreasing maps over effects, and finally, states will be represented as density matrices via the Bush version[10] of the Gleason theorem, or via our state-effect correspondence coming from the preparationally faithfulness of Φ.

8. Infinite Dimension: the C*-Algebra of Transformations

For infinite dimensions we cannot rely on the dimensionality identities in Table 1, and we need an alternative way to derive Quantum Mechanics, such as the construction of a C*-algebra representation of generalized transformations. In order to do that we need to extend the real Banach algebra $\mathfrak{T}_{\mathbb{R}}$ to a complex algebra, and for this we need to derive the *adjoint* of a transformation from the five postulates (we will see that indeed only four of the five postulates are needed). The adjoint is given as the composition of *transposition* and *complex-conjugation* of physical transformations, both maps being introduced operationally on the basis of the existence of a symmetric dynamically faithful state due to Postulate 5. The *complex conjugate* map will be an extension to $\mathfrak{T}_{\mathbb{R}}$ of the involution ς of Section 6. With such an adjoint one then derives a GNS representation[4] for transformations, leading to a C*-algebra.

The transposed transformation. For a symmetric bipartite state that is faithful both dynamically and preparationally, for every transformation on system 1 there always exists a (generalized) transformation on system 2 giving the same operation on that state. This allows us to introduce operationally the notion of *transposed transformation* as follows. For a *faithful* bipartite state Φ, the *transposed transformation* $\mathscr{A}'$ of the transformation $\mathscr{A}$ is the generalized transformation which when applied to the second component system gives the same conditioned state and with the same probability as the transformation $\mathscr{A}$ operating on the first system, namely $(\mathscr{A},\mathscr{I})\Phi = (\mathscr{I},\mathscr{A}')\Phi$ or, equivalently $\Phi(\underline{\mathscr{B}}\circ\mathscr{A},\underline{\mathscr{C}}) = \Phi(\underline{\mathscr{B}},\underline{\mathscr{C}}\circ\mathscr{A}')\ \forall\underline{\mathscr{B}},\underline{\mathscr{C}}\in\mathfrak{P}$.

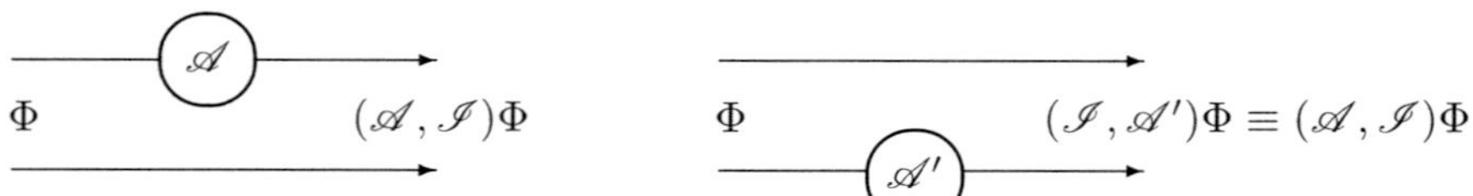

Fig. 2. Illustration of the operational concept of *transposed transformation.*

It is easy to check the axioms of transposition ($(\mathscr{A}+\mathscr{B})' = \mathscr{A}'+\mathscr{B}'$, $(\mathscr{A}')' = \mathscr{A}$, $(\mathscr{A}\circ\mathscr{B})' = \mathscr{B}'\circ\mathscr{A}'$) and that $\mathscr{I}' = \mathscr{I}$. Unicity is implied by faithfulness.

The complex conjugated transformation. Due to the presence of the involution ς, the transposition $\mathscr{A}\to\mathscr{A}'$ does not work as an adjoint for the scalar product $|\Phi|(\underline{\mathscr{A}},\underline{\mathscr{B}})$ (it works as an adjoint for the symmetric bilinear form Φ, which is not positive). In order to introduce an adjoint for generalized transformations (with respect to the scalar product between effects) one needs to extend the involution ς to generalized transformations. With a procedure analogous to that used for effects we introduce the absolute value $|\Phi|$ of the symmetric bilinear form Φ over $\mathfrak{T}_{\mathbb{R}}$, whence extend the scalar product to $\mathfrak{T}_{\mathbb{R}}$. Clearly, since the bilinear form $\Phi(\mathscr{A},\mathscr{B})$ will anyway depend only on the informational equivalence classes $\underline{\mathscr{A}}$ and $\underline{\mathscr{B}}$ of the two transformations, we have many extensions of ς which work equally well. Upon defining $\mathscr{A}^{\varsigma} := \varsigma(\mathscr{A})$, one has $\mathscr{A}^{\varsigma}\in\varsigma(\underline{\mathscr{A}})$, and clearly one has $\varsigma^2(\mathscr{A}) = \varsigma(\mathscr{A}^{\varsigma})\in\underline{\mathscr{A}}$, but generally $\varsigma^2(\mathscr{A})\neq\mathscr{A}$. However, one can always consistently choose the extension such that $\varsigma^2(\mathscr{A}) = \mathscr{A}$. The idea is now that such an involution plays the role of the *complex conjugation*, such that the composition of ς with the transposition provides the adjoint.

The adjoint transformation. Due to the fact that transformations act on effects from the right— i. e. $\underline{\mathscr{B}}\circ\mathscr{A}\in\underline{\mathscr{B}\circ\mathscr{A}}$—in order to keep the usual action on the left in the representation of transformations over generalized effects it is convenient to redefine the scalar product via the bilinear form $\Phi(\mathscr{A}',\mathscr{B}')$ over transposed transformations. Therefore, we define the scalar product between generalized effects as follows

$$ {}_{\Phi}\langle \underline{\mathscr{B}} | \underline{\mathscr{A}} \rangle_{\Phi} := \Phi(\underline{\mathscr{B}}', \varsigma(\underline{\mathscr{A}}')). \tag{1} $$

Notice how in this way one recovers the customary operator-like action of transformations from the left $|\underline{\mathscr{C} \circ \mathscr{A}}\rangle_{\Phi} = |\mathscr{C} \circ \underline{\mathscr{A}}\rangle_{\Phi}$ which follows from ${}_{\Phi}\langle \mathscr{C} \circ \underline{\mathscr{A}} | \mathscr{B} \rangle_{\Phi} = \Phi(\underline{\mathscr{A}}' \circ \mathscr{C}', \varsigma(\underline{\mathscr{B}}'))$. In the following we will equivalently write the entries of the scalar product as generalized transformations or as generalized effects, with ${}_{\Phi}\langle \mathscr{A} | \mathscr{B} \rangle_{\Phi} := {}_{\Phi}\langle \underline{\mathscr{A}} | \underline{\mathscr{B}} \rangle_{\Phi}$, the generalized effects being the actual vectors of the linear factor space of generalized transformations modulo informational equivalence.

For *composition-preserving* involution (i. e. $\varsigma(\mathscr{B} \circ \mathscr{A}) = \mathscr{B}^{\varsigma} \circ \mathscr{A}^{\varsigma}$) one can easily verify[3] that $\mathscr{A}^{\dagger} := \varsigma(\mathscr{A}')$ works as an adjoint for the scalar product, namely

$$ {}_{\Phi}\langle \mathscr{C}^{\dagger} \circ \underline{\mathscr{A}} | \underline{\mathscr{B}} \rangle_{\Phi} = {}_{\Phi}\langle \underline{\mathscr{A}} | \mathscr{C} \circ \underline{\mathscr{B}} \rangle_{\Phi}. \tag{2} $$

In terms of the adjoint the scalar product can also be written as ${}_{\Phi}\langle \mathscr{B} | \mathscr{A} \rangle_{\Phi} = \Phi|_2(\mathscr{A}^{\dagger} \circ \mathscr{B})$. The involution ς is composition-preserving if $\varsigma(\mathfrak{T}) = \mathfrak{T}$ namely if the involution preserves physical transformations. Indeed, for such an involution one can consider its action on transformations induced by the involutive isomorphism $\omega \to \omega^{\varsigma}$ of the convex set of states $\mathfrak{S}$ defined as $\omega^{\varsigma}(\mathscr{A}) := \omega(\varsigma(\mathscr{A}))$, $\forall \omega \in \mathfrak{S}$, $\forall \mathscr{A} \in \mathfrak{T}$. Consistency with state-reduction $\omega^{\varsigma}_{\mathscr{A}}(\mathscr{B}) \equiv \omega_{\mathscr{A}^{\varsigma}}(\mathscr{B}^{\varsigma})$ $\forall \omega \in \mathfrak{S}$, $\forall \mathscr{A}, \mathscr{B} \in \mathfrak{T}$ is then equivalent to $\omega(\varsigma(\mathscr{B} \circ \mathscr{A})) = \omega(\mathscr{B}^{\varsigma} \circ \mathscr{A}^{\varsigma})$ $\forall \omega \in \mathfrak{S}$, $\forall \mathscr{A}, \mathscr{B} \in \mathfrak{T}$. The involution ς of $\mathfrak{S}$ is just the inversion of the principal axes corresponding to negative eigenvalues of the symmetric bilinear form Φ of the faithful state.

The GNS construction and the C*-algebra. By taking complex linear combinations of generalized transformations and defining $\varsigma(c\mathscr{A}) = c^* \varsigma(\mathscr{A})$ for $c \in \mathbb{C}$, we can now extend the adjoint to complex linear combinations of generalized transformations, whose linear space will be denote by $\mathfrak{T}_{\mathbb{C}}$. On the other hand, we can trivially extend the real pre-Hilbert space of generalized effects $\mathfrak{P}_{\mathbb{R}}$ to a complex pre-Hilbert space $\mathfrak{P}_{\mathbb{C}}$ by just considering complex linear combinations of generalized effects. The complex algebra $\mathfrak{T}_{\mathbb{C}}$ (that we will also denote by $\mathcal{A}$) is now a complex Banach algebra of transformations on the Banach space $\mathfrak{P}_{\mathbb{C}}$. We have now a scalar product ${}_{\Phi}\langle \mathscr{A} | \mathscr{B} \rangle_{\Phi}$ between transformations, and an adjoint of transformations with respect to such scalar product. Symmetry and positivity imply the bounding[3] ${}_{\Phi}\langle \mathscr{A} | \mathscr{B} \rangle_{\Phi} \leq \|\mathscr{A}\|_{\Phi} \|\mathscr{B}\|_{\Phi}$, where we introduced the norm induced by the scalar product $\|\mathscr{A}\|^2_{\Phi} \doteq {}_{\Phi}\langle \mathscr{A} | \mathscr{A} \rangle_{\Phi}$. From the bounding for the scalar product it follows that the set $\mathcal{I} \subseteq \mathcal{A}$ of zero norm elements $\mathscr{X} \in \mathcal{A}$ is a left ideal, i. e. it is a linear subspace of $\mathcal{A}$ which is stable under multiplication by any element of $\mathcal{A}$ on the left (i. e. $\mathscr{X} \in \mathcal{I}$, $\mathscr{A} \in \mathcal{A}$ implies $\mathscr{A} \circ \mathscr{X} \in \mathcal{I}$). The set of equivalence classes $\mathcal{A}/\mathcal{I}$ thus becomes a complex pre-Hilbert space equipped with a symmetric scalar product. On the other hand, since the scalar product is strictly positive over generalized effects, the elements of $\mathcal{A}/\mathcal{I}$ are indeed the generalized effects, i. e. $\mathcal{A}/\mathcal{I} \simeq \mathfrak{P}_{\mathbb{C}}$ as linear spaces. Therefore, informationally equivalent transformations $\mathscr{A}$ and $\mathscr{B}$ correspond to the same vector, and there exists a generalized transformation $\mathscr{X}$ with $\|\mathscr{X}\|_{\Phi} = 0$ such that $\mathscr{A} = \mathscr{B} + \mathscr{X}$, and $\|\cdot\|_{\Phi}$, which is a norm on

$\mathfrak{P}_\mathbb{C}$, will be just a semi-norm on $\mathcal{A}$. We can re-define anyway the norm on transformations as $\|\mathscr{A}\|_\Phi := \sup_{\underline{\mathscr{B}}\in\mathfrak{P}_\mathbb{C}, \|\underline{\mathscr{B}}\|_\Phi\leq 1} \|\mathscr{A}\circ\underline{\mathscr{B}}\|_\Phi$. Completion of $\mathcal{A}/\mathcal{I} \simeq \mathfrak{P}_\mathbb{C}$ in the norm topology will give a Hilbert space that we will denote by H_Φ. Such completion also implies that $\mathfrak{T}_\mathbb{C} \simeq \mathcal{A}$ is a complex C*-algebra (i. e. satisfying the identity $\|\mathscr{A}^\dagger\circ\mathscr{A}\| = \|\mathscr{A}\|^2$), as it can be easily proved by standard techniques.[3] The fact that $\mathcal{A}$ is a C*-algebra—whence a Banach algebra—also implies that the domain of definition of $\pi_\Phi(\mathscr{A})$ can be easily extended to the whole H_Φ by continuity.

The product in $\mathcal{A}$ defines the action of $\mathcal{A}$ on the vectors in $\mathcal{A}/\mathcal{I}$, by associating to each element $\mathscr{A}\in\mathcal{A}$ the linear operator $\pi_\Phi(\mathscr{A})$ defined on the dense domain $\mathcal{A}/\mathcal{I}\subseteq\mathsf{H}_\Phi$ as follows

$$\pi_\Phi(\mathscr{A})|\underline{\mathscr{B}}\rangle_\Phi \doteq |\underline{\mathscr{A}\circ\mathscr{B}}\rangle_\Phi. \tag{3}$$

Born rule. From the definition (1) of the scalar product the Born rule rewrites in terms of the pairing $\omega(\underline{\mathscr{A}}) = {}_\Phi\langle\pi_\Phi(\underline{\mathscr{A}})|\pi_\Phi(\omega)\rangle_\Phi$, with representations of states $\pi_\Phi(\omega) = \widetilde{\underline{\mathscr{T}}_\omega} := \underline{\mathscr{T}}'_\omega/\Phi(\mathscr{I},\underline{\mathscr{T}}_\omega)$, and of effects $\pi_\Phi(\underline{\mathscr{A}}) = \underline{\mathscr{A}}'$ (see Ref. 3). Then, the representation of transformations is $\omega(\underline{\mathscr{B}}\circ\mathscr{A}) = {}_\Phi\langle\underline{\mathscr{B}}'|\pi_\Phi(\mathscr{A}^\varsigma)|\pi_\Phi(\omega)\rangle_\Phi$.

References

1. G. M. D'Ariano, *How to Derive the Hilbert-Space Formulation of Quantum Mechanics From Purely Operational Axioms,* work presented at the conference *On the Present Status of Quantum Mechanics* held on 7-9 September 2005, Mali Losinj, Croatia, in celebration of the 70th birthday of Gian Carlo Ghirardi. Also in *quant-ph/0603011.*
2. G. M. D'Ariano, *Quantum Theory, Reconsideration of Foundations - 3, Văxjő, Sweden, 6-11 June 2005*, edited by G. Adenier, A. Y. Khrennikov, and Th. M. Nieuwenhuizen, (American Institute of Physics, Melville, New York, 2006). Also in *quant-ph/0506034.*
3. G. M. D'Ariano, in *Foundations of Probability and Physics-4, Quantum Theory: Reconsideration of Foundations-3, Văxjő, Sweden, 4-9 June 2006*, edited by G. Adenier, A. Y. Khrennikov, and Th. M. Nieuwenhuizen, (American Institute of Physics, Melville, New York, 2007). Also in *quant-ph/0611094.*
4. I. M. Gelfand, and M. A. Neumark, *Mat. Sb.* **12**, 197 (1943).
5. G. Ludwig, *An Axiomatic Basis for Quantum Mechanics I: Derivation of Hilbert Space Structure*, (Springer, SPR:adr, 1985).
6. G. M. D'Ariano, *Tomographic methods for universal estimation in quantum optics*, (IOS Press, Amsterdam, 2002), pp. 385–406, scuola "E. Fermi" on *Experimental Quantum Computation and Information.*
7. G. M. D'Ariano, P. L. Presti, and L. Maccone, *Phys. Rev. Lett.* **93**, 250407 (2004).
8. G. M. D'Ariano, and P. L. Presti, *Phys. Rev. Lett.* **86**, 4195 (2001).
9. G. M. D'Ariano, and P. L. Presti, *Phys. Rev. Lett.* **91**, 047902–1–4 (2003).
10. P. Busch, *Phys. Rev. Lett.* **91** 120403 (2003).

PHASE SPACE DESCRIPTION OF QUANTUM MECHANICS AND NON-COMMUTATIVE GEOMETRY: WINGNER–MOYAL AND BOHM IN A WIDER CONTEXT

B. J. HILEY

TPRU, Birkbeck, University of London,
Malet Street, London WC1E 7HX. UK.
E-mail: b.hiley@bbk.ac.uk

In this paper we approach the question of the existence of a (x, p) phase space in a new way. Rather than abandoning all hope of constructing such a phase-space for quantum phenomena, we take aspects from both the Wigner-Moyal and Bohm approaches and show that although there is no unique phase space, we can form 'shadow' phase spaces. We then argue that this is a consequence of the non-commutative geometry defined by the operator algebra.

Keywords: Quantum mechanics; Wigner–Moyal approach; Bohm model; Poisson deformation algebra; Non–commutative geometry; Shadow manifolds.

1. Introduction

Although Schrödinger's original 'derivation' of his equation[1] originates from a modification of the classical Hamilton-Jacobi equation, the relation between the standard formalism of quantum mechanics and classical mechanics has always been problematical and not without controversy. The success of the Hilbert space formalism and the interpretation of the wave function of a single particle as a probability amplitude has had such predictive success that attempts to find a phase space description of quantum phenomenon have generally been regarded as futile.

In spite of this, physicists often turn to the Wigner-Moyal approach, treating it as a 'semi-classical' approach. Yet when Moyal's origin paper[2] is examined carefully[3] one finds there is nothing semi-classical about it. It turns out that what is called a 'quasi-distribution function' is in fact the density matrix expressed in terms of an $x - p$ variables. We will show that the x and p here are the mean values of a cell constructed in classical phase space, whose size we eventually integrate over.[4] They are not the position and momentum of the particle under consideration. It is important to note that x and p are not operators but ordinary commuting variables.

Fortunately Moyal's work has not been neglected by the mathematicians. They have shown that Moyal's results emerge from a deformed Poisson algebra with Planck's constant being used as the deformation parameter. This algebra reproduces all the expectation values of quantum mechanics exactly from a distribution function

$\rho(x, p, t)$. This algebra has the added advantage that it contains classical mechanics in the limit of an expansion in terms of $\hbar$ to $O(\hbar)$.

There is of course another phase space approach and that is one that I have explored with David Bohm,[56] This goes under the name de Broglie-Bohm although I prefer to call it the Bohm approach because it differs in some significant ways from the way de Broglie originally envisioned. Recently I was surprised to discover[3] that the defining set of equations for the Bohm interpretation are already contained in the appendices of Moyal's original paper.[2] It turns out that in the x-representation, the momentum variable p used in Bohm's model is just the marginal momentum $\overline{\overline{p}}$ in the Moyal approach. One of the task's of this paper is to bring out these connections and to show that at the heart of the phase space approach is the dynamics that depends on two equations, the Liouville equation and a new equation describing the time evolution of the phase. It is this equation that becomes the classical Hamilton-Jacobi equation in the expansion of the deformation algebra to $O(\hbar)$.

The second point I want to bring out is that if we exploit the Heisenberg matrix mechanics so that we have a completely algebraic description of quantum phenomena, we get exactly the operator analogues of the two equation mentioned in the last paragraph. The significance of this result is that we now have a purely algebraic description of the phase space and furthermore this description is non-commutative.

In this algebraic approach we are thus faced with a non-commutative structure. To understand the significance of this for geometry we must recall the work of Gel'fand.[7] He showed that if one has a commutative algebra of functions, we can either start with an underlying continuous manifold with an *a priori* topological and metric structure and then derive the algebra governing the field equations. This is the usual approach adopted in physics. Or one can start with the algebra and deduce the properties of the underlying manifold. Provided the algebra is commutative there is a dual relation between the field algebra and its underlying spatial support.

In quantum mechanics what we do have is a field algebra defined in terms of the operators. Notice further that all the symmetries of the system are carried by the operators. What we do not have is the properties of the manifold, i.e. the phase space that can carry this algebraic structure. Can we obtain the properties of this manifold from the algebra? This is where we hit a snag. If we have a non-commutative structure then there is no single, unique underlying manifold. However one can find shadow manifolds.[7] You can construct these shadow manifolds by projecting the algebra into a sub-space. Doing this is equivalent to projecting into a representation of Hilbert space. What we will show is that the Bohm approach picks out one of these shadow manifolds by choosing the x-representation. This particular choice was a historical contingency, Bohm could have chosen another representation.

What one also finds is that the algebraic time dependent equations do not contain the quantum potential explicitly. Recall it is this strange potential that carries the non-local implications for entangled particles. However the potential re-appears when we project onto the shadow manifold. Thus it is not an intrinsic property of the quantum system but only arises from the projection. In this sense it shares a property with the gravitational potential which arises when the geodesics of curved space-time are projected into a flat space-time. The big difference is that gravitation is universal whereas each quantum potential only applies to its own group of entangled particles. Particles not involved in this entanglement do not see the quantum potential of this particular group.

2. The Moyal Approach

Rather than using Moyal's starting point, let us follow Takasabsi[8] and introduce a two-point quantum density operator $\rho(x_1, x_2, t)$ defined by

$$\rho(x_1, x_2, t) = \psi(x_1, t)\psi^*(x_2, t) = \frac{1}{2\pi}\int \phi(p_1, t)\phi^*(p_2, t)e^{-\frac{i}{\hbar}(x_2p_2 - x_1p_1)}dx_1dx_2 \quad (1)$$

Then let us change co-ordinates, introducing $X = (x_2 + x_1)/2$; $\theta = x_2 - x_1$ and $P = (p_2 + p_1)/2$; $\tau = p_2 - p_1$. Substituting we find

$$\rho(X, \theta, t) = \frac{1}{2\pi}\int \phi^*(P - \hbar\tau/2)e^{\frac{iX\tau}{\hbar}}\phi(P + \hbar\tau/2)e^{\frac{iP\theta}{\hbar}}dPd\tau \quad (2)$$

which we can write in the form

$$\rho(X, \theta, t) = \int f(X, P, t)e^{\frac{iP\theta}{\hbar}}dP. \quad (3)$$

Thus we see that $f(X, P, t)$, which is conventionally treated as a quasi-classical distribution function in the Moyal approach, is actually the Fourier transform of the quantum density matrix. Notice further that the variables X and P are the mean values of a cell in phase space, not the co-ordinates of the single particle we are considering.

Now to link up with quantum mechanics we replace $\hat{A}(\hat{X}, \hat{P})$ by $A(X, P)$ so that mean values can be found from

$$\langle\hat{A}\rangle = \int A(X, P)f(X, P, t)dXdP. \quad (4)$$

Here we have written

$$A(X, P) = \frac{1}{(2\pi)^2}\int \psi^*(X - \hbar\theta/2, t)\hat{A}(\hat{X}, \hat{P})e^{\frac{iP\theta}{\hbar}}\psi(X + \hbar\theta/2, t)d\theta \quad (5)$$

It should be stressed that no approximations have been made in this derivation and consequently we get exactly the expectation values calculated from the standard formalism. However the expression (4) looks like the classical method to derive expectation values if we assume $f(X, P, t)$ is a distribution function. But notice it merely *looks* as if we have a classical theory but it is not a classical theory because $f(X, P, t)$ is the density matrix so it is not surprising to find it takes negative values.[9] Nevertheless we get the correct expectation values, thus the approach is not an approximation to the quantum formalism, it is exact.

3. Relation to the Bohm Approach

In Appendix 1 of Moyal's original paper,[2] a marginal expectation value for the momentum is defined by

$$\rho(X)\overline{\overline{p}} = \int Pf(X,P,t)dP \tag{6}$$

which for a wave function that can be written as $\psi(X) = \rho^{1/2}(X)e^{iS(X)/\hbar}$ gives

$$\overline{\overline{p}}(X) = \frac{\partial S}{\partial X} \tag{7}$$

This is just the expression that Bohm uses for his momentum. But there is more.

In Appendix 4, Moyal considers the transport equation for the probability density ρ and the momentum $\overline{\overline{p}}$. These are found from the time development equation for $f(X,P,t)$ which is derived by Moyal in the form

$$\frac{\partial f(X,P,t)}{\partial t} = H(X,P)\left[\frac{2}{\hbar}\sin\frac{\hbar}{2}\left(\frac{\overleftarrow{\partial}}{\partial X}\frac{\overrightarrow{\partial}}{\partial P} - \frac{\overleftarrow{\partial}}{\partial P}\frac{\overrightarrow{\partial}}{\partial X}\right)\right]f(X,P,t) \tag{8}$$

where the arrows indicate the function which must be differentiated. Let us now consider a particle in three-space and choose the Hamiltonian $H(\mathbf{X},\mathbf{P}) = \frac{1}{2m}\mathbf{P}^2 + V(\mathbf{X},t)$. Integrating equation (8) over $\overline{\overline{\mathbf{p}}}$ we find

$$\frac{\partial \rho}{\partial t} + \sum_i \frac{\partial}{\partial X_i}\left(\frac{\rho}{m}\frac{\partial S}{\partial X_i}\right) = 0 \tag{9}$$

which is just the conservation of probability equation used in the Bohm model.

Let us now work out the transport equation for $\overline{\overline{p}}$. This is obtained by multiplying equation (8) by P_k and integrating over P_k to give

$$\frac{\partial}{\partial t}(\rho\overline{\overline{P}}) + \sum_i \frac{\partial}{\partial X_i}\left(\overline{\overline{\rho P_k\frac{\partial H}{\partial X_i}}}\right) + \rho\overline{\overline{\frac{\partial H}{\partial X_k}}} = 0 \tag{10}$$

If we now substitute from equation (7) and use

$$\overline{\overline{P_i^2}} - \left(\overline{\overline{P}}_i\right)^2 = -\frac{\hbar^2}{4}\frac{\partial^2\rho}{\partial X_i^2} \tag{11}$$

we find

$$\frac{\partial}{\partial X_k}\left[\frac{\partial S}{\partial t} + \overline{\overline{H}} - \frac{\hbar^2}{8m\rho}\sum_i\frac{\partial^2\rho}{\partial X_i^2}\right] = 0 \tag{12}$$

If we write $\rho = R^2$, equation (12) is equivalent to

$$\frac{\partial S}{\partial t} + \frac{1}{2m}(\nabla S)^2 + V - \frac{\hbar^2}{2m}\nabla^2 R/R = 0 \tag{13}$$

which is just the quantum Hamilton-Jacobi used as a basis for the Bohm approach.[6,10] Using $\rho = R^2$ in equation (11) gives

$$\overline{\overline{P_iP_k}} - \overline{\overline{P_iP_k}} = \frac{\hbar^2}{2}\frac{\partial}{\partial X_k^2}\left[\frac{\partial^2 R}{\partial X_i^2}/R\right] \tag{14}$$

Thus we see that the quantum potential $Q = \frac{\partial^2 R}{\partial X_i^2}/R$ arises from the dispersion of the momentum when we use the position representation.

All of this gives a different insight into the the Bohm model showing that it is an intrinsic part of the quantum formalism even when we approach the formalism from the Moyal point of view.

4. The Star-Product and Its Properties

The algebra behind the Moyal structure is known as the Poisson deformation algebra (See Chari and Pressley[11]), an algebra based on new star-product defined by

$$A(x,p) \star B(x,p) := A(x,p) \exp\left[\frac{i\hbar}{2}\left(\frac{\overleftarrow{\partial}}{\partial X}\frac{\overrightarrow{\partial}}{\partial P} - \frac{\overleftarrow{\partial}}{\partial P}\frac{\overrightarrow{\partial}}{\partial X}\right)\right] B(x,p) \tag{15}$$

This Moyal product is non-commutative so that we can from two brackets that play a key role in the Wigner-Moyal approach. The first is the anti-symmetric Moyal bracket defined by

$$\{A(x,p), B(x,p)\}_{MB} = \frac{A \star B - B \star A}{i\hbar} \tag{16}$$

and the other is the symmetric Baker bracket[12]

$$\{A(x,p), B(x,p)\}_{BB} = \frac{A \star B + B \star A}{2} \tag{17}$$

These brackets are the analogues of the commutator and the anti-commutator, better known in mathematical circles as the Jordon product. We can write these brackets as

$$\{A, B\}_{MB} = A(x,p)\left[\frac{2}{\hbar}\sin\frac{\hbar}{2}\left(\frac{\overleftarrow{\partial}}{\partial x}\frac{\overrightarrow{\partial}}{\partial p} - \frac{\overleftarrow{\partial}}{\partial p}\frac{\overrightarrow{\partial}}{\partial x}\right)\right] B(x,p) \tag{18}$$

and

$$\{A, B\}_{BB} = A(x,p)\left[\cos\frac{\hbar}{2}\left(\frac{\overleftarrow{\partial}}{\partial x}\frac{\overrightarrow{\partial}}{\partial p} - \frac{\overleftarrow{\partial}}{\partial p}\frac{\overrightarrow{\partial}}{\partial x}\right)\right] B(x,p) \tag{19}$$

Expanding in powers of $\hbar$ to order $\hbar^2$ we find

$$\{A, B\}_{MB} = \{A, B\}_{PB} + O(\hbar^2) \approx \left[\frac{\partial A}{\partial x}\frac{\partial B}{\partial p} - \frac{\partial A}{\partial p}\frac{\partial B}{\partial x}\right] \tag{20}$$

where the bracket on the RHS is the Poisson bracket. Similarly expanding (17) we find

$$\{A, B\}_{BB} = AB + O(\hbar^2) \tag{21}$$

Thus we see that the Moyal approach has a very natural classical limit. There is no need to appeal to decoherence to reach the classical limit. Processes with high values of action naturally behave classically.

5. Time Dependent Equations in the Moyal Approach

Let us now examine the time evolution from a different point of view from that used by Moyal.[2] We take as our starting point the work of Fairlie and Manogue[13] and write down the analogues to the Schrödinger equation and its complex conjugate. These are

$$H \star f = \frac{i}{2\pi}\int d\theta e^{-\frac{ip\theta}{\hbar}}\left[\psi^*(x-\hbar\theta/2,t)\frac{\partial\psi(x+\hbar\theta/2,t)}{\partial t}\right] \tag{22}$$

and

$$f \star H = \frac{i}{2\pi}\int d\theta e^{-\frac{i\theta p}{\hbar}}\left[\frac{\partial\psi^*(x-\hbar\theta/2,t)}{\partial t}\psi(x+\hbar\theta/2,t)\right] \tag{23}$$

If we take the difference between these two equations we find

$$\frac{\partial f}{\partial t} + \{f, H\}_{MB} = 0 \tag{24}$$

which is just equation (8). In the limit of $O(\hbar^2)$, this equation becomes the usual classical Liouville equation

$$\frac{\partial f}{\partial t} + \{f, H\}_{PB} = 0 \tag{25}$$

where we find the Poisson bracket has replaced the Moyal bracket.

We can now explore the result of adding the two equations. We now obtain the equation

$$\begin{aligned}\{H, f\}_{BB} = H \star f + f \star H = \\ \frac{i}{2\pi}\int d\theta e^{-\frac{i\eta p}{\hbar}}\left[\psi^*(x-\hbar\theta/2,t)\frac{\partial\psi(x+\hbar\theta/2,t)}{\partial t}\right] \\ -\frac{i}{2\pi}\int d\theta e^{-\frac{i\theta p}{\hbar}}\left[-\frac{\partial\psi^*(x-\hbar\theta/2,t)}{\partial t}\psi(x+\hbar\theta/2,t)\right]\end{aligned} \tag{26}$$

If we try to simplify by writing $\psi = Re^{iS/\hbar}$ we find the RHS of (26) can be written as

$$\begin{aligned}&\left[\psi^*\frac{\partial\psi}{\partial t} - \frac{\partial\psi^*}{\partial t}\psi\right] \\ &= \left[\frac{1}{R(x+\hbar\theta/2)}\frac{\partial R(x+\hbar\theta/2)}{\partial t} - \frac{1}{R(x-\hbar\theta/2)}\frac{\partial R(x-\hbar\theta/2)}{\partial t}\right]\psi^*\psi \\ &+\frac{i}{\hbar}\left[\frac{1}{S(x+\hbar\theta/2)}\frac{\partial S(x+\hbar\theta/2)}{\partial t} - \frac{1}{S(x-\hbar\theta/2)}\frac{\partial S(x-\hbar\theta/2)}{\partial t}\right]\psi^*\psi\end{aligned} \tag{27}$$

This still looks a very messy result but if we go to the limit $O(\hbar^2)$, we find

$$\left[\psi^*\frac{\partial\psi}{\partial t} - \frac{\partial\psi^*}{\partial t}\psi\right] \to -\frac{\partial S}{\partial t}f + O(\hbar^2) = 0 \tag{28}$$

and since the Baker bracket becomes the product, we have

$$\frac{\partial S}{\partial t} + H = 0 \tag{29}$$

which is just the classical Hamilton-Jacobi equation. Thus we see that the Poisson deformation algebra contains both the classical Liouville equation and the classical Hamilton-Jacobi equations as limiting equations of the Schrödinger equation and its dual.

6. The Operator Formalism

If we examine the quantum Hamilton-Jacobi equation (8) which forms one of the defining equations of the Bohm approach, we see that we obtain the classical limit by putting the quantum potential to zero. This suggests that there must be some relation between equation (8) and equation (26). To bring this connection out further let us recall that it is the operator equations that have a similar form to the classical equations. For example, Hamilton's equations of motion can be written in the form

$$\dot{x} = \{x, H\}_{PB} \qquad\qquad \dot{p} = \{p, H\}_{PB} \tag{30}$$

while the corresponding equation of motion in the Heisenberg picture are expressed in terms of the commutator, viz

$$i\hbar\dot{\hat{X}} = [\hat{X}, \hat{H}] \qquad\qquad i\hbar\dot{\hat{P}} = [\hat{P}, \hat{H}] \tag{31}$$

so it is necessary to look for operator equations which are of the same form as equations (24) and (26). These equations will be expressed in terms of the commutator and the anti-commutator. In fact these equations have been found in Brown and Hiley,[14] and Hiley.[3] These equations are

$$i\hbar\frac{\partial\hat{\rho}}{\partial t} + \left[\hat{\rho}, \hat{H}\right]_{-} = 0 \qquad\qquad \hat{\rho}\frac{\partial\hat{S}}{\partial t} + \frac{1}{2}\left[\hat{\rho}, \hat{H}\right]_{+} = 0 \tag{32}$$

Here $\hat{\rho}$ is the density operator and $\hat{S}$ is the phase operator. To obtain these equations the wave function have been replaced by an element of the left (right) ideal so that the wave function is replaced by a 'wave operator'[a].

The first equation in (32) is just the quantum Liouville equation giving rise to the conservation of probability equation. The second equation is the operator form of the equation involving the Baker bracket (17). It should be noticed that this equation should be the operator analogue of of the quantum Hamilton-Jacobi equation (13) but notice there is no explicit expression for the quantum potential. The quantum arises when we project the equation into a sub-space.

Suppose we chose the projection operator $P_x = |x\rangle\langle x|$ then the first equation, the quantum Liouville equation, becomes

$$\frac{\partial P}{\partial t} + \nabla\left(P\frac{\nabla S}{m}\right) = 0 \tag{33}$$

---

[a]This wave operator is expressed in terms of a polar decomposition[3]

which is just the conservation of probability equation (9). The second equation becomes

$$\frac{\partial S}{\partial t} + \frac{1}{2m}(\nabla S)^2 + V - \frac{\hbar^2}{2mR}\nabla^2 R \tag{34}$$

This is just the quantum Hamilton-Jacobi equation (13) where the quantum potential appears explicitly.

In section 3 we showed that these equations were obtained from a marginal distribution found by averaging over the momentum P. In the operator approach they are obtained by projecting into the x-representation. We could equally have taken the marginal distribution by integrating over the position X. It can be shown that this is equivalent to a projection into the p-representation. In this case we get a corresponding formalism in p-space including a quantum Hamilton-Jacobi equation with a corresponding quantum potential. Thus we have complete $x - p$ symmetry in the Bohm model, removing one of Heisenberg's original objections[15] to the model, namely, the apparent lack of this symmetry in Bohm's original approach.[10]

Thus we have a complete correspondence between the equations of motion derived from the Poisson deformation algebra and those obtained from the operator algebra. Furthermore we see that there is an additional equation to the Liouville equation which appears to have received scant attention[13]

7. Conclusion

In this paper we have brought out some of the deeper connections between the Wigner-Moyal approach, the Bohm approach and the Heisenberg-type algebra of operators. These approaches are, in fact, different aspects of the same mathematical structure, each emphasising a different features of this structure.

The Moyal algebra (Poisson deformation algebra) constructs a $X - P$ phase space where X and P are mean co-ordinates of a pair of points in classical phase space. In this sense it is essentially a bilocal model. This fits comfortably with the notion of quantum cells or 'quantum blobs' in phase space suggested by several authors including de Gosson.[4] The Bohm model appears in the Moyal algebra as a result of forming a marginal distribution which produces a mean momentum, $\overline{\overline{p}}$, which can be identified with the Bohm momentum. This marginal distribution produces what looks like a classical phase space provided it is supplemented with the quantum potential.

We also show how a purely algebraic approach, exploiting the algebra of operators, produces dynamical equations of motion in terms of operators which have the same form as those produced in the Moyal algebra. In this operator algebra approach, the Bohm model arises from a projection into an algebraic sub-space, which seem to be playing a role similar to the marginal distributions in the Moyal theory.

Finally I want to suggest that we can perhaps make more sense of all this if we adopt what I call the Gel'fand approach. Here rather than starting with an *a priori* given phase space manifold, we use the algebra of operators to abstract the underlying manifold. This works beautifully when the algebra is commutative but not when in is non-commutative. In this case there is no unique underlying manifold, but we are forced to introduce the concept of "shadow manifolds". This seems to be exactly what is happening when we project the algebraic form of the dynamical equations into a sub-algebra, the choice of the projection determining which shadow manifold is being chosen. All this supports the notion that the geometry underlying quantum mechanics is a non-commutative geometry.[16] It fits neatly into the philosophical framework of the implicate/explicate order introduced by Bohm.[17]

References

1. E. Schrödinger, *Ann. der Physik*, **79**, 361 (1926). English Translation in *Collective Papers on Wave Mechanics* (Blackie and Son, London, 1928).
2. J. E. Moyal, *Quantum Mechanics as a Statistical Theory, Proc. Camb. Phil. Soc.* **45**, 99 (1949).
3. B. J. Hiley, *Phase Space Descriptions of Quantum Phenomena, Proc. Int. Conf. Quantum Theory: Reconsideration of Foundations*, **2**, 267-86 (ed. Khrennikov, A., Växjö University Press, Växjö, Sweden, 2003).
4. M. de Gosson, *Phys. Lett.* **A 330**, 161 (2004).
5. D. Bohm and B. J. Hiley, *Phys. Reports* **144**, 323 (1987).
6. D. Bohm and B. J. Hiley, *The Undivided Universe: an Ontological Interpretation of Quantum Theory*, (Routledge, London 1993).
7. J. Demaret, M. Heller and D. Lambert, *Foundations of Science* **2**, 137 (1997).
8. T. Takabayasi, *Prog. of Theor. Phys.* **11**, 341 (1954).
9. R. P. Feynman, in *Quantum Implications: Essays in Honour of David Bohm*, (Ed. Hiley, B. J. and Peat, F. D., pp. 235-248, Routledge & Kegan Paul, London 1987).
10. D. Bohm, *Phys. Rev.* **85**, 166 (1952); and *Phys. Rev.* **85**, 180 (1952).
11. V. Chari, and A. Pressley, *A Guide to Quantum Groups* (Cambridge, England, 1995).
12. Jn. G. A. Baker, *Phys. Rev.* **109**, 2198 (1958).
13. D. B. Farlie and C. A. Manogue, *J. Phys. A: Maths. Gen.* **24**, 3807 (1991).
14. M. R. Brown and B. J. Hiley, 2000, *Schrödinger revisited: an algebraic approach*, quant-ph/0005026.
15. W. Heisenberg, *Physics and Philosophy: the revolution in modern science*, (George Allen and Unwin, London, 1958).
16. A. Connes, *Noncommutative Geometry* (Academic Press, San Diego, 1990).
17. D. Bohm, *Wholeness and the Implicate Order* (Routledge, London, 1980).

QUANTUM MECHANICS AS SIMPLE ALGORITHM FOR APPROXIMATION OF CLASSICAL INTEGRALS

A. Yu. KHRENNIKOV

Center for Mathematical Modelling in Physics and Cognitive Sciences, University of Växjö, S-35195, Sweden
E-mail: Andrei.Khrennikov@msi.vxu.se

In previous publications we constructed a kind of theory with hidden variables – Pre-quantum Classical Statistical Field Theory (PCSFT). The role of hidden variables was played by classical fields. Since the corresponding phase space of classical fields has the infinite dimension, the model was quite complicated from the mathematical viewpoint. It was based on integration over the infinite dimensional Hilbert space. In this note we consider a finite dimensional illustration for PCSFT. Now phase space has a finite dimension, all integrals are usual Gaussian integrals. It becomes completely evident that quantum mechanics can be considered as just a simple algorithm for approximation of Gaussian integrals.

Keywords: Pre–quantum classical statistical field theory; Taylor expansion; Small parameter; Approximation; Gaussian integrals; Finite dimensional quantum mechanics.

1. Introduction

During last one hundred years people have been extremely curious why there exist two kinds of physical phenomena, so called classical and quantum, which are described by so different mathematical models, namely, by classical statistical mechanics[1] and quantum mechanics.[2] The first model is heuristically very natural. It does not induce the feeling of mysteries. One need not change philosophy and even logic. The second one is really mysterious (especially, for the Copenhagen or many worlds interpretations). It generated new "quantum philosophy"[3,4] and even new logic – quantum logic. Some people like mysteries, some like idealistic philosophy. For them quantum mechanics was the great domain of presentation of their views. For instance, N. Bohr and W. Heisenberg were under influence of idealistic philosophy long before starting quantum investigations. They were happy that quantum formalism provided so nice possibility to express their very special philosophic views. On the other hand, not everybody was lucky to share views of Bohr[5] and Heisenberg.[6] A rather large group of scientists rejected the Copenhagen interpretation. They were permanently looking for a better explanation of relation between classical and quantum mechanics. We can mention Einstein, De Broglie, Schrödinger, Lande, Bohm, Nelson, Accardi and at the present conference – Santos, Hiley, 't Hooft, Cetto, Cole, de la Peña, Hess, Nieuwenhuizen, Khrennikov and some others. During many years the strategy of the Copenhagen party was based on proving various "no-go"theorems. The first one was von Neumann's theorem.[2] Now days the most popular is Bell's theorem. There is no problem if one interpret the "no-go"-activity as an attempt to support idealistic philosophy by using mathematics. Unfortunately, mathematical "no-go" theorems are interpreted as kinds of physical statements about nature. In[7] we emphasized that the "no-go"-activity is more or less meaningless, because nobody knows how a possible coupling between classical and quantum models should look out. Every author (starting with von Neumann) first proposed his own picture of classical-quantum relation and then proved that his picture contradicted to the mathematical formalism of quantum mechanics. Each new generation of "no-go"-researchers started with sharp critique of rules for classical-quantum relation which were established by the previous generation. For instance, Bell strongly criticized assumptions which were done by von Neumann, Gleason, Jauch and Piron. After distraction of the previous pictures of classical-quantum correspondence, J. Bell created his own picture (which was also criticized by many authors, see, e.g.[8–16]). We can conclude that, in spite of all "no-go" theorems (those that have already been proved and those that will be proved in future), we can expect creation of prequantum classical-like models. On one hand, such models would demystify quantum mechanics. On the other hand, (and this is more important) they would give us the possibility to go beyond quantum mechanics and give new experimental predictions which would violate quantum mechanical laws. Of course, a possibility of violation of quantum laws is considered as nonsense by the majority of quantum community. Typically it is pointed out that these laws have been confirmed by many experiments during the last hundred years. As a counterargument, I can say that laws of Newtonian mechanics were confirmed by the huge number of experiments (in particular, in our common day life) and during three hundred years nobody could even imagine that these laws could be violated. I think that as any mathematical model quantum mechanics is an approximative model of reality and soon or later we shall approach a technological level giving the possibility to go beyond the quantum approximation.

Of course, we all know that quantum mechanics is a nonrelativistic approximation of quantum field theory (it is well known that relativistic quantum mechanics is not a well possed model). But we do not discuss at the moment relation relativistic/nonrelativistic. We discuss relation between quantum and classical probabilistic laws. We would like to say that there is only one randomness - the *classical ensemble randomness* and the quantum random behavior might be reduced to the classical one (in contrast to views of e.g. von Neumann[2]).

We are not saying that something is wrong in quantum mechanics by itself. It works well in its domain of application. The problem is that it pretends to be a *complete theory* and that it is forbidden go beyond it. Any theory should be quantized, i.e., to have its quantum version at the top-stage of its development. For example, huge population of physicists are looking for such a thing as quantum gravity. However, if one rejects pretensions on completeness of quantum mechanics, it becomes quite doubtful that quantum gravity (as a physical theory and not a mathematical model) should exist at all. If gravity by itself provides a deeper level of the mathematical description of nature than quantum mechanics, then there are no reasons to quantize gravity.

We can mention a few prequantum classical-like models: De Broglie double solution model,[19] Bohmian mechanics,[20] Bohm-Vigier stochastic mechanics,[20] Nelson's stochastic mechanics,[21] SED[22–,25] 't Hooft's deterministic prequantum mechanics.[26] We emphasize that all *stochastic models are of the random field type.*

In previous publications[7,17,18] we constructed a kind of theory with hidden variables – Prequantum Classical Statistical Field Theory (PCSFT). The role of hidden variables was played by *classical fields.* As well as mentioned models, PCSFT is a random field model. Since the corresponding phase space of classical fields has the infinite dimension, the model was quite complicated from the mathematical viewpoint. It was based on integration over the infinite dimensional Hilbert space with respect to Gaussian measures. In this note we consider a finite dimensional illustration for PCSFT. To simplify model even more, we start our considerations with a toy model of quantum mechanics over the reals and only then we proceed to complex (finite dimensional) quantum mechanics. Now phase space has a finite dimension, all integrals are usual Gaussian integrals. It becomes completely evident that quantum mechanics can be considered as just a simple algorithm for approximation of Gaussian integrals. Interpretational questions for PCSFT were discussed in[7,17,.18]

2. Finite Dimensional Quantum Mechanics

2.1. *Quantum mechanics over real numbers*

We shall use a toy model of quantum mechanics which based on the real space. Statistical features of the correspondence between a prequantum classical statistical model and quantum mechanics are more evident for this toy model. Denote the algebra of all $(m \times m)$ real matrices by the symbol $M^{(r)}(m)$. We denote by $\mathbf{D}^{(r)}(m)$ the class of nonegative symmetric trace-one matrices $\rho \in M^{(r)}(m)$. We call them "density operators." We denote by $L_s^{(r)}(m)$ the class of all symmetric matrices. In the quantum model (for the m-dimensional real space) statistical states (describing ensembles of systems prepared for measurement) are represented by density matrices and quantum observables by matrices belonging $L_s^{(r)}(m)$. The *quantum average* of an observable $A \in L_s^{(r)}(m)$ with respect to a statistical state $\rho \in \mathbf{D}^{(r)}(m)$ is given by the von Neumann trace class formula:[2]

$$< A >_\rho = \text{Tr } \rho \text{A}. \tag{1}$$

In the operator representation observables and density matrices are corresponding classes of $\mathbf{R}$-linear operators. Denote the quantum model by $N_{\text{quant}}^{(r)} = (\mathbf{D}^{(r)}(m), L_s^{(r)}(m))$.

If $m = 1$, then quantum observables are given by real numbers (operators of multiplication by real numbers on the real line) and there is only one statistical state $\rho = 1$. Here $< A >_\rho = \rho A = A$.

2.2. *Quantum mechanics over complex numbers*

Denote the algebra of all $(m \times m)$ complex matrices by the symbol $M^{(c)}(m)$. We denote by $\mathbf{D}^{(c)}(m)$ the class of nonegative symmetric trace-one matrices $\rho \in M^{(c)}(m)$. We call them "density operators." We denote by $L_s^{(c)}(m)$ the class of all symmetric matrices. In the quantum model (for the m-dimensional complex space) statistical states (describing ensembles of systems prepared for measurement) are represented by density matrices and quantum observables by matrices belonging $L_s^{(c)}(m)$. The *quantum average* is given[2] by (1). In the operator representation observables and density matrices are corresponding classes of $\mathbf{C}$-linear operators. Denote the quantum model by $N_{\text{quant}}^{(c)} = (\mathbf{D}^{(c)}(m), L_s^{(c)}(m))$.

If $m = 1$, then quantum observables are given by real numbers (operators of multiplication by real numbers on the complex plane) and and there is only one statistical state $\rho = 1$. Here $< A >_\rho = \rho A = A$.

3. Illustration for the von Neumann Trace–Class Formula in the One Dimensional Case

States of systems are represented by real numbers, $q \in Q = \mathbf{R}$. Ensembles of such systems are described by probability measures on the real line, *statistical states.* We consider a special class of preparation procedures. They produce ensembles of systems described by Gaussian probability distributions on Q having the zero mean value and dispersion

$$\sigma^2(\mu) = \alpha + O(\alpha^2), \tag{2}$$

where as always $|O(\alpha^2)| \leq C\alpha^2$ for some constant C and a sufficiently small α. The crucial point is that α is *a small parameter* of our model. Denote this class of probability distributions by the symbol $S_G^\alpha(Q)$.

For a probability $\mu \in S_G^\alpha(Q)$, we have: $d\mu(q) = \frac{e^{\frac{-q^2}{2(\alpha+O(\alpha^2))}}dq}{\sqrt{2\pi(\alpha+O(\alpha^2))}}$. We recall that, for a probability with the zero mean value, its dispersion is given by

$$\sigma^2(\mu) = \frac{1}{\sqrt{2\pi(\alpha + O(\alpha^2))}} \int_{-\infty}^{\infty} q^2 e^{\frac{-q^2}{2(\alpha+O(\alpha^2))}} dq. \tag{3}$$

As was already pointed out, we consider α as a small parameter. Therefore Gaussian probability distributions are very *sharply concentrated around the point* $q_0 = 0$. By using the terminology of functional analysis we say that $\{\mu \equiv \mu(\alpha)\}$ is a δ-family: $\lim_{\alpha \to 0} \mu(\alpha) = \delta$ in the sense of theory of distributions.

In the approximation $\alpha = 0$ all systems are located at a single point, namely, q_0. However, a finer description (in that α can not be neglected) provides the picture of Gaussian bells concentrated nearby q_0. We remark that in average a system cannot go far away from q_0. By using the Chebyshov inequality one obtain for any $C > 0$:

$\mu\{q : |q| > C\} \leq \frac{\alpha+O(\alpha^2)}{C^2} \to O, \alpha \to 0$. But the probabilistic inequality does not exclude the possibility that some system could move far from q_0 (of course, with a small probability).

We also introduce a class of *physical variables* in the classical statistical model under consideration: a) $f \in C^\infty(\mathbf{R})$, a smooth function; b) $f(0) = 0$; c) $|f^{(4)}(q)| \leq c_f e^{r_f|q|}, c_f, r_f \geq 0$. Denote this functional space by the symbol $\mathbf{V}(Q), Q = \mathbf{R}$. [a]

We defined the following classical statistical model on the real line: 1) *states* of systems are real numbers; 2) *statistical states* (ensembles of systems) are represented by Gaussian probabilities having zero average and dispersion $\sigma^2(\mu) = \alpha + O(\alpha^2), \ \alpha \to 0$; 3) *physical variables* are smooth functions with exponentially growing fourth derivative which map zero into itself. We denote this model by $N^\alpha_{\text{class}} = (S^\alpha_G(Q), \mathbf{V}(Q))$.

As always in classical statistical physics, the average of a physical variable $f \in \mathbf{V}(Q)$ with respect to an ensemble of systems which is described by a probability $\mu \in S^\alpha_G(Q)$ is given by the integral:

$$< f >_\mu = \frac{1}{\sqrt{2\pi(\alpha + O(\alpha^2))}} \int_{-\infty}^{\infty} f(q) e^{\frac{-q^2}{2(\alpha+O(\alpha^2))}} dq. \tag{4}$$

Since α is a parameter of the model, we can consider averages as functions of $\alpha :< f >_\mu \equiv < f >_\mu (\alpha)$. We are interested in the asymptotic expansion of averages when $\alpha \to 0$. In particular, such an asymptotic expansion will give us the possibility to calculate averages approximately. Let $f \in \mathbf{V}(Q)$ and let $\mu \in S^\alpha_G(Q)$. Then

$$< f >_\mu (\alpha) = \frac{\alpha}{2} f''(0) + O(\alpha^2). \tag{5}$$

To prove this asymptotic equality, we first make the scaling of the state variable:

$$q = \sigma(\mu) x \tag{6}$$

We have:

$$< f >_\mu (\alpha) = \frac{1}{\sqrt{2\pi}} \int_{-\infty}^{\infty} f(\sigma(\mu)x) e^{\frac{-x^2}{2}} dx. \tag{7}$$

We now expand $f(\sigma(\mu)x)$ by using the fourth order Taylor formula with the integral remainder,

$$< f >_\mu (\alpha) = \frac{\sigma^2(\mu)}{2} f''(0) \tag{8}$$

$$+ \frac{\sigma^4(\mu)}{4!\sqrt{2\pi}} \int_{-\infty}^{\infty} x^4 \Big(\int_0^1 (1-\theta)^3 f^{(4)}(\sigma(\mu)x\theta) d\theta \Big) e^{\frac{-x^2}{2}} dx.$$

[a] The restriction to the growth of the fourth derivative will be used when we shall consider the Taylor expansion of f up two the fourth term. The exponential growth implies integrability with respect to any Gaussian measure.

We recall that for a Gaussian measure with zero mean value all odd momenta are equal to zero. This is an important point of our considerations. This imply that the first nonzero contribution to the classical average is given by the second derivative – quadratic term. Disappearance of the third order term implies the asymptotics $O(\alpha^2)$. We now estimate the remainder to obtain this asymptotics:

$$|R(f,\mu)| \leq \frac{C\sigma^4(\mu)}{4!\sqrt{2\pi}} \int_{-\infty}^{\infty} x^4 \Big(\int_0^1 (1-\theta)^3 e^{r\sigma(\mu)|x|\theta} d\theta \Big) e^{\frac{-x^2}{2}} dx.$$

Since we consider α as a small parameter, we can assume that $|\sigma(\mu)| \leq 1$ in the exponential function. Thus: $|R(f,\mu)| \leq C'\sigma^4(\mu) \int_{-\infty}^{\infty} x^4 e^{r|x|-\frac{x^2}{2}} dx$. Since $\sigma^2(\mu) = \alpha + O(\alpha^2)$, we have that $R(f,\mu) = O(\alpha^2), \alpha \to 0$.

We consider the dispersion $\sigma^2(\mu)$ as the *intensity of fluctuations* in the ensemble of systems. We define the *relative average* with respect to the intensity of fluctuations by normalizing the average by the main term – namely, α – in the intensity of fluctuations: $\langle f \rangle_\mu = \frac{<f>_\mu}{\alpha}$. Then $\langle f \rangle_\mu = \frac{f''(0)}{2} + O(\alpha)$, and in particular, $\lim_{\alpha \to 0} \langle f \rangle_\mu(\alpha) = \frac{f''(0)}{2}$.

We have shown that $\frac{f''(0)}{2}$ gives the approximation of the (classical) relative average. The precision of such an approximation is α. If the level of development of measurement technology is such that all contributions of the magnitude α are neglected in measurements, then averages can be calculated by using the following simple rule:

$$\langle f \rangle_\mu^{\text{approx}} = \Big[\frac{< f >_\mu}{\sigma^2(\mu)} \Big]^{\text{approx}} = \frac{f''(0)}{2}. \tag{9}$$

At the first sight such averages have nothing to do with classical averages given by integrals. There could be even presented an interpretation of physics claiming that rules of classical probability theory are violated and relating the exotic rule (9) for calculating of averages to special features of systems under consideration (and not to a special approximation procedure for averages).

Finally, we remark that calculation of averages by (9) is essentially simpler than classical probabilistic averages given by Lebesgue integrals.

4. Finite Dimensional Quantum Mechanics Over Reals as an Approximative Theory

States are vectors $q \in Q = \mathbf{R}^m$; statistical states are Gaussian distributions with the zero mean value and dispersion $\sigma^2(\mu) = \alpha + O(\alpha^2)$. Denote this class of probabilities by the symbol $S_G^\alpha(Q)$. We introduce the scalar product and norm on $Q : (\xi, q) = \sum_{j=1}^m \xi_j q_j$ and $\|q\|^2 = \sum_{j=1}^m q_j^2$. If a Gaussian measure μ is nondegenerate (so the measure of any open set is positive), then

$$d\mu(q) = \frac{e^{-\frac{1}{2}(B^{-1}q,q)}\, dq}{\sqrt{(2\pi)^m \det B}},$$

where B is a positive operator (we consider everywhere only Gaussian measures with zero mean values). If $\mu \in S_G^\alpha(Q)$ and nondegenerate, then

$$\sigma^2(\mu) = \frac{1}{\sqrt{(2\pi)^m \det B}} \int_{\mathbf{R}^m} \|q\|^2 \, e^{-\frac{1}{2}(B^{-1}q,q)} \, dq = \alpha + O(\alpha^2).$$

In the general case the easiest way to define a Gaussian measure is to use its Fourier transform:

$$\tilde{\mu}(\xi) = \int_{\mathbf{R}^m} e^{i(\xi,q)} \, d\mu(q) = e^{-\frac{1}{2}(Bq,q)},$$

where $B = \text{cov}\ \mu$ is the *covariation operator:*

$$(B\xi_1, \xi_2) = \int_{\mathbf{R}^m} (\xi_1, q) \, (\xi_2, q) \, d\mu(q).$$

We remark that by definition a covariation operator is *positively defined and symmetric.* Let μ be a Gaussian measure with the zero mean value and let A be a symmetric operator. Then

$$\int_{\mathbf{R}^m} (Aq, q) \, d\mu(q) = \text{Tr BA},$$

where $B = \text{cov}\ \mu$. To obtain this formula, we just expand the quadratic form (Aq, q) with respect to an orthonormal basis. In particular, we have (by choosing $A = I$) that the dispersion can be represented in the trace-form:

$$\sigma^2(\mu) = \int_{\mathbf{R}^m} \|q\|^2 \, d\mu(q) = \text{Tr B}.$$

Thus, for $\mu \in S_G^\alpha(Q)$, Tr cov $\mu = \alpha + \text{O}(\alpha^2)$. We now define a class of physical variables – $\mathbf{V}(Q)$: a) $f \in C^\infty(\mathbf{R}^m)$; b) $f(0) = 0$; c) $\|f^{(4)}(q)\| \leq c_f e^{r_f \|q\|}, c_f, r_f \geq 0$. Thus we have defined the following classical statistical model:

$$N_{\text{class}} = (S_G^\alpha(Q), \mathbf{V}(Q)).$$

As in the one dimensional case by using scaling and the Taylor formula, we obtain:

$$< f >_\mu (\alpha) = \frac{\sigma^2(\mu)}{2} \text{Tr}\ \rho \text{f}''(0) \tag{10}$$

$$+ \frac{\sigma^4(\mu)}{4!} \int_{\mathbf{R}^m} \Big(\int_0^1 (1-\theta)^3 \, f^{(4)}(\sigma(\mu)x\theta)(q,q,q,q) \, d\theta \Big) \, d\mu_{\text{scal}}(x),$$

where μ_{scal} is a normalized Gaussian measure – the image of μ under the scaling (6). This implies:

$$< f >_\mu (\alpha) \equiv \int_{\mathbf{R}^m} f(q) \, d\mu(q) = \frac{\alpha}{2} \text{Tr}\ \rho \text{f}''(0) + \text{O}(\alpha^2), \tag{11}$$

where ρ is a density operator; in fact, $\rho = \text{cov}\ \mu_{\text{scal}}$.

As in the one-dimensional case, we introduce the relative average:

$$\langle f \rangle_\mu \equiv \frac{< f >_\mu}{\alpha}.$$

Then

$$\langle f \rangle_\mu = \frac{1}{2} \text{Tr } \rho \text{ f}''(0) + \text{O}(\alpha). \tag{12}$$

Thus if one neglects by terms of the magnitude α, it is possible to use the following approximative calculus of averages: $\langle f \rangle_\mu^{\text{approx}} = \frac{1}{2} \text{Tr } \rho \text{A}$, where $A = f''(0)$ and $\rho = \text{cov } \mu_{\text{scal}}$. This is nothing else than the von Neumann trace formula for quantum averages, see (1). To proceed more formally, we consider maps:

$$T : S_G^\alpha(Q) \to \mathbf{D}^{(r)}(m), \rho = T(\mu) = \text{cov } \mu_{\text{scal}}; \tag{13}$$

$$T : \mathbf{V}(Q)) \to L_s^{(r)}(m)), A = T(f) = f''(0) \tag{14}$$

(we recall that Hessian is always a symmetric matrix). The maps (13), (14) project the classical statistical model $N_{\text{class}} = (S_G^\alpha(Q), \mathbf{V}(Q))$ onto the quantum model $N_{\text{quant}}^{(r)} = (\mathbf{D}^{(r)}(m), L_s^{(r)}(m))$ in such a way that classical and quantum averages are coupled by the asymptotic equality:

$$\langle f \rangle_\mu = \frac{1}{2} < T(f) >_{T(\mu)} + O(\alpha). \tag{15}$$

5. Prequantum Phase Space – the Two Dimensional Case

In previous sections we considered the prequantum toy model in that the phase space structure was not taken into account. The corresponding quantum model was over the reals. On the other hand, physical reality is described by the classical phase space mechanics and the complex quantum mechanics. We shall see that it is possible to create a prequantum phase space model reproducing the complex quantum mechanics. The crucial point is that classical variables and statistical states – functions and measures on phase space – should be *invariant with respect to a special group of transformations of phase space.*

This *fundamental prequantum group* is very simple – the *special orthogonal group* $SO(2)$, the group of rotations of phase space.

States of systems are now represented by points $\psi = (q, p) \in \Omega = Q \times P$, where $Q = P = \mathbf{R}$. Here the q is the position and the p is momentum, so Ω denotes phase space. Statistical states are represented by Gaussian $SO(2)$-invariant measures having zero mean value and dispersion

$$\sigma^2(\mu) = 2\alpha + O(\alpha^2); \tag{16}$$

physical variables are by $SO(2)$-invariant maps, $f : \Omega \to \mathbf{R}$, which satisfy conditions a), b), c) specifying variables in the real case. Denote these classes of measures and functions, respectively, $S_G^\alpha(\Omega|SO(2))$ and $\mathbf{V}(\Omega|SO(2))$.

The appearance of the factor 2 has the following motivation: there are two contributions into fluctuations – fluctuations of positions and momenta. We shall see that they are equally distributed. Therefore it is natural to consider as a small parameter of the model the dispersion of e.g. the q-fluctuations (which equals to the dispersion of the p-fluctuations).

We consider the classical model $N_{\rm class} = (S_G^\alpha(\Omega|SO(2)), \mathbf{V}(\Omega|SO(2))$. As in the real case, we can obtain the asymptotic expansion of the classical averages, see (12). However, in quantum mechanics we consider the complex structure. We would like to recover it in our classical model. To do this, we shall study in more detail properties of classical probabilities and variables.

A measure μ is invariant if for any $u \in SO(2) : \int_{\mathbf{R}^2} f(uq)d\mu(q) = \int_{\mathbf{R}^2} f(q)d\mu(q)$. For a Gaussian measure μ with the covariation matrix B, this is equivalent to the condition: $[u, B] = 0, \ u \in SO(2)$.

Let f be a two times differentiable invariant map, so $f(u\psi) = f(\psi)$, for any $u \in SO(2)$. By representing $u = u_\theta = \begin{pmatrix} \cos\theta & -\sin\theta \\ \sin\theta & \cos\theta \end{pmatrix}$, we have that $f(\cos\theta q - \sin\theta p, \sin\theta q + \cos\theta p) = f(q,p)$. This is a rather strong constraint determining a very special class of maps. In particular, we obtain: $u^*\nabla f(u\psi) = \nabla f(\psi)$ and $u^* f''(u\psi)u = f''(\psi)$. Hence $u^*\nabla f(0) = \nabla f(0)$ for any rotation, and thus $\nabla f(0) = 0$ and $[f''(0), u] = 0, \ u \in SO(2)$.

It is convenient to introduce the *commutator* of the set $SO(2)$ in the algebra of all two by two matrices $M^{(r)}(2)$:

$$SO'(2) = \{A \in M^{(r)}(2) : [A, u] = 0, u \in SO(2)\}$$

We remark that a generator of $SO(2)$ can be chosen as the symplectic operator: $J = \begin{pmatrix} 0 & 1 \\ -1 & 0 \end{pmatrix}$. Therefore the commutator of $SO'(2)$ coincides with the commutator of $J : \{J\}' = \{A \in M^{(r)}(2) : [A, J] = 0\}$.

Let $\mu \in S_G^\alpha(\Omega|SO(2))$and let $f \in \mathbf{V}(\Omega|SO(2))$. Then $B = \text{cov}\ \mu$ and $A = f''(0)$ belong to $SO'(2)$.

We note that a matrix A belongs to the commutator $SO'(2)$ iff $A = \begin{pmatrix} R & -S \\ S & R \end{pmatrix}$. If A is also symmetric, then it is diagonal: $A = \begin{pmatrix} R & 0 \\ 0 & R \end{pmatrix}$. In particular, its trace is given by TrA = 2R.

Thus if $\mu \in S_G^\alpha(\Omega|SO(2))$, then its covariation matrix is diagonal $B = \begin{pmatrix} b & 0 \\ 0 & b \end{pmatrix}$, where $2b = \alpha + O(\alpha^2)$. Fluctuations of the coordinate q and the momentum p are independent and equally distributed: $d\mu(q) = \frac{1}{2\pi b} \exp\{-\frac{q^2+p^2}{2b}\}dq$. Denote the marginal distributions of μ by the symbols μ_q and μ_p, respectively. Then

$$\sigma^2(\mu_q) = \frac{1}{\sqrt{2\pi b}} \int_{-\infty}^{+\infty} q^2 \, e^{-\frac{q^2}{2b}} dq = \sigma^2(\mu_p) = \frac{1}{\sqrt{2\pi b}} \int_{-\infty}^{+\infty} p^2 \, e^{-\frac{p^2}{2b}} dp.$$

Hence $\sigma^2(\mu_q) = \sigma^2(\mu_p) = \frac{1}{2}\sigma^2(\mu) = \alpha + O(\alpha^2)$.

We are now completely ready to recover the complex structure of quantum mechanics. Any matrix belonging the commutator $SO'(2)$ can be represented in the form: $A = R\, I\ +\ S\ (-J)$. By mapping I into 1 and $(-J)$ into i we obtain a map of the commutator $SO'(2)$ onto the set of complex numbers $\mathbf{C}$:

$$j : SO'(2) \to \mathbf{C}, z = j(A) = R + iS. \tag{17}$$

This is the isomorphism of two fields. In particular, a symmetric matrix $A = \begin{pmatrix} R & 0 \\ 0 & R \end{pmatrix}$ is represented by the real number $j(A) = R$. This is the operator of multiplication by R. The trace of this operator in the one dimensional complex space $\mathbf{C}$ (with the scalar product, $(z, w) = z\bar{w}$) equals R. We have:

$$\text{Tr A} = 2\text{Tr j(A)}, \tag{18}$$

where at the left-hand side we have the real trace and at the right-hand side – the complex trace. Now we can write the basic asymptotic equality for averages in the complex form. In the funny way the Taylor factor $\frac{1}{2}$ disappears through the transition from the real to complex structure, see (18).

Let $f \in \mathbf{V}(\Omega|SO(2))$ and let $\mu \in S_G^\alpha(\Omega|SO(2))$. Then

$$< f >_\mu (\alpha) \equiv \int_{\mathbf{R}^2} f(q,p)\, d\mu(q,p) = \alpha\, j(f''(0)) + O(\alpha^2). \tag{19}$$

To prove this asymptotic equality, we make the scaling of the state variable:

$$\psi = \frac{\sigma(\mu)}{\sqrt{2}} \Psi \tag{20}$$

Then the image of μ is again a Gaussian measure, say μ_{scal}, having the dispersion $\sigma^2(\mu_{\text{scal}}) = 2$. Set $D = \text{cov}\ \mu_{\text{scal}}$. In the two dimensional case $D = \begin{pmatrix} 1 & 0 \\ 0 & 1 \end{pmatrix}$ and $\text{Tr D} = 2$. We now have: $< f >_\mu (\alpha) = \frac{\sigma^2(\mu)}{4} \text{Tr D f}''(0) + \text{O}(\alpha^2)$. Thus $< f >_\mu (\alpha) = \frac{\sigma^2(\mu)}{2}\, j(D)\, j(f''(0)) + O(\alpha^2)$. Finally, we note that in the two dimensional case: $j(D) = 1$. Thus we obtain: $< f >_\mu (\alpha) = \frac{\sigma^2(\mu)}{2}\, j(f''(0)) + O(\alpha^2)$, and hence (19).

We recall that in the one dimensional quantum mechanics there is just one "density matrix", namely, $\rho = 1 \in \mathbf{R}$. It is convenient to consider the renormalization of averages by the main term in the intensities of fluctuations of the coordinate and momenta: $\langle f \rangle_\mu = \frac{<f>_\mu}{\alpha}$. Then we get: $\langle f \rangle_\mu(\alpha) = j(f''(0)) + O(\alpha)$.

6. Prequantum Phase Space for Finite Dimensional Quantum Mechanics

States of systems are now represented by points $\psi = (q,p) \in \Omega = Q \times P$, where $Q = P = \mathbf{R}^m$. Here the $q = (q_1, ..., q_n)$ is the position and the $p = (p_1, ..., p_n)$ is momentum, so Ω denotes phase space. Let us consider the canonical representation of the group $SO(2)$ in the phase space $\Omega = Q \times P : u = u_\theta = \begin{pmatrix} \cos\theta\, I & -\sin\theta\, I \\ \sin\theta\, I & \cos\theta\, I \end{pmatrix}$, where I is the unit matrix from $M^{(r)}(m)$. The corresponding group of $\mathbf{R}$-linear operators (or $2m \times 2m$ matrices) we denote by the symbol $SO_m(2)$.

The classical model $N_{\text{class}} = (S_G^\alpha(\Omega|SO_m(2))$ and $\mathbf{V}(\Omega|SO_m(2))$ in defined in the same way as in the two dimensional case. A Gaussian measure is invariant iff its covariation operator belongs to the commutator $SO'_m(2) = \{A \in M^{(r)}(2m) : [A, u] = 0, u \in SO_m(2)\}$. If a smooth function f is invariant then all its odd derivatives equal to zero and the second derivative belong to the $SO'_m(2)$. A matrix $A \in SO'_m(2)$ if it has the same block structure as was considered in section 5: here $R, S \in M^{(r)}(m)$. In contrast to the two dimensional case a symmetric matrix from $SO'_m(2)$ can be nondiagonal: here $R^* = R$ and $S^* = -S$.

There is a natural map (generalizing the map $j : SO'(2) \to \mathbf{C}$) of the commutator $SO'_m(2)$ onto the set of complex matrices $M^{(c)}(m)$:

$$j : SO'_m(2) \to M^{(c)}(m), z = j(A) = R + iS. \tag{21}$$

This is the isomorphism of two rings. Symmetric matrices are mapped onto symmetric matrices. Let us denote real and complex conjugations by $*$ and $\star$, respectively. We have $(R + iS)^\star = R^* - iS^* = R + iS$.

Let $f \in \mathbf{V}(\Omega|SO_m(2))$ and let $\mu \in S_G^\alpha(\Omega|SO_m(2))$. Then

$$< f >_\mu (\alpha) = \alpha \ \mathrm{Tr}\rho\, \mathrm{j}(\mathrm{f}''(0)) + \mathrm{O}(\alpha^2), \tag{22}$$

where $\rho \in \mathbf{D}^{(c)}(m)$. We now modify the classical$\to$ quantum projections, (13), (14), to make them consistent with the complex structure:

$$T : S_G^\alpha(\Omega) \to \mathbf{D}^{(c)}(m), \rho = T(\mu) = j(\text{cov } \mu_{\text{scal}}) \tag{23}$$

(here the measure μ_{scal} is the image of μ under scaling (20));

$$T : \mathbf{V}(Q)) \to L_s^{(c)}(m)), A = T(f) = j(f'')(0) \tag{24}$$

The maps (23), (24) project the classical statistical model $N_{\text{class}} = (S_G^\alpha(\Omega|SO_m(2))$ and $\mathbf{V}(\Omega|SO_m(2))$ onto the quantum model $N_{\text{quant}}^{(c)} = (\mathbf{D}^{(c)}(m), L_s^{(c)}(m))$ in such a way that classical and quantum averages are coupled by the asymptotic equality: $\langle f \rangle_\mu =< T(f) >_{T(\mu)} +O(\alpha)$.

Conclusion. *Quantum mechanics with finite dimensional state space can be considered as a Taylor approximation of classical statistical mechanics on phase space.*

References

1. L. Boltzman, *Lectures on gas theory* (Dover Publications, Inc., New York, 1995).
2. J. von Neumann, *Mathematical foundations of quantum mechanics* (Princeton Univ. Press, Princeton, N.J., 1955).

3. A. Plotnitsky, Reading Bohr: Complementarity, Epistemology, Entanglement, and Decoherence, in *NATO Workshop Decoherence and its Implications for Quantum Computations*, A. Gonis and P. Turchi,Eds. (IOS Press, Amsterdam, 2001), pp.3–37.
4. A. Plotnitsky, *The knowable and unknowable (Modern science, nonclassical thought, and the "two cultures"* (Univ. Michigan Press, 2002).
5. N. Bohr, *The philosophical writings of Niels Bohr*, 3 vols., (Woodbridge, Conn., Ox Bow Press, 1987).
6. W. Heisenberg, *Physics and philosophy (the revolution in modern science)* (Prometheus Books, New York, 1999).
7. A. Yu. Khrennikov, A pre-quantum classical statistical model with infinite-dimensional phase space. *J. Phys. A: Math. Gen.*, **38**, 9051 (2005).
8. W. M. De Muynck, Interpretations of quantum mechanics, and interpretations of violations of Bell's inequality. In *Foundations of Probability and Physics*, ed. by A. Khrennikov, *Q. Prob. White Noise Anal.* **13**, 95 (2001).
9. W. M. De Muynck, *Foundations of quantum mechanics, an empiricists approach* (Kluwer, Dordrecht, 2002).
10. W. De Baere, *Lett. Nuovo Cimento* **39**, 234 (1984); **40**, 488(1984); *Advances in electronics and electron physics* **68**, 245 (1986).
11. E. Santos, *Found. Phys.* **21**, 221 (1991).
12. E. Santos, *Phys. Rev. Lett.* **66**, 1388 (1991).
13. A. Yu. Khrennikov, *Interpretations of Probability* (VSP Int. Sc. Publishers, Utrecht/Tokyo, 1999; second edition, 2004).
14. L. Accardi, Could one now convince Einstein? In *Quantum theory: reconsideration of foundations—3*, ed. by A. Khrennikov, AIP Conf. Proc. Ser., **810** (Amer. Inst. Phys., Melville, NY, 2006), pp. 3-18.
15. G. Adenier and A. Yu. Khrennikov, Anomalies in the EPR-Bell experiments. In *Quantum theory: reconsideration of foundations—3*, ed. by A. Khrennikov, AIP Conf. Proc. Ser., **810** (Amer. Inst. Phys., Melville, NY, 2006), pp. 283-293.
16. K. Hess and W. Philipp, *Proc. Nat. Acad. Sc.* **98**, 14224 (2001); **98**, 14227(2001); **101**, 1799 (2004); *Europhys. Lett.* **57**, 775 (2002).
17. A. Yu. Khrennikov, *Doklady Mathematics,* **73**, 445 (2006)
18. A. Yu. Khrennikov, *Physics Letters A*, **357**, 171 (2006).
19. L. De Broglie, *The current interpretation of wave mechanics, critical study* (Elsevier Publ., Amsterdam-London-New York, 1964).
20. D. Bohm and B. Hiley, *The undivided universe: an ontological interpretation of quantum mechanics,* Routledge and Kegan Paul, London, 1993.
21. E. Nelson, *Quantum fluctuation* (Princeton Univ. Press, Princeton, 1985).
22. L. de la Pena and A. M. Cetto, *The Quantum Dice: An Introduction to Stochastic Electrodynamics Kluwer* (Dordrecht, 1996).
23. L. de la Pena and A. M. Cetto, Recent development in linear stochastic electrodynamics. In *Quantum theory: reconsideration of foundations—3*, ed. by A. Khrennikov, AIP Conf. Proc. Ser., **810** (Amer. Inst. Phys., Melville, NY, 2006), pp. 131-140.
24. Th. M. Nieuwenhuizen, Classical phase space density for relativistic electron. In *Quantum theory: reconsideration of foundations—3*, ed. by A. Khrennikov, AIP Conf. Proc. Ser., **810** (Amer. Inst. Phys., Melville, NY, 2006), pp. 198-210.
25. D. Cole, Simulation results related to stochastic electrodynamics. In *Quantum theory: reconsideration of foundations—3*, ed. by A. Khrennikov, AIP Conf. Proc. Ser., **810** (Amer. Inst. Phys., Melville, NY, 2006), pp. 99-113.
26. G. 't Hooft, quant-ph/0212095; hep-th/0105105; quant-ph/0604008.

NONCOMMUTATIVE QUANTUM MECHANICS VIEWED FROM FEYNMAN FORMALISM

J. LAGES

Institut UTINAM, équipe Dynamique des Structures Complexes, UMR du CNRS 6213, Université de Franche-Comté, Campus La Bouloie, 25030 Besançon Cedex, France
E-mail: jose.lages@utinam.cnrs.fr

A. BERARD, H. MOHRBACH and Y. GRANDATI

Laboratoire de Physique Moléculaire et des Collisions, ICPMB, IF CNRS N° 2843, Université Paul Verlaine, Institut de Physique, Bd Arago, 57078 Metz Cedex 3, France

P. GOSSELIN

Institut Fourier, Université de Grenoble I, UMR 5582 CNRS-UJF, UFR de Mathématiques, BP74, 38402 St Martin d'Hères, Cedex France

Dyson published in 1990 a proof due to Feynman of the Maxwell equations. This proof is based on the assumption of simple commutation relations between position and velocity. We first study a nonrelativistic particle using Feynman formalism. We show that Poincaré's magnetic angular momentum and Dirac magnetic monopole are the direct consequences of the structure of the SO(3) Lie algebra in Feynman formalism. Then we show how to extend this formalism to the dual momentum space with the aim of introducing Non-commutative quantum mechanics which was recently the subject of a wide range of works from particle physics to condensed matter physics.

Keywords: Non-commutative quantum mechanics; Feynman formalism; Dirac magnetic monopole; Poincaré momentum; Maxwell's equations; Angular algebra symmetry; Dirac quantization; Berry curvature.

1. Introduction

Feynman's ideas[1] were exposed by Dyson in an elegant publication. Initially Feynman's motivation was to develop a quantization procedure without resort to a Lagrangian or a Hamiltonian. Assuming minimal commutation relations between position and velocity and using Newton's second law, Feynman derived the source-less set of Maxwell's equations which are galileo invariant. The interpretation of the Feynman's derivation of the Maxwell's equations has aroused[2–11] a great interest among physicists. In particular Tanimura[2] has generalized the Feynman's derivation in a Lorentz covariant form with a scalar time evolution parameter. An extension of the Tanimura's approach has been achieved[3] in using the Hodge duality to derive the two groups of Maxwell's equations with a magnetic monopole in a flat and in a curved spaces. In Ref. 4 the descriptions of relativistic and non relativistic particles in an electromagnetic field was studied, whereas in Ref. 5 a dynamical equation for spinning particles was proposed. A rigorous mathematical interpretation of Feynman's derivation connected to the inverse problem for the Poisson dynamic has been formulated in Ref. 6. Also in Refs. 7 and 8 the Feynman's derivation is considered in the frame of the Helmholtz's inverse problem for the calculus of variations. Other works[9–11] have provided new looks on the Feynman's derivation of the Maxwell's equations. More recently,[12] some of the authors embedded Feynman's derivation of the Maxwell's equation in the framework of noncommutative geometry. As Feynman's brackets can be interpreted as a deformation of Poisson brackets we showed that the Feynman brackets can be viewed as a generalization of the Moyal brackets defined over the tangent bundle space.[12]

The mathematical foundations of Feynman's formalism is presented in Section 2 and is used to review the Feynman's derivation of the Maxwell's equation in Section 3. It is well known that velocities do not commute in the presence of an electromagnetic field. For this reason the angular algebra symmetry, *e.g.* the SO(3) symmetry in the Euclidean case, is broken. In Section 4 we show how to restore such a symmetry and we point out in this context the necessity of adding the Poincaré momentum $\mathbf{M}$ to the simple angular momentum $\mathbf{L}$. The direct consequence of this restoration is then the generation of a Dirac magnetic monopole. A natural extension of Feynman's formalism is to consider the dual momentum space. In Section 5 we embed then our work in the natural generalization of quantum mechanics involving noncommutative coordinates. This generalization was originally introduced by Snyder[13] as a short distance regularization to improve the problem of infinite self energies inherent in a Quantum Field Theory. Due to the advent of the renormalization theory this idea was not very popular[a] until Connes[14] analyzed Yang Mills theories on noncommutative space. More recently a correspondence between a noncommutative gauge theory and a conventional gauge theory was introduced by Seiberg and Witten.[15] Non-commutative gauge theories were also found as being naturally related to string and M-theory.[16] Applications of noncommutative theories were also found in condensed matter physics, for instance in the Quantum Hall effect[17] and in the noncommutative Landau problem,[18,19] the name of noncommutative quantum mechanics started then to be used.[19–21] In section 5 we resume our paper[22] which shows that in our model a quantum particle in a harmonic potential has a behavior similar to a particle in a constant magnetic field θ in standard quantum mechanics, since a paramagnetic term appears in the Hamiltonian. Moreover, the particle acquires an effective dual mass in the same way as an electron moving in a periodic potential in solid state physics. Again, the angular algebra symmetry is naturally broken and the restoration of this symmetry gives then a dual Dirac monopole in momentum space field configuration.

[a]See also contribution of van Huele in this volume.

2. Mathematical Foundations of Feynman's Formalism

Let a particle with a mass m and an electrical charge q be described by the vector $\mathbf{x} = \{x^i\}_{i=1,\dots,N}$ which defines its position on the manifold $\mathcal{M}$. Let the manifold $\mathcal{M}$ be a N-dimensional vectorial manifold diffeomorphic to $\mathbb{R}^N$. Let τ be the parameter of the group of diffeomorphisms $\mathcal{G} : \mathbb{R} \times \mathcal{M} \to \mathcal{M}$ such as $\mathcal{G}(\tau, \mathbf{x}) = \mathcal{G}^\tau \mathbf{x} = \mathbf{x}(\tau)$. Then taking τ as the time parameter of our physical system, we are able to define a velocity vector $\dot{\mathbf{x}} \in \mathcal{M}$ as $\dot{\mathbf{x}} = \dfrac{d\mathbf{x}}{d\tau} = \mathcal{G}^\tau \mathbf{x} = \{\dot{x}^i(\tau)\}_{i=1,\dots,N}$. Let $T(\mathcal{M})$ be the tangent bundle space associated with the manifold $\mathcal{M}$, a point on $T(\mathcal{M})$ is described then by a $2N$ dimensional vector $\boldsymbol{\xi} = \{\mathbf{x}, \dot{\mathbf{x}}\}$. Let $A^0(T(\mathcal{M})) = C^\infty(T(\mathcal{M}), \mathbb{R})$ be the algebra of differential functions defined on the manifold $T(\mathcal{M})$. We define a Poisson structure on $T(\mathcal{M})$ which is an internal skew-symmetric bilinear multiplicative law on $A^0(T(\mathcal{M}))$ denoted $(f, g) \to [f, g]$ and satisfying the Leibnitz rule

$$[f, gh] = [f, g]h + [f, h]g \tag{1}$$

and the Jacobi identity

$$J(f, g, h) = [f, [g, h]] + [g, [h, f]] + [h, [f, g]] = 0. \tag{2}$$

The manifold $T(\mathcal{M})$ with such a Poisson structure is called a Poisson manifold. We define a dynamical system on the Poisson manifold $T(\mathcal{M})$ by the following differential equation

$$\frac{df}{d\tau} = [f, H] \tag{3}$$

where $H \in A^0(T(\mathcal{M}))$ is the Hamiltonian of the dynamical system.

With such definitions we derive the following important relations for functions belonging to $A^0(T(\mathcal{M}))$

$$\begin{aligned}[f(\boldsymbol{\xi}), h(\boldsymbol{\xi})] &= \{f(\boldsymbol{\xi}), h(\boldsymbol{\xi})\} \\ &\quad + \left[x^i, x^j\right] \frac{\partial f(\boldsymbol{\xi})}{\partial x^i} \frac{\partial h(\boldsymbol{\xi})}{\partial x^j} + \left[\dot{x}^i, \dot{x}^j\right] \frac{\partial f(\boldsymbol{\xi})}{\partial \dot{x}^i} \frac{\partial h(\boldsymbol{\xi})}{\partial \dot{x}^j},\end{aligned} \tag{4}$$

where we have introduced Poisson-like brackets defined by

$$\{f(\boldsymbol{\xi}), g(\boldsymbol{\xi})\} = \left[x^i, \dot{x}^j\right] \left(\frac{\partial f(\boldsymbol{\xi})}{\partial x^i} \frac{\partial h(\boldsymbol{\xi})}{\partial \dot{x}^j} - \frac{\partial f(\boldsymbol{\xi})}{\partial \dot{x}^i} \frac{\partial h(\boldsymbol{\xi})}{\partial x^j} \right). \tag{5}$$

We can see the relation (4) as the simple deformation of the Poisson-like brackets introduced in (5). It is obvious that the tensors $\left[x^i, x^j\right]$ and $\left[\dot{x}^i, \dot{x}^j\right]$ are skew symmetric. We introduce then the following notations

$$\left[x^i, x^j\right] = \frac{q}{m^2}\, \theta^{ij}(\boldsymbol{\xi}), \quad \left[x^i, \dot{x}^j\right] = \frac{1}{m}\, g^{ij}(\boldsymbol{\xi}), \quad \left[\dot{x}^i, \dot{x}^j\right] = \frac{q}{m^2}\, F^{ij}(\boldsymbol{\xi})$$

where $g^{ij}(\boldsymbol{\xi})$ is the $N \times N$ metric tensor, and where $\theta^{ij}(\boldsymbol{\xi})$ and $F^{ij}(\boldsymbol{\xi})$ are two $N \times N$ skew symmetric tensors, $F^{ij}(\boldsymbol{\xi})$ being related to the electromagnetic tensor introduced in a preceding paper.[23]

3. Maxwell's Equations

In a three dimensional flat space we have $g^{ij}(\mathbf{x}) = \delta^{ij}$ and the Hamiltonian of the Poisson structure reads then

$$H = \frac{1}{2}m\dot{x}^i\dot{x}_i + f(\mathbf{x}), \tag{6}$$

where

$$x_i = g_{ij}\, x^j. \tag{7}$$

The Jacobi identity (2) involving position and velocity components

$$J(x^i, \dot{x}^j, \dot{x}^k) \propto \frac{\partial F^{jk}(\boldsymbol{\xi})}{\partial \dot{x}^i} = 0. \tag{8}$$

shows that the gauge curvature is velocity independent, $F^{ij}(\boldsymbol{\xi}) \equiv F^{ij}(\mathbf{x})$. From the Jacobi identity (2) involving only velocities components we derive the Bianchi equation

$$J(\dot{x}^i, \dot{x}^j, \dot{x}^k) \propto \varepsilon^k{}_{ji}\frac{\partial F^{ij}(\mathbf{x})}{\partial x^k} = 0 \tag{9}$$

which, if we set $F^{ij}(\mathbf{x}) = \varepsilon^{ji}{}_k B^k(\mathbf{x})$, gives the following Maxwell's equation

$$\nabla \cdot \mathbf{B} = 0. \tag{10}$$

Now using the dynamical equation (3) we obtain the following equation of motion

$$m\ddot{x}^i = m\left[\dot{x}^i, H\right] = qF^{ij}(\mathbf{x})\dot{x}_j + qE^i\left(\mathbf{x}\right) \tag{11}$$

where

$$qE^i\left(\mathbf{x}\right) = -\frac{\partial f(\mathbf{x})}{\partial x_i}. \tag{12}$$

We have then a particle of mass m and electrical charge q moving in flat space where a magnetostatic and an electrostatic external field are present. We are able now to derive the other Maxwell's equation of the first group. With the dynamical equation (3) we express the time derivative of the magnetic field

$$\frac{dB^i}{dt} = \frac{1}{2}\varepsilon^i{}_{jk}\left[F^{jk}, H\right] = \frac{m^2}{2q}\varepsilon^i{}_{jk}\left[\left[\dot{x}^j, \dot{x}^k\right], H\right] \tag{13}$$

and we use the Jacobi identy (2) to rewrite the last term of the last equation. After some calculus we obtain

$$\frac{dB^i}{dt} = -\dot{x}^i\nabla\cdot\mathbf{B} + \frac{\partial B^i}{\partial x_j}\dot{x}_j + \varepsilon^i{}_{jk}\frac{\partial E^j}{\partial x_k} \tag{14}$$

which using (10) gives the second Maxwell's equation

$$\frac{\partial \mathbf{B}}{\partial t} - \nabla \times \mathbf{E} = \mathbf{0} \tag{15}$$

for static fields and electric fields deriving from any potential $f(\mathbf{x})$ (12).

As the two other Maxwell's equations are not Galileo invariant they cannot be deduced from the formalism and can be merely seen as a definition of the charge density and the current density. Nevertheless, as shown in Ref. 12 the complete set of the Maxwell's equations can be deduced in the relativistic generalization.

4. SO(3) Algebra and Poincaré Momentum

One of the most important symmetry in physics is naturally the spherical symmetry corresponding to the isotropy of the physical space. This symmetry is related to the SO(3) algebra. In the following we show that this symmetry is broken when an electromagnetic field is applied. In order to study the symmetry breaking of the SO(3) algebra we use the usual angular momentum $L^i = m\varepsilon^i{}_{jk}x^j\dot{x}^k$ which is a constant of motion in absence of gauge field. In fact, no electromagnetic field implies $F^{ij}(\mathbf{x}) = [\dot{x}^i, \dot{x}^j] = 0$, and the expression of the SO(3) Lie algebra with our brackets (4) gives then the standard algebra defined in terms of the Poisson brackets (5)

$$\left\{\begin{array}{l} [x^i, L^j] = \{x^i, L^j\} = \varepsilon^{ij}{}_k x^k, \\ [\dot{x}^i, L^j] = \{\dot{x}^i, L^j\} = \varepsilon^{ij}{}_k \dot{x}^k, \\ [L^i, L^j] = \{L^i, L^j\} = \varepsilon^{ij}{}_k L^k. \end{array}\right. \tag{16}$$

When the electromagnetic field is turned on this algebra is broken in the following manner

$$\left\{\begin{array}{lll} [x^i, L^j] & = \{x^i, L^j\} & = \varepsilon^{ij}{}_k x^k, \\ [\dot{x}^i, L^j] & = \{\dot{x}^i, L^j\} & + \frac{q}{m}\varepsilon^j{}_{kl}x^k F^{il}(\mathbf{x}) \\ & = \varepsilon^{ij}{}_k\dot{x}^k & + \frac{q}{m}\varepsilon^j{}_{kl}x^k F^{il}(\mathbf{x}), \\ [L^i, L^j] & = \{L^i, L^j\} & + q\varepsilon^i{}_{kl}\varepsilon^j{}_{ms}x^k x^m F^{ls}(\mathbf{x}) \\ & = \varepsilon^{ij}{}_k L^k & + q\varepsilon^i{}_{kl}\varepsilon^j{}_{ms}x^k x^m F^{ls}(\mathbf{x}). \end{array}\right. \tag{17}$$

In order to restore the SO(3) algebra we introduce a new angular momentum $M^i(\boldsymbol{\xi})$ which is *a priori* position and velocity dependent. We consider then the following transformation law

$$L^i(\boldsymbol{\xi}) \rightarrow \mathcal{L}^i(\boldsymbol{\xi}) = L^i(\boldsymbol{\xi}) + M^i(\boldsymbol{\xi}), \tag{18}$$

and we require that this new angular momentum $\mathcal{L}^i$ verifies the usual SO(3) algebra

$$\left\{\begin{array}{l} [x^i, \mathcal{L}^j] = \{x^i, \mathcal{L}^j\} = \varepsilon^{ij}{}_k x^k, \\ [\dot{x}^i, \mathcal{L}^j] = \{\dot{x}^i, \mathcal{L}^j\} = \varepsilon^{ij}{}_k \dot{x}^k, \\ [\mathcal{L}^i, \mathcal{L}^j] = \{\mathcal{L}^i, \mathcal{L}^j\} = \varepsilon^{ij}{}_k \mathcal{L}^k. \end{array}\right. \tag{19}$$

These equations (19) gives then three constrains on the expression of the angular momentum $\mathcal{L}^i$. From the first relation in (19) we easily deduce that M^i is velocity independent $M^i(\boldsymbol{\xi}) = M^i(\mathbf{x})$, from the second relation we obtain

$$[\dot{x}^i, M^j] = -\frac{1}{m}\frac{\partial M^j(\mathbf{x})}{\partial x_i} = -\frac{q}{m}\varepsilon^j{}_{kl}x^k F^{il}(\mathbf{x}) \tag{20}$$

and finally the third relation gives

$$M^i = \frac{1}{2} q \varepsilon_{jkl} x^i x^k F^{jl}(\mathbf{x}) = -q\,(\mathbf{x} \cdot \mathbf{B})\, x^i. \tag{21}$$

Equations (20) and (21) are compatible only if the magnetic field $\mathbf{B}$ is the Dirac magnetic monopole field

$$\mathbf{B} = \frac{g}{4\pi} \frac{\mathbf{x}}{\|\mathbf{x}\|^3}. \tag{22}$$

The vector $\mathbf{M}$ allowing us to restore the SO(3) symmetry (19) is then the Poincaré momentum[24]

$$\mathbf{M} = -\frac{qg}{4\pi} \frac{\mathbf{x}}{\|\mathbf{x}\|}.$$

already found in a preceding paper[25].[26] The total angular momentum is then

$$\boldsymbol{\mathcal{L}} = \mathbf{L} - \frac{qg}{4\pi} \frac{\mathbf{x}}{\|\mathbf{x}\|}. \tag{23}$$

This expression was initially found by Poincaré in a different context.[24] Actually he was looking for a new angular momentum that would be a constant of motion. In our framework this property is trivially verified by using the dynamical relation (3). This procedure of symmetry restoration has also been performed for Lorentz algebra in a curved space.[12] An other generalization of this formalism can be find in a recent interesting work where the study of the Lorentz generators in N-dimensional Minkowski space is proposed.[27,28]

Let us now discuss an important point. As the Dirac magnetic monopole is located at the origin we have

$$J(\dot{x}^i, \dot{x}^j, \dot{x}^k) = \nabla \cdot \mathbf{B} = g\delta^3(\mathbf{x}). \tag{24}$$

The preservation of the SO(3) symmetry in the presence of a gauge field is then incompatible with the requirement of the Jacobi identity at the origin of the coordinates and we have to exclude the origin from the manifold $\mathcal{M}$. As the Jacobi identity is the infinitesimal statement of associativity in the composition law of the translation group,[29] the breakdown of the Jacobi identity (24) when $\nabla \cdot \mathbf{B} \neq 0$ implies that finite translations do not associate. In usual quantum mechanics non-associativity between operators acting on the Hilbert space can not be tolerated, one has to use the Dirac's quantization procedure to save associativity (24).

In order to consider quantum mechanics within our framework, we have to quantify as usual the total angular momentum $\mathcal{L}$. Considering the rest frame of the particle we have the following Dirac quantization

$$\frac{gq}{4\pi} = \frac{n}{2}\hbar. \tag{25}$$

5. Noncommutative Quantum Mechanics

Let now the momentum vector $\mathbf{p}$ replace the velocity vector $\dot{\mathbf{x}}$ in the Feynman formalism presented in Sec. 2. Consider a quantum particle of mass m whose coordinates satisfy the deformed Heisenberg algebra

$$[x^i, x^j] = i\hbar q_\theta \theta^{ij}(\mathbf{x},\ \mathbf{p}), \quad [x^i, p^j] = i\hbar\delta^{ij}, \quad [p^i, p^j] = 0,$$

where θ is a field which is *a priori* position and momentum dependent and q_θ is a charge characterizing the intensity of the interaction of the particle and the θ field. The commutation of the momentum implies that there is no external magnetic field. It is well known that these commutation relations can be obtained from the deformation of the Poisson algebra of classical observable with a provided Weyl-Wigner-Moyal product[30] expanded at the first order in θ.

5.1. *Jacobi Identities*

The Jacobi identity $J(p^i, x^j, x^k) = 0$ implies the important property that the θ field is position independent $\theta^{jk} = \theta^{jk}(\mathbf{p})$. Then one can see the θ field like a dual of a magnetic field and q_θ like a dual of an electric charge. The fact that the field is homogeneous in space is an essential property for the vacuum. In addition, one easily sees that a particle in this field moves freely, that is, the vacuum field does not act on the motion of the particle in the absence of an external potential. The effect of the θ field is manifest only in presence of a position dependent potential. To look further at the properties of the θ field consider the other Jacobi identity $J(x^i, x^j, x^k) = 0$ giving the equation of motion of the field

$$\frac{\partial \theta^{jk}(\mathbf{p})}{\partial p^i} + \frac{\partial \theta^{ki}(\mathbf{p})}{\partial p^j} + \frac{\partial \theta^{ij}(\mathbf{p})}{\partial p^k} = 0, \tag{26}$$

which is the dual equation of the Maxwell equation $\nabla \cdot \mathbf{B} = 0$. As we will see later, equation (26) is not satisfied in the presence of a monopole and this will have important consequences.

5.2. *Position Transformation*

Now consider the position transformation $X^i = x^i + q_\theta a^i_\theta(\mathbf{x}, \mathbf{p})$, where a_θ is *a priori* position and momentum dependent, that restores the usual canonical Heisenberg algebra

$$[X^i, X^j] = 0, \quad [X^i, p^j] = i\hbar\delta^{ij}, \quad [p^i, p^j] = 0.$$

The second commutation relation implies that a_θ is position independent, while the commutation relation of the positions leads to the following expression of θ in terms of the dual gauge field a_θ

$$\theta^{ij}(\mathbf{p}) = \frac{\partial a^i_\theta(\mathbf{p})}{\partial p^j} - \frac{\partial a^j_\theta(\mathbf{p})}{\partial p^i}, \tag{27}$$

which is dual to the standard electromagnetic relation in position space.

5.3. *Field Properties*

In order to examine more in detail the properties of this new field, let us consider initially the case of a constant field what is usual in noncommutative quantum mechanics. In the case of an harmonic oscillator expressed in terms of the original $\{\mathbf{x}, \mathbf{p}\}$ coordinates the Hamiltonian reads

$$H_\theta(\mathbf{x}, \mathbf{p}) = \frac{\mathbf{p}^2}{2} + \frac{k}{2}\mathbf{x}^2 \tag{28}$$

from which we have $p^i = m\dot{x}^i - kq_\theta\theta^{ij}x_j$, $\dot{p}^i = -kx^i$ and the equation of motion $m\ddot{x}^i = kq_\theta\theta^{ij}\dot{x}_j - kx^i$ which corresponds formally to a particle in a harmonic oscillator submitted to an external constant magnetic field. From equation (27) we deduce that $a^i_\theta(\mathbf{p}) = q_\theta\ \theta^{ij}p_j$, so $X^i = x^i + \frac{1}{2}q_\theta\theta^{ij}p_j$, and the Hamiltonian can then be written

$$H_\theta(\mathbf{X}, \mathbf{p}) = \frac{\left(m_*^{-1}\right)^{ij}p_ip_j}{2} + \frac{k}{2}\mathbf{X}^2 - k\frac{q_\theta}{2m}\mathbf{\Theta}\cdot\mathcal{L}, \tag{29}$$

with $\theta^{ij} = \varepsilon^{ijk}\Theta_k$, $\mathcal{L}^i(\mathbf{X}, \mathbf{p}) = \frac{1}{2}\varepsilon^i{}_{jk}\left(X^jp^k + p^kX^j\right)$ and $\sigma^{ij} = \delta^{ij}\mathbf{\Theta}^2 - \Theta^i\Theta^j$, the dual tensor of the Maxwell constraint tensor. Note that the interaction with the field θ is due to the presence of the position dependent harmonic potential and leads to a dual paramagnetic interaction which could be experimentally observable. Like an electron in the effective periodic potential of ions, the particle in the θ field acquires an effective mass tensor $m_*^{ij} = m\left(\delta^{ij} + \frac{\hbar^2kq_\theta^2}{4}\sigma^{ij}\right)^{-1}$ which breaks the homogeneity of space. This strong analogy with the vacuum of the solid state leads us to consider this field as a property of the vacuum.

5.4. *Angular Momentum*

Consider now the problem of angular momentum. It is obvious that the angular momentum expressed according to the canonical coordinates satisfies the angular momentum algebra however it is not conserved

$$\frac{d}{dt}\boldsymbol{\mathcal{L}}(\mathbf{X}, \mathbf{p}) = kq_\theta\boldsymbol{\mathcal{L}}\wedge\mathbf{\Theta}. \tag{30}$$

In the original $\{x, p\}$ space the usual angular momentum $L^i(\mathbf{x}, \mathbf{p}) = \varepsilon^i{}_{jk}x^jp^k$, does not satisfy this algebra. So it seems that there are no rotation generators in the $\{x, p\}$ space. We will now prove that a true angular momentum can be defined only if θ is a non constant field. From the definition of the angular momentum we deduce the following commutation relations

$$[x^i, L^j] = i\hbar\varepsilon^{ijk}x_k + i\hbar q_\theta\varepsilon^j{}_{kl}p^l\theta^{ik}(\mathbf{p}), \quad [p^i, L^j] = i\hbar\varepsilon^{ijk}p_k, \tag{31}$$

$$[L^i, L^j] = i\hbar\varepsilon^{ij}{}_kL^k + i\hbar q_\theta\varepsilon^i{}_{kl}\varepsilon^j{}_{mn}p^lp^n\theta^{km}(\mathbf{p}), \tag{32}$$

showing in particular that the SO(3) Algebra is broken. To restore the angular momentum algebra consider the transformation law

$$L^i \to \mathbb{L}^i = L^i + M^i_\theta(\mathbf{x}, \mathbf{p}), \tag{33}$$

and require the usual algebra

$$[x^i, \mathbb{L}^j] = i\hbar\varepsilon^{ijk}x_k, \quad [p^i, \mathbb{L}^j] = i\hbar\varepsilon^{ijk}p_k, \quad [\mathbb{L}^i, \mathbb{L}^j] = i\hbar\varepsilon^{ijk}\mathbb{L}_k. \tag{34}$$

The second equation in (34) implies the position independent property $M_\theta^j(\mathbf{x}, \mathbf{p}) = M_\theta^j(\mathbf{p})$, while the third equation leads to

$$M_\theta^i(\mathbf{p}) = \frac{1}{2}q_\theta\varepsilon_{jkl}p^i p^l \theta^{kj}(\mathbf{p}). \tag{35}$$

Using this equation we rewrite the third equation in (34) and we obtain a dual Dirac monopole[31] defined in momentum space

$$\mathbf{\Theta}(\mathbf{p}) = \frac{g_\theta}{4\pi}\frac{\mathbf{p}}{\|\mathbf{p}\|^3}. \tag{36}$$

We have introduced the dual magnetic charge g_θ associated to the Θ field. Consequently we have

$$\mathbf{M}_\theta(\mathbf{p}) = -\frac{q_\theta g_\theta}{4\pi}\frac{\mathbf{p}}{\|\mathbf{p}\|} \tag{37}$$

which is the dual of the famous Poincaré momentum introduced in positions space.[23,24] Then the generalized angular momentum

$$\mathbb{L} = \mathbf{r} \wedge \mathbf{p} - \frac{q_\theta g_\theta}{4\pi}\frac{\mathbf{p}}{\|\mathbf{p}\|}. \tag{38}$$

is a genuine angular momentum satisfying the usual algebra. It is the summation of the angular momentum of the particle and of the dual monopole field. One can check that it is a conserved quantity.

The duality between the monopole in momentum space and the Dirac monopole is due to the symmetry of the commutation relations in noncommutative quantum mechanics where $[x^i, x^j] = i\hbar q_\theta\varepsilon^{ijk}\Theta_k(\mathbf{p})$ and the usual quantum mechanics in a magnetic field where $[v^i, v^j] = i\hbar q\varepsilon^{ijk}B_k(\mathbf{x})$. Therefore the two gauge fields $\Theta(\mathbf{p})$ and $B(\mathbf{x})$ are dual to each other. Note that in the presence of the dual monopole the Jacobi identity fails

$$J(p^i, x^j, x^k) = -q_\theta\hbar^2\frac{\partial\Theta^i(\mathbf{p})}{\partial p_i} = -4\pi q_\theta\hbar^2 g_\theta\delta^3(\mathbf{p}) \neq 0. \tag{39}$$

This term is responsible for the violation of the associativity which is only restored if the following quantification equation is satisfied

$$\int d^3p\frac{\partial\Theta^i}{\partial p_i} = \frac{2\pi n\hbar}{q_\theta} \tag{40}$$

leading to $q_\theta g_\theta = \frac{n\hbar}{2}$, in complete analogy with Dirac's quantization.[29]

5.5. *Physical Realization*

A physical realization of our theory was found very recently in the context of the anomalous Hall effect in a ferromagnetic crystal.[32] The main point is the consideration of the Berry phase[33] $a_n^\mu(\mathbf{k}) = i \langle u_{n\mathbf{k}} | d_k | u_{n\mathbf{k}} \rangle$ where the wave functions $u_{n\mathbf{k}}$ are the periodic part of the Bloch waves. In their work, the authors introduced a gauge covariant position operator of the wave packet associated to an electron in the n band $x^\mu = i\frac{\partial}{\partial k_\mu} - a_n^\mu(\mathbf{k})$, whose commutator is given by

$$[x^\mu, x^\nu] = \frac{\partial a_n^\nu(\mathbf{k})}{\partial k^\mu} - \frac{\partial a_n^\mu(\mathbf{k})}{\partial k^\nu} = -iF^{\mu\nu}(\mathbf{k}) \tag{41}$$

where $F^{\mu\nu}(\mathbf{k})$ is the Berry curvature in momentum space.

The connection with our noncommutative quantum mechanics theory is then clearly apparent. The $\theta(\mathbf{p})$ field corresponds to the Berry curvature $F(k)$ and $a_\theta(\mathbf{p})$ is associated to the Berry phase $a_n(k)$. This shows that physical situations with a Berry phase living in momentum space could be expressed in the context of a noncommutative quantum mechanics. It is essential to mention that the monopole in momentum space, that we deduced from general symmetry considerations applied to the noncommutative quantum mechanics, was highlighted in very beautiful experiments of Fang *et al.*[32]

6. Conclusion

Starting from the derivation of Maxwell's equations, we reviewed the Feynman formalism. The angular algebra symmetry is naturally broken in the presence of a magnetic field, we showed within the framework of the Feynman formalism how to restore this symmetry. The restoration generates then a Dirac magnetic monopole and implies in addition to the usual angular momentum an associated Poincaré momentum. In Ref. 12 this restoration has been performed also in curved space and a direct application to gravitoelectromagnetism has been given.

Going from the tangent bundle space to the cotangent bundle space and requiring the restoration of the Heisenberg algebra, we have shown that Non-commutative quantum mechanics can be viewed from Feynman formalism. In order to maintain the SO(3) algebra a dual monopole in momentum space is generated. This monopole is responsible for the violation of the Jacobi identity and implies the non associativity of the law of addition of the momentum. To restore associativity a Dirac's quantization of the dual charges is necessary. As a natural physical realization of our theory we can see the $\theta(\mathbf{p})$ field like a Berry curvature associated to a Berry phase expressed in momentum space. The monopole in momentum space predicted by our generalization of noncommutative quantum mechanics was found recently[32] in condensed matter physics experiments.

References

1. F. Dyson, *Am. J. Phys.* **58**, p. 209 (1990).

2. S. Tanimura, *Ann. Phys.* **220**, p. 229 (1992).
3. A. Bérard, Y. Grandati and H. Mohrbach, *Jour. Math. Phys.* **40**, p. 3732 (1999).
4. C. R. Lee, *Phys. Lett. A* **148**, p. 146 (1990).
5. C. H. Chou, *Phys. Lett. B* **323**, p. 147 (1994).
6. J. F. Carinena, L. A. Ibort, G. Marmo and A. Stern, *Phys. Rep.* **263**, p. 153 (1995).
7. S. A. Hojman and L. C. Shepley, *J. Math. Phys.* **32**, p. 142 (1991).
8. R. J. Hughes, *Am. Jour. Phys.* **60**, p. 301 (1992).
9. M. Montesinos and A. Perez-Lorentzana, *Int. Jour. Theor. Phys.* **38**, p. 901 (1999).
10. P. Singh and N. Dadhich, *Mod. Phys. Lett. A* **16**, p. 83 (2001).
11. Z. K. Siligadze, *Annales Fond. Broglie* **27**, p. 241 (2002).
12. A. Bérard, J. Lages and H. Mohrbach, *Eur. Phys. J. C* **35**, p. 373 (2004).
13. H. Snyder, *Phys. Rev.* **71**, p. 38 (1947).
14. A. Connes, *Noncommutative Geometry* (Academic press, San Diego, 1994).
15. N. Seiberg and E. Witten, *J. High Energy Phys.* **09**, p. 32 (1999).
16. A. Konechny and A. Schwarz, *Phys. Rep.* **360**, p. 353 (2002).
17. J. Bellissard, H. Schulz-Baldes and A. van Elst, *J. Math. Phys.* **35**, p. 5373 (1994).
18. G. Dunne, R. Jackiw and C. Trugenberg, *Phys. Rev. D* **41**, p. 661 (1990).
19. C. Duval and P. Horvathy, *J. Phys. A* **34**, p. 10097 (2001).
20. J. Gamboa, F. Méndez, M. Loewe and J. C. Rojas, *Mod. Phys. Lett. A* **16**, p. 2075 (2001).
21. V. P. Nair and A. P. Polychronakos, *Phys. Lett. B* **505**, p. 267 (2001).
22. A. Bérard and H. Mohrbach, *Phys. Rev. D* **69**, p. 127701 (2004).
23. A. Bérard, H. Mohrbach and P. Gosselin, *Int. Jour. Theor. Phys.* **39**, p. 1051 (2000).
24. H. Poincaré, *C. R. Acad. Sci. Paris* **123**, p. 530 (1896).
25. A. Bérard, Y. Grandati and H. Mohrbach, *Phys. Lett. A* **254**, p. 133 (1999).
26. A. Bérard and H. Mohrbach, *Int. Jour. Theor. Phys.* **39**, p. 2633 (2000).
27. M. Land, hep-th/0603055, (2006).
28. M. Land, hep-th/0606046, (2006).
29. R. Jackiw, *Phys. Rev. Lett.* **54**, p. 159 (1985).
30. J. E. Moyal, *Proc. Cam. Phil. Soc.* **45**, p. 99 (1949).
31. P. A. M. Dirac, *Proc. R. Soc. A* **113**, p. 60 (1931).
32. Z. Fang, N. Nagaosa, K. S. Takahashi, A. Asamitsu, R. Mathieu, T. Ogasawara, H. Yamada, M. Kawasaki, Y. Tokura, and K. Terakura, *Science* **302**, p. 92 (2003).
33. M. V. Berry, *Proc. R. Soc. London A* **392**, p. 45 (1984).

BEYOND THE QUANTUM IN SNYDER SPACE

*J. F. S. van HUELE and M. K. TRANSTRUM

Department of Physics and Astronomy, Brigham Young University, Provo, UT 84602, USA

**E-mail: vanhuele@byu.edu*

Snyder Space is a modified formulation of quantum mechanics, proposed by Snyder in Phys. Rev. **71**, 38-41 (1947) to circumvent infinities in high-energy physics. Snyder space proposes discrete space, continuous time, continuous momentum and continuous energy. It also leads to modified commutation relations for position and momentum operators. Snyder's algebra is therefore related to current theories beyond the standard model, such as noncommutative quantum mechanics, minimal length uncertainty relations, and dynamical quantization. Snyder obtains his algebra by postulating an underlying space $\mathcal{U}$ on which both position and momentum operators are defined and by requiring Lorentz invariance of the theory. When solving problems in Snyder space one can in principle study the equations as they appear in coordinate space, momentum space, or in the underlying space. Alternatively one can transform Snyder space problems into modified problems in standard quantum mechanics using operator methods. In the limit that the spatial lattice parameter a vanishes, Snyder space reduces to standard quantum mechanics, thereby giving a quantitative meaning to the preposition "beyond" in "beyond the quantum". We give a brief introduction to Snyder space and quote results for some simple systems. Finally we speculate on the connection with other approaches including some that assume the existence of a structure underlying quantum mechanics.

Keywords: Snyder; Noncommutative structures; Commutation relations; Minimal length; Representations; Underlying space.

1. Why Go Beyond the Quantum?

Motivations for searching beyond the quantum are multiple. Let us list a few that come to mind. In the first place the need to modify quantum theory may arise from the discovery of a discrepancy between a specific prediction of the theory and experimental results. Only slightly less urgent is the realization that the theory fails to give a specific prediction for the outcome of an experiment that has been performed or will be performed one day. Continuing down this path, a dissatisfaction with the type of predictions that the theory makes, or with the type of explanation that the theory gives, or even more generally with the ontological status of the theory might be a call for action to supplement or modify the existing theory. In some extreme cases, and depending on the level of dissatisfaction, a total overhaul of the theory is called for.

There are certainly also motivations to go beyond the quantum that have little to do with a dissatisfaction with the theory, but rather paradoxically originate from our satisfaction with it. On the one hand, there is probably no better way to grasp the current theory than to step out and to circumscribe it from the outside. It should be mentioned here that, contrary to what is intended, the theory beyond the quantum might turn out to be less satisfactory than the original! By viewing what happens when a specific aspect of the theory is relaxed, one gains understanding in its particular role. On the other hand, if what is possible is also necessary, then charting random explorations beyond the quantum becomes an exercise into discovering the physics of the future. And even if the possible is not necessary, these explorations provide a way for highly and narrowly trained physicists to enlarge their favorite playground as new material becomes available for study with techniques of theoretical and mathematical physics.

There is no doubt that the different contributions to a conference entitled "Beyond the Quantum" each build on a different mix of motivations. It is quite unlikely however that any approach beyond the quantum, once it is reasonably worked out, can be reduced to only one such motivation, for even a project started with a single purpose is likely to overlap with many others. That is certainly the case for one particular approach beyond the quantum, entitled quantum mechanics in Snyder space, and the subject of this paper.

2. Go Beyond What Quantum?

Before one can tackle what lies beyond it, one needs to specify the quantum. The old quantum theory, quantum mechanics, quantum field theory, and quantum gravity are all quantum theories: they all claim the "quantum". Do they also share it? Certainly each one of these efforts goes beyond the previous one in a way that does not leave the quantum unaffected. In fact each successive step can be seen as an attempt to fix a problem with the previous "quantum". Are we to invent the next term in this series of theories in which the amount of quantumness actually seems to increase? Or is something more radical needed that does away with the quantum altogether? Is there any way to quantify the level of "quantumness" in any existing quantum theory or the lack of quantumness in any proposed theory beyond the quantum?

As an example, a possible attitude is to claim that problems associated with measurement were introduced in quantum mechanics and were not removed by the introduction of quantum field theory. The goal of reaching beyond the quantum now becomes the resolution of the measurement problem, maybe through the introduction of an underlying structure, a view represented at the meeting[a]. Another goal might be the simultaneous description of quantum theory and relativity and the ensuing concern of how much quantum needs to be removed to make the two theories compatible.

[a]See the contribution of A. Allahverdyan, R. Balian and Th. M. Nieuwenhuizen.

Hartland Snyder's problem in 1947 was different. Infinities were creeping into high-energy physics calculations and to avoid them he sought a modification of our quantum way to look at space and time.

3. Snyder's Goes Beyond the Quantum

Because Snyder did his work just before the development of renormalization in quantum electrodynamics by Feynman, Schwinger, and Tomonaga, he started by going beyond quantum mechanics[1] before addressing fields.[2] It is likely that the success of QED discouraged him from pursuing this line of work since the original motivation disappeared. After laying mostly dormant for almost fifty years, Snyder's contribution has reappeared and now figures prominently in the references of papers in the current field of noncommutative theories, minimal length as discretization of space, extensions of the standard model, particle physics, string theory, and others. Still, except for highlighting the fact that Snyder's approach preserves Lorentz covariance, Snyder is usually cited as a precursor, and, it seems, mostly for historical reasons. Here we follow a different approach and take Snyder's work at face value. We consider Snyder's alternate form of quantum mechanics, quote some results that can be derived from it, and investigate what can be learned about going beyond the quantum from following Snyder's particular way of doing it.

4. Snyder Space Essentials

To start with, Snyder defines a five-dimensional de Sitter space $\mathcal{U}(\eta_0, \eta_1, \eta_2, \eta_3, \eta_4)$ with η_i real such that

$$-\eta^2 = \eta_0^2 - \eta_1^2 - \eta_2^2 - \eta_3^2 - \eta_4^2. \tag{1}$$

Note that $\mathcal{U}$ is a projective space of four-dimensional space. This underlying space has no direct physical reality but allows us to define physical variables on it. Position and time are defined as operators on this configuration space

$$\hat{x}_i = ia(\eta_4 \frac{\partial}{\partial \eta_i} - \eta_i \frac{\partial}{\partial \eta_4}) \tag{2}$$

$$\hat{x}_0 = ia(\eta_4 \frac{\partial}{\partial \eta_0} + \eta_0 \frac{\partial}{\partial \eta_4}) \tag{3}$$

where a is a fundamental length or lattice spacing. Until then, it was believed that the introduction of a fundamental length would break the Lorentz invariance of the theory. The momentum and energy operators

$$\hat{p}_i = (\hbar/a)(\frac{\eta_i}{\eta_4}) \tag{4}$$

$$\hat{p}_0 = (\hbar/a)(\frac{\eta_0}{\eta_4}) \tag{5}$$

also depend on the fundamental length a. A Lorentz transformation in η-space induces a corresponding transformation of the space-time and momentum-energy operators, leaving the spectra invariant. Snyder then shows that the following definition of angular momenta and boost generators

$$\hat{L}_1 = i\hbar(\eta_3 \frac{\partial}{\partial \eta_2} - \eta_2 \frac{\partial}{\partial \eta_3}) \tag{6}$$

$$\hat{M}_i = i\hbar(\eta_0 \frac{\partial}{\partial \eta_i} + \eta_i \frac{\partial}{\partial \eta_i}), \tag{7}$$

leads to the usual relations

$$\hat{L}_k = \quad \epsilon_{ijk}\ \hat{x}_i \hat{p}_j \tag{8}$$

$$\hat{M}_i = \hat{x}_i \hat{p}_0 + \hat{x}_0 \hat{p}_i \tag{9}$$

These definitions allow one to find the commutation relations between all these operators. Defining the parameter $\alpha = \frac{a^2}{\hbar^2}$, which expresses the lattice spacing in inverse momentum squared, one gets

$$[\hat{x}_i, \hat{p}_i] = i\hbar(1 + \alpha \hat{p}_i^2) \tag{10}$$

$$[\hat{x}_0, \hat{p}_0] = i\hbar(1 - \alpha \hat{p}_0^2) \tag{11}$$

$$[\hat{x}_i, \hat{p}_j] = i\hbar\alpha \hat{p}_i \hat{p}_j \tag{12}$$

$$[\hat{x}_j, \hat{p}_i] = i\hbar\alpha \hat{p}_i \hat{p}_j \tag{13}$$

$$[\hat{x}_i, \hat{p}_0] = i\hbar\alpha \hat{p}_i \hat{p}_0 \tag{14}$$

$$[\hat{p}_i, \hat{x}_0] = i\hbar\alpha \hat{p}_i \hat{p}_0, \tag{15}$$

as well as

$$[\hat{x}_i, \hat{x}_j] = i\hbar\alpha\epsilon_{ijk} \hat{L}_k \tag{16}$$

$$[\hat{x}_i, \hat{x}_0] = i\hbar\alpha \hat{M}_i, \tag{17}$$

showing that small modifications can have profound consequences! Snyder's particular definition of the the physics operating on the underlying space opens the door to a world where positions along different axes do not commute. As to how noticeable the physics beyond the quantum is, let us remember that the magnitude of this new effect is of proportional to α and that other commutators remain unchanged as

$$[\hat{p}_i, \hat{p}_j] = 0 \tag{18}$$

$$[\hat{L}_i, \hat{L}_j] = \quad i\hbar\epsilon_{ijk} \hat{L}_k \tag{19}$$

$$[\hat{M}_i, \hat{M}_j] = -i\hbar\epsilon_{ijk} \hat{L}_k. \tag{20}$$

Before moving on to solutions in Snyder space, we want to mention an unexpected result. Snyder space as defined above depends on two parameters, $\hbar$, the usual "quantum" parameter, and a or α, the new, or "beyond" parameter. This ability to quantify the "beyond" the quantum is most useful both conceptually and experimentally, but there is more. Taking the limit of vanishing a, usual quantum theory is recovered. One is then free to consider a classical limit by removing $\hbar$. Once Snyder space is established however, the order in which the limits are taken can be switched. If the quantum parameter vanishes as fast as a, leaving α constant, the classical limit of the quantum operators still don't commute, leading to a non-commutative classical structure. Beyond the quantum in Snyder space has opened the door to beyond-more-than-just-the-quantum! There is no reason why such a possibility should not be considered, at least in principle, by any theory that aims beyond the quantum.

5. Snyder Space Solutions

We are now ready to consider new possibilities offered by solving simple problems in Snyder space rather than in quantum mechanics. We restrict our discussion to systems in one spatial dimension, although, as previously mentioned, multiple dimensions can lead to considerably more interesting situations. In one dimension the operators $\hat{x}, \hat{p}$ in the polar representation $\eta_1 = \eta \sin(\phi), \eta_4 = \eta \cos(\phi)$ take the form:

$$\hat{x} = ia\frac{\partial}{\partial \phi} \tag{21}$$

$$\hat{p} = \frac{\hbar}{a}\tan(\phi) \tag{22}$$

and we notice that η, the curvature of the de Sitter space, has become irrelevant! Single-valuedness of the eigenfunctions imposes a discrete spectrum for the position operator. The normalized eigenstates $|m>$ correspond to eigenfunctions with eigenvalues ma in the ϕ representation

$$< \phi|m >= \frac{1}{2\pi}e^{-im\phi}, \tag{23}$$

where m ranges over all integers. A Fourier decomposition of a wave function results in the representation in the position eigenfunction basis. Since momentum is a multiplicative operator on the underlying space, as well as in momentum space, the relation between the two can be obtained through a simple change of variables

$$\hat{p} = \frac{\hbar}{a}\tan(\phi) = p \tag{24}$$

and the position operator in momentum space can then be found by applying this transformation on the position operator in the underlying space

$$\hat{x} = i\hbar(1 + \alpha p^2)\frac{\partial}{\partial p}, \tag{25}$$

all expressions that reduce to the standard momentum representation expressions when α becomes negligible.

Let us first illustrate how this can be used to obtain exact solutions directly in the case of the nonrelativistic free particle. A solution exists in both the underlying space and the momentum space. The solution can also be expanded in terms of the position eigenfunctions.

In the underlying space the equation for the free particle with mass μ is

$$\langle\phi|\hat{H}|\psi\rangle \;=\; \langle\phi|\frac{\hat{p}^2}{2\mu}|\psi\rangle \;=\; \frac{\hbar^2}{2a^2\mu}\tan^2(\phi)\langle\phi|\psi\rangle \;=\; E\langle\phi|\psi\rangle, \tag{26}$$

or

$$\left(\frac{\hbar^2}{2a^2\mu}\tan^2(\phi) - E\right)\langle\phi|\psi\rangle = 0, \tag{27}$$

from which it immediately follows that

$$\langle\phi|\psi\rangle \;=\; \delta\left(\phi - \tan^{-1}\frac{a}{\hbar}\sqrt{2\mu E}\right). \tag{28}$$

The result is very similar in momentum space

$$\langle p|\psi\rangle \;=\; \delta\left(p - \sqrt{2\mu E}\right), \tag{29}$$

providing an example where Snyder space gives the same answer as standard quantum mechanics.

We can express this result in terms of the position eigenfunctions using the relation

$$\delta(\phi) = \sum_{m=-\infty}^{\infty}\frac{1}{\sqrt{2\pi}}e^{im\phi} = \sum_{m=-\infty}^{\infty}\langle\phi|m\rangle \tag{30}$$

where $|m\rangle$ is the eigenstate of the position operator with eigenvalue ma. Thus, we have,

$$|\psi\rangle = \sum_{m=-\infty}^{\infty} e^{-i\,\tan^{-1}\frac{a}{\hbar}\sqrt{2\mu E}}|m\rangle \tag{31}$$

As a second example we illustrate a different solution technique in Snyder space. This technique, the operator transformation method,[3] allows one to relate an exact problem in Snyder space, in this example the simple harmonic oscillator (SHO), to an approximate problem in standard quantum mechanics, in casu a perturbed SHO. It actually turns out that in this case an exact solutions can also be found for the SHO in Snyder space along the lines of the free particle example above and that correspondingly, the perturbation problem converges in quantum mechanics.

Using a subscript S to denote the Snyder space operators, we look for a transformation

$$x_S = x_S(x,p) \tag{32}$$

$$p_S = p_S(x,p), \tag{33}$$

where x and p obey the canonical commutation relations whereas x_S and p_S obey the commutation relation given in section 4.

An important point to observe is that x and p are not the standard operators of quantum mechanics, although they may appear to be very similar. One reason for making this distinction, is that x and p are no longer a priori required to be Hermitian, and as such they do not correspond to physical observables. On the other hand,

x_S and p_S are Hermitian and correspond to the observable position and momentum. In fact, hermiticity is used to select an appropriate operator transformation so that the well-established techniques from standard quantum mechanics might be applied.

One may verify that

$$x_S = x + \alpha p\, x\, p \tag{34}$$

$$p_S = p \tag{35}$$

allow all operators to satisfy their prescribed commutation relations. Another acceptable transformation is given by

$$x_S = x \tag{36}$$

$$p_S = \frac{1}{\sqrt{\alpha}} \tan\left(\sqrt{\alpha} p\right). \tag{37}$$

When substituting the transformed position and momentum operators into the SHO Hamiltonian, additional anharmonic terms show up, and, performing the perturbation theory, one can show that the energy spectrum is modified to

$$E_n = \hbar\omega \left\{ \sqrt{1 + \left(\frac{\beta}{2}\right)^2} \left(n + \frac{1}{2}\right) + \frac{\beta}{2} \left[\left(n + \frac{1}{2}\right)^2 + \frac{1}{4} \right] \right\}, \tag{38}$$

where $\beta = \alpha m \omega \hbar$. Not surprisingly, results for systems in more than one dimension are harder to obtain but can be found for free particles, simple harmonic oscillators and particles in a rigid box.[4]

6. Snyder Space Overlapping Beyond the Quantum

As previously mentioned, the particular commutation relations proposed by Snyder have been recovered in many different contexts. Dynamical quantization, first introduced[5] as a method to describe quark interactions modifies the Heisenberg algebra by adding an extra term

$$[x_i, p_j] = \left(i\hbar + i\frac{l}{c}H\right)\delta_{ij}, \tag{39}$$

where H is the Hamiltonian of the system being considered, l is some fundamental unit of length, and c is the speed of light. All other commutation relations are unmodified. For a free-particle Hamiltonian H, the commutation relation reduces to the position-momentum relation of Snyder space.

The field of noncommutative quantum mechanics arising from the low-energy limit prediction of string theory modifies the usual commutation algebra among position operators of different dimension, x_i and x_j, to

$$[x_i, x_j] = i\Theta_{ij}, \tag{40}$$

where Θ_{ij} is a real (constant) antisymmetric matrix with dimensions of length squared. This Space-Space algebra is by far the most common algebra studied in noncommutative quantum mechanics, but space-momentum and momentum-momentum algebras also occur.

Finally, minimal length uncertainty relation considerations[6,7] suggest that the uncertainty in position Δx has a minimal value x_0. This can be obtained if the uncertainty relation is given by

$$\frac{\Delta x}{x_0} \geq \frac{1}{2}\left(\frac{p_0}{\Delta p} + \frac{\Delta p}{p_0}\right), \tag{41}$$

from which it follows that Δx has a minimum value of x_0 at $\Delta p = p_0$. By relating the uncertainty relation to the commutator and making the approximation

$$\langle p \rangle \approx 0, \tag{42}$$

then

$$\Delta p^2 = \langle p^2 \rangle - \langle p \rangle^2 \approx \langle p^2 \rangle, \tag{43}$$

and it follows that the commutation relation

$$\left[\frac{x}{x_0}, \frac{p}{p_0}\right] = i\left(1 + \frac{p^2}{p_0^2}\right) \tag{44}$$

implies the minimal uncertainty relation. By requiring that $[x, p] \approx i\hbar$ in the low-energy limit, it follows that $x_0 p_0 = \hbar$. Finally, since x_0 is the minimal length that can be resolved, we can set $x_0 = a$, i.e. the fundamental length of Snyder space, and we have reproduced one-dimensional Snyder space.

7. Lessons Learned from Snyder Space Beyond the Quantum

Let us conclude by briefly contrasting possible limitations of Snyder space with some of the lessons learned. One might wonder to what extent Snyder space really reaches beyond the quantum as opposed to just modifying a few rules in quantum mechanics. To the extent that our current theories form one coherent, although possibly incomplete entity, even tweaking a few "details" is likely to affect the whole edifice in a subtle way. Although Snyder space may seem tailored to address problems at very high energies, it is formally applicable at any scale, including the currently accessible scale or that of the next experiment.[8] It was mentioned at the conference that quantum gravity is like a big rock in the garden of quantum mechanics. Shouldn't the shape of the rock help determine how to rearrange the garden and shouldn't more time spent studying it? The Snyder space results quoted here can be related to a modification of the time-independent Schrödinger equation. No real time evolution was considered. Still, the Snyder space formalism gives an operational definition of space and time operators that lies closer to what both relativity and experiment demand than what quantum mechanics can currently provide. Also, the underlying space is hardly reminiscent of a hidden variable manifold or of a parallel world, but then new directions are indeed needed and Snyder space does allow new representations that may prove useful like, say, the Wigner distribution in quantum optics. Finally, the ability to quantify the degree to which Snyder space trespasses quantum mechanics is a useful concept. At the very least, Snyder space physics can be used as a benchmark and a source of inspiration on how to tread carefully beyond the quantum.

References

1. H. S. Snyder, *Phys.Rev.* **71**, 38 (1947).
2. H. S. Snyder, *Phys.Rev.* **72**, 68 (1947).
3. H. O. Girotti, *Am.J.Phys.* **72**,608 (2004).
4. M. K. Transtrum, Thesis, (Brigham Young University, 2006).
5. I. Saavedra and C. Utreras, *Phys Lett B* **98**, 74 (1981).
6. A. Kempf,*J.Phys. A*, **30**,2093 (1997).
7. A. Kempf, *J.Math.Phys.* **38**,1347 (1997).
8. D. Meschini,arXiv:gr-qc0601097 (2006).

PART E

Stochastic Electrodynamics

SOME QUANTUM EXPERIMENTS FROM THE POINT OF VIEW OF STOCHASTIC ELECTRODYNAMICS

V. ŠPIČKA*, J. J. MAREŠ, P. HUBÍK, and J. KRIŠTOFIK

Institute of Physics Academy of Sciences of the Czech Republic
Cukrovarnická 10, Praha 6, The Czech Republic
**E-mail: spicka@fzu.cz*

Stochastic electrodynamics (SED), an alternative theory to quantum phenomena based on laws of classical physics is shortly reviewed and compared with quantum electrodynamics. Experiments supporting the existence of zero-point fluctuating radiation field, the key concept of SED, are discussed. Relation between measurements of the black-body radiation spectrum and noise is analysed to define conditions under which the zero-point component of radiation or noise can be observed. Further, it is shown that stability of weakly localized orbits, measured in disordered solid state systems, can be explained by the presence of zero-point fluctuations of vacuum.

Keywords: Stochastic electrodynamics; Zero-point energy; Black-body radiation; Noise measurements; Weak localization.

1. Introduction

Stochastic electrodynamics (SED) can be understood as a theory the main aim of which is to explain quantum behaviour of matter within a conceptual frame of laws of classical physics. It is based on the enlargement of a set of concepts creating description of the world in classical physics by one essential additional assumption: a randomly fluctuating electromagnetic field is permanently present even at the absolute zero temperature at every point of the universe. The idea is that this all pervasive, so called zero-point radiation (ZPR), causes all quantum effects we observe.

Even if not so strongly as quantum mechanics and its field generalization, quantum electrodynamics (QED), SED changed (in comparison with a commonly accepted picture of world of classical physics of the 19th century) essentially our ideas about the universe and its observation in two aspects: it introduced fluctuations into the realm of classical physics and changed the meaning of vacuum. Up to the end of the 19th century, the development of classical physics was based on concepts of continuous space, time and energy and perception of the concept of vacuum as a completely empty space. Movements of bodies were governed by Newton equations with continuous changes of bodies energies. Electromagnetic field was described by Maxwell's equations with zero solution without source terms. The vacuum was thus understood as a state where not only any particles but even no electromagnetic field exist. This concept of vacuum was based on the absolutely extraordinary, seemingly convincing, boundary condition for electromagnetic field in vacuum, i.e. that at $T = 0$ no fields are presented at all. Space from which both ordinary matter and electromagnetic fields are removed was identified with vacuum. Assumption was that there is simply just nothing in vacuum. Realisation of such an operation of emptying space completely in some confined region was assumed to be possible, at least in principle in abstract sense of gedanken experiments. This way of thinking was traditionally used from the times of Bartoli and Boltzmann and culminated at the beginning of the 20th century when even the idea of ether was abandoned. As generally known, really severe troubles with these concepts emerged, however, when physicists began to deal with the explanation of the black-body radiation spectrum. Due to the conflict with basic concepts as for space and energy then generally accepted, Planck had unpleasant time during his work on the derivation of what is nowadays called Planck's law. Conceptual troubles (and changes in commonly accepted views on behaviour of "known" universe) related to the explanation of the black-body radiation spectrum were mirroring some of essential problems we have with the new concepts introduced by formulation of quantum theory up to nowadays. Black-body radiation spectrum measurement "forced" Planck to adopt discrete energy levels to explain the measured curves. His and other physicists considerations about black-body radiation spectrum led not only to the idea of discrete levels. They later caused also change of our perception of structure and behaviour of vacuum and to emergence of stochastic approach for understanding experiments. As for explanation of experiments oversuccessful theory, the enigma called quantum mechanics and its generalizations to quantum theories of fields, were gradually established.

Development of description of quantum phenomena followed mainly the way, nowadays presented in most textbooks, that the only possible solution of black-body radiation problem, is just Planck's solution which admits only discrete sets of allowed states corresponding to universally quantized action. On the other hand, such a collapse of idea of continuum which is one of the cornerstones of mathematical thinking together with deny of the positivistic logic of classical physics continued to be a nightmare not only for many founders of quantum mechanics, followed by their students, but it also led again and again, just from the beginning of quantum theory, to attempts to explain quantum phenomena within some more "reasonable" generalization of classical physics concepts. One of these attempts is SED. Since this is a theory dealing with electromagnetic field and its presence at the absolute zero temperature, it is closely related to the black-body radiation problem. In fact, with Planck's discovery of zero-point energy of harmonic oscillator, door was open not only for the creation of quantum mechanics and QED, but also for SED. Essential conceptual changes of our view of world started by attempts to explain black-body radiation spectrum.

1.1. *Contents*

Basic concepts of quantum mechanics and QED and their relation to the emergence of stochastic electrodynamics as an alternative to quantum description are discussed in Section 2. Emphasis is put on the role of black-body radiation, zero-point energy and problems with understanding of the role of vacuum fluctuations in QED, stochastic approach to quantum phenomena and formulation of SED. Phenomena, such as Casimir forces, which are supporting at least indirectly the reality of ZPR and related vacuum field fluctuations are mentioned. Section 3 is devoted to the analysis of black-body radiation spectrum. Properties of zero-point radiation spectral distribution (its Lorentz invariance and its divergence and the existence of the natural cut-off energy) are discussed. Especially, possibilities to detect ZPR spectrum via black-body radiation measurements are analysed. The following section deals with the role of Zitterbewegung in SED. Section 5 introduces periodic reactions as systems in which diffusive processes can be explained via stochastic approach to quantum dynamics; quantum diffusion and its relation to SED is considered. The last section provides new insight into the problem of the stability of matter within SED: weak localization phenomenon, which is traditionally treated as quantum interference effect, is explained in terms of SED. At the end of the section, the possibility to explain superconductivity in some systems, which are exhibiting weak localization behaviour, with the help of SED is mentioned.

2. Development of concepts: QED and SED

In this section we briefly comment on the development of quantum theory and problems of its field counterpart, QED. This is essential for the understanding of motivation and reasons, why SED was established as a complementary theory to the very successful QED. We will start with a short discussion of history of the black-body radiation problem since this is, for this purpose, the most important from all quantum phenomena observed and explained during the early stages of history of quantum mechanics. After this we will characterize both QED and SED.

2.1. *Notes to history of quantum mechanics, QED and SED*

The very first non-classical ingredient appeared in the first Planck's theory of black-body radiation[1] via his formulas for mean energy of oscillator and related energy density spectrum of black-body radiation. It is, however, very important to realize, that this was only the second theory of Planck[2] which brought for the first time zero-point energy on the scene (even if as a zero-point term of average energy of only harmonic oscillator in this case), which plays a central role in possible differences between QED and SED. Planck had severe problems with conceptual questions related to the black-body radiation spectrum, which prevented him from arriving to the zero point energy also for energy density spectrum of the black-body radiation - he just did not equal oscillators and electromagnetic field from this point of view arguing that only emission process relating to oscillators can be quantized. Einstein and Stern[3] a year later obtained twice larger contribution to the zero-point energy of oscillator and they did not want to accept zero-point energy of electromagnetic field, either.

It was Nernst[4] who first introduced the zero-point energy for both, oscillators as well as electromagnetic fields, coming with zero-point term for the spectral energy density of the black-body radiation. In addition, he admitted the possibility that zero-point fluctuating electromagnetic field with the corresponding zero-point spectral energy density is a really existing field in vacuum. He also immediately realized some of consequences of its possible existence: he was the first who wrote about the possibility to extract a useful work from it. On the other hand, he was also aware of conceptual trouble which the zero-point energy represents due to its divergence when the total energy by integrating over energies is calculated. He raised a question of consequences of such a huge energy of vacuum on gravitation effects in the universe: his consideration thus preceded nowadays vacuum catastrophe scenarios and problems with cosmological constant of general relativity. The most important from the point of view of SED is to realize that Nernst[4] came with an attempt to re-establish the classical theory of black-body radiation by assuming the existence of background zero-point radiation and continuous distribution of states in energy. This could lead already before QED to establishment of SED, but in the time when Nernst came with his ideas of zero-point fields, it was too late since attention was almost completely turned to the development of quantum mechanics followed by QED mathematical formalism. This attempt to explain quantum effects within classical physics was overlooked and forgotten for some time. Therefore, from historical reasons, basis of the description of quantum effects by quantum mechanics and QED were established during the twenties of the 20th century and preferably used up to now. SED was formulated with a delay, partly as a reaction to lasting conceptual troubles of quantum mechanics and QED, which are, however, from the practical point of view, very successful theories.

Quantum mechanics and QED created a very advanced formalism which enables us up to now to describe all observable quantum effects with sufficient accuracy. They are based on less intuitive concepts than classical physics and include some inconsistencies related mainly to its understanding of vacuum: QED adopted the idea of zero-point field but, as a virtual, in reality not existing phantom, the consequences of which are, nevertheless, sometimes observable and so there are mathematical procedures within QED, which are "materializing" the influence of this phantom. But this is exactly one of points where it is reasonable to consider intuitively a more accessible alternative of QED. Before we are coming to this point, let us discuss how QED deals with zero-point energy and related problems in more detail.

2.1.1. *Essential problems of QED: infinities and the structure of vacuum*

At the end of the twenties QED was formulated. Emergence of a problematic ZPR contribution to quantum electromagnetic field was "suppressed" with the help of formalism. Quantized electromagnetic field fluctuates and zero-point radiation modes are present, which is the mirror of the fact that annihilation and creation operators does not commute (consequently electromagnetic field has a dispersion according to uncertainty principle and therefore it fluctuates). It seemed first that problems with zero-point modes could be avoided formally by relating observables only with normal ordering of creation and annihilation operators mimicking thus in formalism considerations that only differences in energies matter and in this way infinite energies can be subtracted. The solution of the problem, however, cannot be so simple since this has two serious drawbacks: first, there are effects which are observed, as Casimir effect[5–7] or Lamb shift,[6–10] and they are based on the fluctuations of vacuum fields. Second, even more serious, relates to the gravitation where the idea of a subtraction of energies just does not work: the total energy counts, not only energy differences, according to general theory of relativity. QED has been a very successful theory, but it has always had, just from its very beginning, severe problems with infinities and related questions of structure and behaviour of vacuum. Even if many troubles of QED were overcome by formal treatment via mass and charges renormalization procedures, the most severe problem of fluctuations of vacuum and zero-point energy persists up to now.

The problem with the interpretation of structure of vacuum is reappearing again and again up to now, starting already in early stages of QED, e.g. in relation with understanding of phenomena of spontaneous emission. There are two basic explanations (interpretations) of spontaneous emission. The first one is based on the idea of radiation reaction while the second one directly deals with vacuum fluctuations (spontaneous emission can be viewed as stimulated emission by the zero-point fluctuations). Both these approaches are complementary and they are related to each other by fluctuation-dissipation theorem. Stochastic approach to problems of electromagnetic field naturally emerged also with explanations of Lamb shift[6–10] and Casimir forces.[5–7] Discussions about real or virtual (not conserving energy during virtual processes) vacuum fluctuations of field, which re-emerged on these occasions, continue up to present days. Paradoxically, questions of non-trivial boundary conditions for electromagnetic field in vacuum reappeared in QED when explaining Casimir forces, too. For the explanation of this effect it is necessary to take into account that boundaries in the space change the spectrum of zero-point radiation, and therefore there is a non-zero finite vacuum energy $E_V = E_{V,boundary} - E_{V,without\ boundary}$. It means that this energy depends essentially on the geometry and the corresponding Casimir force exists only due to boundaries and it is crucially influenced by the geometry of these boundaries for modes of vacuum zero-point field (e.g. change from quantization of the field in the infinite space to the field in the space bounded by parallel plates is related to the observed Casimir force between these plates).

Ideas of ZPR, fluctuations of vacuum and related non-trivial boundary conditions for (electromagnetic) fields were thus forced by nature to reappear in QED. By this development, QED indirectly supported revival of old ideas of Nernst and formulation and later on simultaneous development of stochastic electrodynamics (SED), which offers complementary views on the laws of nature.

SED is, however, also challenged by the second essential problem of QED, i.e. by the relation between energy content of vacuum and gravitation mentioned already by Nernst. As it is known and it will be discussed later on in this article, the integral over energy from the zero-point energy density diverges. The divergence in the total zero-point energy of vacuum becomes really important (even from the point of view of very basic concepts on which QED and SED are based) since the corresponding fields (quantum fields of QED or classical fields of SED) are naturally supposed to be coupled to gravity: any energy density is a source for gravitational field, and a uniform, divergent vacuum energy should lead, according to estimates, to much larger gravitational phenomena than they are observed. The conflict between the predicted energy content of vacuum and the real size of gravitational effects is sometimes called vacuum catastrophe[11] and it is closely related to the problem of the size of cosmological constant in the general theory of relativity.[12,13] This leads again even to considerations that vacuum fields are not real at all and cannot be so coupled to gravity. On the other hand, other phenomena than gravitational, such as Casimir forces and Lamb shift, which are related to vacuum fields, are observed. We will not deal with the problem of gravity in this article since it is too complicated. This is the essential problem of all recent field theories.

2.1.2. *Summary of the early development of quantum mechanics and QED*

Shortage of insight from the point of view of classical physics, led to an introduction of quantum mechanics and later on quantum electrodynamics to solve urgent problems of physics. In the time of early development of quantum mechanics it seemed that the only possibility how to explain black-body radiation spectrum and stability of matter lies in the formalism of quantum mechanics. This, however, later on did not turn to be true assumption. In fact, problems which emerged at the beginning of the 20th century with data from the black-body radiation spectrum, question of stability of matter and so on could be more naturally explained fully within the realm of classical physics, at least at a level of accuracy which experiments then provided. With a delay, ideas of stochastic electrodynamics envisaged by Nernst as a complementary explanation of quantum phenomena within classical physics were returning to physics. Classical physics of the 19th century, was, however, essentially enriched by new ideas of stochasticity and by different perception (understanding) of vacuum and its role in formulation of laws of nature.

2.1.3. *Randomness and discussion of Earnshaw's theorem: a way to SED*

The way towards better understanding of problems with quantum mechanics, QED and their interpretations and maybe even towards partial solutions of these problems is shown not only by Nernst's ideas, but also by considerations related to Earnshaw's theorem: a system which consists of classical charged particles cannot exist in static, stable equilibrium only due to electric fields.[14] The observed stability of matter (i.e. system in stable equilibrium) consisting of charged particles can be explained by self-consistent influences of charged particles and electromagnetic field: random movement of charged particles cause changes of the electromagnetic field which in the same time modifies paths of particles. In other words, stochastic character of both movement of particles and electromagnetic field can lead to the stability of observed matter. Therefore stochastic approach to classical physics may provide a reasonable solution of observed quantum phenomena.

2.2. *Stochastic approach to quantum phenomena*

Basic idea is to explain quantum behaviour of systems in terms of classical physics. This does not mean completely deterministic approach of Newtonian physics since fluctuations are involved and are supposed to be an innate part of the world in this approach. Classical trajectories thus fluctuate.

Stochastic approach to quantum mechanics addresses a question of how quantum phenomena emerge from the classical physics world. Two possible streams of stochastic, but classical approach to quantum phenomena, exists: 1. stochastic (random) quantum mechanics (SQM),[15–23] 2. stochastic (random) electrodynamics (SED).[6,7]

SQM and SED have in common that observed quantum behaviour of systems is assumed to be caused by fluctuating behaviour of particles trajectories (properties of these fluctuations bring Planck's constant $\hbar$ into the game) governed by laws of classical physics. They, however, differ in their approach to the source of fluctuations.

2.2.1. *Stochastic quantum mechanics*

This approach started by Schrödinger who noticed a formal similarity between Schrödinger[24] and diffusion equations. This idea was discussed in more detail by Fürth.[25] In this approach quantum phenomena are caused by some kind of Brownian motion of particles. At the beginning Brownian motion was viewed as appropriate and at the end Nelson developed Markovian theory of quantum processes.[16] This approach is useful but it is only approximate. The assumption that stochastic processes behind quantum phenomena have Markovian character was later on criticized since non-Markovian processes are needed to describe details of quantum behaviour.[7]

Stochastic quantum mechanics does not care about the cause of stochasticity behind quantum processes. This aim is targeted by the stochastic electrodynamics.

2.2.2. *Stochastic electrodynamics*

Stochastic electrodynamics (SED) is based on the assumption of random fluctuations of phases of real classical electromagnetic field in vacuum. These vacuum fluctuations and their interaction with charges in matter cause behaviour of matter known in terms of quantum mechanics as quantum phenomena. Classical particles as well as their movement are fully governed by laws of classical physics, the world of SED consists of Newton's equation of motion and Maxwell equations. Planck constant $\hbar$ plays in SED just a role of a constant to fit experimental data.

Due to the presence of electromagnetic field, SED is the field theory alternative to QED. Its basic idea of real field fluctuating in vacuum can be, however, generalized to other fields being thus an alternative to quantum field theories in general.

SED can be also understood as a hidden variable theory since it is based on classical trajectories and aims at explaining quantum phenomena. From this point of view the formulation and investigation of consequences of SED offers possible alternative to quantum mechanical description of matter, but also better understanding of quantum mechanics itself. This is may be the main reason, together with its ability to explain even complicated real quantum effects, why SED has attracted attention of excellent physicists.

Surprisingly, SED explained several non-trivial quantum effects (e.g. ground state of hydrogen, stability of matter, black-body radiation spectrum spontaneous emission, van der Waals and Casimir forces, Lamb shift, Davies-Unruh effect, see[7]) which are traditionally considered as purely quantum ones, some of them even being phenomena which forced physicists to formulate quantum mechanics in contradiction with some of cornerstone concepts of classical physics. Within SED, all quantum effects are considered as "classical" ones induced by real (not virtual) fluctuations of zero-point radiation of vacuum.

Initial ideas of SED can be traced to considerations of Nernst,[4] Planck,[1,2] Einstein[3,26] and Bohr.[5] Later on, important hints for its formulation appeared in works of Welton[27] and Casimir.[5] The real development of SED as a consistently formulated theory started in the 1960's in independent studies of Marshall[28–30] and Boyer[31–37] who both published a lot of key studies. Their work was followed by many authors. Some of them, as Cole,[38–48] de la Pena and Cetto,[49–54] Franca,[55–59] Puthoff,[60–65] Rueda and Haisch,[66–71] Santos[72,73] worked on problems of SED intensively over decades. They applied SED for explanation of many quantum phenomena and established thus SED as a serious, even if still less far used and developed, alternative of QED. Of course, also other important contributions to understanding of quantum systems within SED were published, e.g. review article by Guerra;[74] some of them are listed in references.[75–94] Milloni discussed problems related to SED, e.g. Casimir forces, in several articles,[95–104] a review article[105] and a monograph[6] and commented the results of SED as viewed from the position of QED. A monograph written by de La Pena and Cetto[7,106] provides an exhaustive review of key concepts and history of SED up to approximately 1995. The very recent development of SED (articles published from 2005 up to 2006) can be found in.[107–113]

2.2.3. *Quantum phenomena explained by SED*

In this section we will briefly mention some examples of quantum phenomena which first led to a formulation of stochastic electrodynamics (SED) and then were successfully treated by classical physics in terms of SED independently of quantum mechanics.

First, we have to mention black-body radiation spectrum and stability of matter. These effects are traditionally considered as phenomena which cannot be explained by classical physics. Both, black-body radiation spectrum and stability of matter, however, strongly motivated not only the formulation of QED, but also SED. At the end they were explained also within classical physics by SED, even if with a delay in comparison with QED; in some sense the historical reasons for the introduction of quantum mechanics appear artificial. Due to their extraordinary importance for many ideas related to SED, both of these phenomena will be discussed in more detail in the next sections and they will be related to experiments which have not been yet treated by SED up to now.

Development of SED has been also strongly supported by ideas behind the explanation of Casimir, van der Waals forces and Lamb shift. On the other hand, the existence of these effects is also considered as one of indirect proofs of the reality of the ZPR.

Inspiration for explanation of the Casimir effect within SED can be found already in Le Sage's theory of gravitation (1784); his idea of shielding of isotropic flux of "particles" corresponds to the explanation of different density of ZPR inside and outside of two parallel planes, which was the system first considered by Casimir when he was deriving what is nowadays known as Casimir forces. In the case of parallel planes Casimir forces are attractive. This even lead to consideration to use Casimir forces for gaining useful work from contraction of parallel plates.

Casimir forces of balls inspired Casimir to consider theory of classical electron based on this effect. At the beginning this model was not, however, working since in the first rough (long distances) perturbation scheme calculations, Casimir force between balls emerged as repulsive for all distances. Only later on, more detailed calculations showed that spontaneous shrinkage of the sphere can take place for very small diameters. Repulsive forces for large radius lead to expansion; conducting spherical shell placed solely in physical vacuum represents a macroscopic quantum system of well defined geometrical extent revealing the tendency to expand spontaneously at the expense of practically inexhaustible non-localized energy of ZPR. There are numerous articles about problems of Casimir forces,[114–119] including excellent reviews[120–125] and a monograph.[126]

3. Black-body radiation spectrum and stochastic electrodynamics

During the period of establishing both competitive theories, QED and SED, the existence of zero-point energy (at absolute zero temperature) and related idea of fluctuations of vacuum played the essential role: analysis of the theory of black-body radiation is therefore the most important from all quantum effects discussed in the first period of their development. All essential conceptual changes of view of our world leading to QED or SED are related to attempts to explain black-body radiation spectrum.

3.1. *Black-body radiation spectrum and quantum mechanics*

Black-body radiation spectrum motivated creation of both quantum and stochastic electrodynamics. As already said, very surprisingly non-classical quantum picture was historically established as the first one.

In his paper from 1912[2] Planck presented his second version of black-body radiation spectrum (where oscillator energies were assumed discrete in agreement with what will be later on known as quantum theory contrary to classical physics expectations). In this theory mean equilibrium energy of oscillator per degree of freedom (mode) is given by well known formula:

$$E(\omega, T) = \frac{1}{2}\hbar\omega + \frac{\hbar\omega}{\exp\frac{\hbar\omega}{kT} - 1}, \tag{1}$$

where the first term on the right is the mean zero-point energy of the oscillator ground state. This term, for the first time in history appearing in this formula, is completely independent of temperature, and plays an essential role in troubles as for our understanding of nature. Expression (1), when integrated over energy, gives infinite contribution, which is the cause of on the one side many problems and on the other side useful ideas discussed later on in this article. To discuss them, let us introduce the spectral energy density $\rho(\omega, T)$ of oscillator, which is given by

$$\rho(\omega, T) = N(\omega)E(\omega, T), \tag{2}$$

where the total number N of modes of electromagnetic field in the ground state in large (infinite) space region with energy between ω and $d\omega$ per unit volume is given by:

$$N d\omega = 2\frac{1}{(2\pi)^3}\frac{4\pi\omega^2 d\omega}{c^3}. \tag{3}$$

Additional factor 2 on the right hand side appears there to account for two polarizations.

From (2) and (3):

$$\rho(\omega, T) = \frac{\omega^2}{\pi^2 c^3}E(\omega, T) = \frac{\hbar\omega^3}{2\pi^2 c^3} + \frac{1}{(\pi^2 c^3)}\left(\frac{\hbar\omega^3}{\exp\frac{\hbar\omega}{kT} - 1}\right). \tag{4}$$

To describe the black-body radiation spectrum, Planck originally considered all these relations only for oscillators which were in his theory on the one hand representing electromagnetic field, but on the other hand he did not admit full equivalence between oscillators and electromagnetic field in the cavity. It is worth mentioning that this was the reason why Planck in his derivation of the spectral energy density did not use (2) and (3) and he obtained formula (4) without the first term on the right, i.e. without the zero-point energy contribution. Later on Einstein and Nernst "postulated" the same relations also for other fields, especially electromagnetic and the zero-point contribution emerged. We note that the neglect of ZPR contribution in the black-body radiation spectrum by Planck did not matter from the point of view of experiment due to to the fact that this contribution was not measured not only in experiments available to Planck, but neither in any other similar experiments, see the text later on.

The result of Planck was surprising mainly from two reasons: 1. It introduces the concept of discrete levels of energy, 2. even at the absolute zero temperature, which was traditionally related to the ground (vacuum= empty) state of matter, the mean energy of oscillator (as found later on even energy density of electromagnetic field) has non-zero value of energy, so called zero-point energy (ZPR).

3.2. *Zero-point radiation*

Zero-point energy is the energy corresponding to the spectral energy density for zero temperature:

$$\rho_{ZP}(\omega) \equiv \rho(\omega, T = 0) = \frac{\omega^2}{\pi^2 c^3}\frac{\hbar\omega}{2} = \frac{\hbar\omega^3}{2\pi^2 c^3}. \tag{5}$$

Both discrete energies and ZPR were concepts considered incompatible with the principles of classical physics. We can see, however, immediately that ZPR is in no contradiction with Maxwell equations since to include ZPR into the description only means to change the boundary conditions for Maxwell equations. Instead we take the solution of Maxwell equations without sources as zero, we admit the existence of zero electromagnetic field background which is not related to any source terms in Maxwell equations.

The ZPR plays the essential role for the formulation of stochastic electrodynamics as it was envisaged already by Nernst: in this case this contribution represents zero-point energy of randomly fluctuating (average is zero) radiation field which is real and all the time present in the whole space. Planck's constant in this case sets only the scale of electromagnetic ZPR, it does not have any other meaning. The possibility to use these facts to construct classical theory of quantum phenomena was not, however, realized by founders of quantum theory.

Zero-point spectrum of electromagnetic field is homogeneous and isotropic in the space. Contrary to the temperature dependent contribution to the spectral density energy, ρ_{ZP} is Lorentz invariant. This property is very essential for the formulation of SED.

3.2.1. *Lorentz invariance of ZPR spectral distribution*

Boyer was the first who realized that the Lorentz invariance of ZPR spectrum could be used to derive its shape independently of assumptions related to quantum theory. To point out this turn in approach to the black-body radiation, we will follow Boyer's ideas and present here largely simplified (in one dimension only), derivation of ZPR spectrum based on the requirement of the Lorentz invariance of the ZPR spectrum: The energy spectrum of ZPR observed in different inertial systems must be, not to violate principle of special relativity, isotropic and of the same spectral composition in all these inertial systems.

Let us now apply these requirements to the case where an inertial system $S^{'}$ is approaching the stationary one S along the common x-axis with the constant speed $v < c$. Observing in $S^{'}$ the light beams sent from S along the x-axis in the direction opposite to the movement of $S^{'}$, the frequency has to transform as $\omega^{'} = K\omega$ $(d\omega^{'} = K d\omega)$ and the spectral density of modes as $\eta(\omega) \to \eta^{'}(\omega) = \eta(\omega/K)$. Here, $K > 1$ is so called Doppler's factor which is in this particular case given by

$$K = \frac{1+\beta}{(1-\beta^2)^{1/2}}, \tag{6}$$

where $\beta = v/c$. Taking into account the Einstein-Planck's relativistic requirement according to which energy flux U in a light beam is directly proportional to its spectral density of modes η, i.e.

$$U(\omega)d\omega \propto \hbar\omega\eta(\omega)d\omega, \tag{7}$$

we can express the energy flux put through the band pass filter A-B in S by relation $\int_A^B c\hbar\omega\eta(\omega)d\omega$. As a result, flux measurements (using the same filter) must be the same in both S and $S^{'}$ and the following integral has inevitably to cancel

$$\int_A^B c\hbar\omega\left[\eta(\omega) - K^2\eta(\omega/K)\right] d\omega \equiv 0 \tag{8}$$

and consequently, the following functional equation must be valid

$$\eta(\omega) = K^2\eta(\omega/K). \tag{9}$$

Its solution[127] is given by formula $\eta(\omega) = const \times \omega^2$, where $const$ is certain Lorentz invariant. From it we immediately obtain important formula for spectral density of energy flux of ZPR

$$U(\omega)d\omega = const \times \omega^3 d\omega. \tag{10}$$

This may be completed easily by inserting there Weyl's density of states in a large cavity and polarisation degeneracy. If transformed to spectral energy density it leads immediately to formula (5).

3.2.2. *Divergence of ZPR spectrum and natural cut-off*

The integral over all energies from energy density spectrum of ZPR is divergent:

$$\omega \to \infty \quad E = \frac{\hbar}{2\pi^2 c^3} \int_0^\infty \omega^3 d\omega \to \infty. \tag{11}$$

The problem of this divergence can be solved when we suppose that the nature offers us some natural cut-off ω_{max} of available frequencies

$$E_{tot} = \frac{\hbar}{2\pi^2 c^3} \int_0^{\omega_{max}} \omega^3 d\omega = \frac{\hbar \omega_{max}^4}{8\pi^2 c^3}. \tag{12}$$

To determine the natural cut-off frequency it is reasonable to consider behaviour of particles occurring in nature.[128] We are dealing with electromagnetic fields, therefore it is natural to limit us to charged particles, e.g. electrons.

We can solve the problem of this divergence assuming that a charged particle interacts with zero-point fluctuations of background electromagnetic field. Since a charged particle cannot follow the electromagnetic vibrations with the velocity exceeding the speed of light over an appreciable amplitude, the overall response of a certain type of particles (e. g. electrons) has its natural upper frequency limit given approximately by its Compton frequency ω_C, all frequencies $\omega > \omega_C = mc^2/\hbar$ are unaccessible to any measurement process related with these charged particles, i.e. we can suppose they do not play any role and they do not contribute to observable phenomena, therefore there is no problem with the divergence. Obviously, such a response cut-off frequency must be Lorentz invariant. In an opposite case, the measurement of cut-off frequency would make it possible to determine, in contradiction to requirements of relativity principle, the absolute movement of a given inertial system. On the other hand, assuming that the ZPR is generated at a distant Universe in various inertial systems by the same type of particles (i.e. electrons) the resulting spectrum should be smeared in the vicinity of ω_C (for a possible shape of such a smearing see e.g.[4]).

3.2.3. *Black-body radiation spectrum in SED*

Formula (10) provides the density of ZPR radiation independently of any assumption of quantum theory. This fact was used as a motivation to introduce SED as a theory, independent of quantum mechanics and QED, to describe quantum behaviour of systems. To proceed on the way to SED we need to have a possibility to derive also the temperature dependent part of the black-body radiation without postulates of quantum theory.

Indeed, the whole, even the temperature dependent part, black-body radiation spectrum may be derived completely independently of any quantum theory if the existence of classical homogeneous fluctuating field with a Lorentz invariant spectrum is assumed at the absolute zero of temperature. This was shown for the first time by Marshall and Boyer and later on followed by others. No discrete energy levels are needed, but, of course, the constant $\hbar$ plays a different role now: it is just a parameter to be fitted to explain experiments. This step opens a possibility to introduce stochastic electrodynamics as a theory based on completely classical description of movement of charged particles, which interact with electromagnetic field. This field has non-zero fluctuations in vacuum. Energy of each mode in vacuum has a value $\hbar/2\omega$ and spectrum of this ZPR is of Lorentz invariant form.

3.3. *Measurement of black-body radiation and ZPR spectrum*

We have seen that the idea of ZPR existence is essential for the formulation of SED and therefore we need to discuss possible experiments supporting the existence of ZPR. Let us start with discussion of a possibility to measure ZPR just from black-body radiation spectrum.

Among the most direct experiments yielding the spectral distribution of electromagnetic radiation in equilibrium with ordinary matter belong measurements of high temperature spectrum of black-body radiation and of low-temperature noise spectrum in electronic circuits. We would like to stress here that there is a close connection between these two types of experiment, in fact it is the same experiment performed in another spectral range and using quite different technical means. For example, the arrangement of isothermal noise circuit and an illustration adapted from original Kirchhoff's paper[129] on black-body theory are depicted in Figure 1 for comparison. The analogical function of semi-permeable filter P and LC cir-

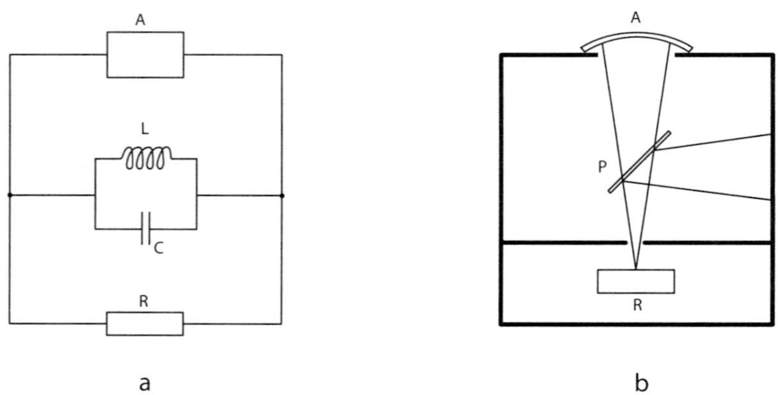

Fig. 1. Comparison of a tuned noise generating circuit (a) with an arrangement used by Kirchhoff in his considerations (b). A ... power absorber, R ... emitter, P ... optical band-pass filter.

cuit and other parts (emitter R, absorber A) of both arrangements is apparent at first glance. Furthermore, Kirchhoff's theorem claiming the universality of ratio between absorption and emission coefficients for the light resembles the very content of fluctuation-dissipation theorem expressed, in a special case of electronic circuits, as the equivalence between resistor and noise generator. No wonder that the process of emission and absorption of electromagnetic signal has to be described in both cases by the same formula (4). There is, however, a point, which is of primary importance, and simultaneously, which is almost completely ignored in the literature. This is the question of under which conditions the ZPR term in formula (4) can (or cannot) be observed in the experiment.

The thought experiment similar to that of Nyquist used for the derivation of his famous formula[130] can be exploited. Let us take then a circuit made of two separated, otherwise identical resistors kept at different temperatures connected with a common tuned non-resistive LC circuit defining the eigenfrequency $\Omega \sim 1/\sqrt{(}LC)$ of the system and the band half-width $\Delta \sim 1/RC$ as well. Taking further into account Weber's theorem,[131] according to which the resistor is in the circuit equivalent to the noise voltage generator producing the voltage per mode and is given by formula (akin to formula (1))

$$\langle V^2 \rangle = \frac{1}{C}\left(\frac{\hbar\Omega}{2} + \frac{\hbar\Omega}{\exp(\frac{\hbar\Omega}{kT}) - 1}\right). \tag{13}$$

The resulting superposition of mean square voltages due to both noise generators replacing resistors involves the ZPR components or, strictly speaking their average value $\hbar\Omega/2C$. If we, however, compute energy transfer from one resistor to another, the ZPR components cancel, and the net energy flux is given only by the difference of temperature dependent term in (4).

Hence, in an experiment where the absorber absorbs the radiation or, generally, the noise signal, the ZPR term cannot be observed in principle. This is evidently the case of classical investigation of black-body spectrum by Lummer and Pringsheim[132] where the radiation from open aperture passes monochromator (band filter) and it is, as far as possible, captured by a bolometer (wide band detector). On the other hand, the mean square amplitude of integral of electric field vector $\langle [\int \vec{E} d\vec{l}]^2 \rangle$ taken around the circuit can be measured provided that no power is extracted from the system. These two types of essentially different and in some sense complementary measurements (i.e. with and without power extraction) should be carefully distinguished. Unfortunately, the conditions encountered in practical realizations of noise experiments[78–80,113] are very often mixed, so that the analysis of resulting data is, as a rule, not trivial.

4. Zitterbewegung and SED

Assume, in accordance with the fundamental Ansatz of SED, a permanent interaction of the electron with the ZP background radiation. Accordingly, this must result in its chaotic motion or jiggling (called even here Zitterbewegung traditionally) and eventually, in the statistical uncertainty in its position. Indeed, mean square of a coordinate corresponding to the harmonic response of electron on the external ZPR field may be written as:

$$\langle q^2 \rangle = \int_{\omega_1}^{\omega_2} \langle q_\omega^2 \rangle d\omega = \frac{2\alpha}{\pi}\left(\frac{\hbar}{mc}\right)^2 \int_{\omega_1}^{\omega_2} \frac{d\omega}{\omega} = \frac{2\alpha}{\pi}\left(\frac{\hbar}{mc}\right)^2 \ln(\frac{\omega_2}{\omega_1}), \tag{14}$$

where the logarithm is known as Bethe's integral and $\alpha = e^2/4\pi\epsilon_0\hbar c$ is the fine structure constant (coupling parameter). Computing the lower limit of Bethe's integral, e.g. for the grounded environment of radius $a = 1$ m, we obtain an estimate $\omega_1 = 2.74\ c/a = 8.2 \times 10^8\ \mathrm{s}^{-1}$, the mean square root amplitude of Zitterbewegung will be $\sim \sqrt{\langle q^2 \rangle} \approx 7.2 \times 10^{-13}$ m, which is a value $200\times$ larger than the classical electron radius and comparable with Compton radius of electron. Interestingly, this picture resembles the Dirac model of electron, i.e. point charge oscillating in a sphere of Compton radius with frequency ranging up to ω_C.

Notice, the way of argumentation in the frame of quantum mechanics is quite opposite.[133] The starting Ansatz in QED is Heisenberg's uncertainty relation from which one can derive that an electron has to have inevitably a rest energy bound to the jiggling equal to $\sim \hbar\omega/2$ at frequency ω. This energy is, surprisingly, not interpreted as a real free energy of the electron, but only as a virtual quantity intrinsic to every "quantum" particle which is necessary for keeping the commutation properties of corresponding operators. This difficult logic is, as we believe, attainable only because of the microscopic nature of Zitterbewegung and related quantum process in general. The minuteness of effects which are of "quantum nature" is, however, not a universal property. (The black-body radiation of the Sun is a good counter-example!)

5. Periodic reactions and quantum diffusion

There are, moreover, even unaided eye observations revealing quantum effects on the laboratory scale, namely so called periodic self-organised diffusion-controlled reactions (e.g. Belousov-Zhabotinsky's waves, Liesegang's rings[134]) playing extraordinary role especially in biological structures. Curiously, in some cases the effective diffusion action corresponding to these reactions formally satisfies "quantum mechanical" relationship

$$Mv\lambda \approx \hbar, \tag{15}$$

where M is the molecular weight of precipitate, v is the speed of spreading of the reaction front, and λ is a length parameter characterizing macroscopic reaction patterns. The wave-like nature of these process, revealing e.g. interference and diffraction effects in which the parameter λ plays the role quite analogous to that of the de Broglie wavelength in quantum mechanics, is very impressive. As has been recently shown[135,136] such a somewhat enigmatic behaviour can be accounted for by the classical theory of Brownian motion just completed by an assumption that the stochastic process behind is controlled prevailingly by the interaction between diffusing molecules and ZP radiation. Formally the transition from classical to quantum regime may be achieved if $D \to D_Q$, where D is the diffusion constant and D_Q is so called Fürth's parameter defined as

$$D_Q = \frac{\hbar}{2M}. \tag{16}$$

As experiments showed, relation (15) is valid in case where empirical value of D approaches the quantity D_Q.

5.1. *Quantum diffusion and stochastic electrodynamics*

The problem just treated is closely related to attempts to derive Schrödinger equation directly from the classical theory of Brownian motion and/or diffusion. The striking analogy between diffusion equation (generalized Fick's law) and Schrödinger equation was already observed in the 1930's by Schrödinger[24] and Fürth.[25] The latter showed that formal substitution of diffusion constant by iD_Q (i is the imaginary unit and D_Q is defined above by equation (16)) in equations describing diffusion and Brownian motion transforms them into Schrödinger equation. The epistemological problem arising from the presence of imaginary unit in the transformation is very likely due to the phase-space based formalism used both in classical analytical and quantum mechanics.[16] In configuration space, namely, the quantum motion and diffusive motion of a particle are indistinguishable having the same Hausdorff's dimensions there.[137] The only difference between these two cases is the difference in sources of stochastic behaviour; i.e. molecular collisions for Brownian diffusion and interaction with random ZPR for quantum motion. From this point of view de Broglie wave and wave-like properties of particles may be interpreted as a direct consequence of interaction of the particle with real fluctuating ZPR background.

6. Stability of matter and weak localization

In this section we will first deal with phenomenon of stability of matter which has often been discussed in both QED and SED. Motivated by related considerations we will then discuss an experimentally observed phenomenon, weak localization of electrons, which is commonly treated by quantum mechanics, but up to our knowledge it has never been discussed from the point of view of SED. Within the standard approach this phenomenon is understood as to be caused by quantum interference.[138] We will see that the effect of weak localization can be interpreted in terms of SED, too. We will also briefly deal with a possibility to describe superconductivity within SED. There is a class of superconducting materials the mechanism of superconductivity of which may be based on the effect of weak localization.

6.1. *Stability of matter and SED*

It is an enormously fruitful idea developed in the frame of SED that the moving charged particle, electron for example, can be kept on a stationary orbit in consequence of dynamical equilibrium between absorbed ZPR and emitted recoil radiation.[7] In quantum mechanics the stationary states being defined as solutions, i.e. eigenfunctions, of Schrödinger's equation may be mathematically represented by a stationary wave-function pattern. Such a pattern is, speaking more physically, a result of constructive interference of electron waves which is compatible with boundary conditions characterizing a given system. The essential features of just mentioned duality between SED and QED approach to the stationary states (ZPR-recoil radiation equilibrium in SED versus constructive interference in QED) may be illustrated by the following simple model of an electron trapped on a closed orbit.

Because of changes in direction of its movement an electron moving in an environment where it suffers only elastic collisions has to emit recoil radiation of power P_E given by Larmor's formula[139]

$$P_E = \frac{e^2}{6\pi\epsilon_0 c^3} v^2 \Omega^2, \qquad \Omega = \theta\frac{v}{\lambda}, \tag{17}$$

where v is the classical speed of the electron, λ is its free path, and the change of direction characterized by angle θ. Ω is thus the effective angular frequency, see Figure 2. Zero-point radiation P_A absorbed simultaneously by the electron within

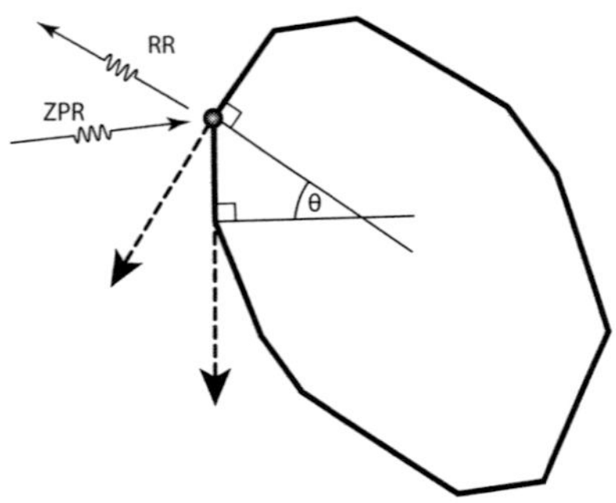

Fig. 2. Illustration of a movement of electron on a closed orbit. Electron is maintained in a stationary state by an equilibrium between ZPR and recoil radiation (RR).

the frequency range from 0 to Ω can be computed by the following formula[140]

$$P_A = \frac{\hbar^2 \lambda \Omega^4}{3\pi^2 m c^3}. \tag{18}$$

Equating (17) and (18), we immediately obtain a relation putting the "quantum" conditions on the possible shape of the orbit

$$P_A = P_E \quad \Rightarrow \quad \lambda\left(\frac{\Omega}{v}\right)^2 = \frac{mc^2}{2\pi^2\epsilon_0}. \tag{19}$$

Remarkably, for the case of electron in hydrogen atom, where $\theta = 2\pi$ and $\lambda = 2\pi a_B$ "quantum" condition (19) reads

$$a_B = \frac{4\pi\epsilon_0 \hbar^2}{me^2}, \tag{20}$$

giving thus its first Bohr's radius a_B. Based on this result, we can conclude that the stability of matter in general can be, at least in principle, accounted for by the very presence of ZPR in electronic processes.

6.2. *Weak localization: an interference phenomenon explained by SED*

Obviously, formula (19) can be used for the description of more general class of closed orbits than are atomic orbits. As an example, the so called weakly localized orbits can be given. These relatively very large (typically $\sim 10^{-7} - 10^{-6}$ m) and stable orbits exist in disordered metals and semiconductors where they are responsible for a plenty of interesting transport effects. The weakly localized orbits are within the frame of quantum mechanics explained as large interference patterns constructively composed of partial electron waves split on imperfections in conductor.

The corresponding SED description of weakly localized orbit which is, as was mentioned above, based on the balance of ZPR and recoil radiation is, however, very convenient for treatment of cases where this balance is disturbed by some external mean. For example, SED theory enables us to compute very simply the effect of a conducting (reflecting) cavity limiting the access of ZPR to the electronic system enclosed. Figure 3 depicts the temperature dependence of "quantum correction to

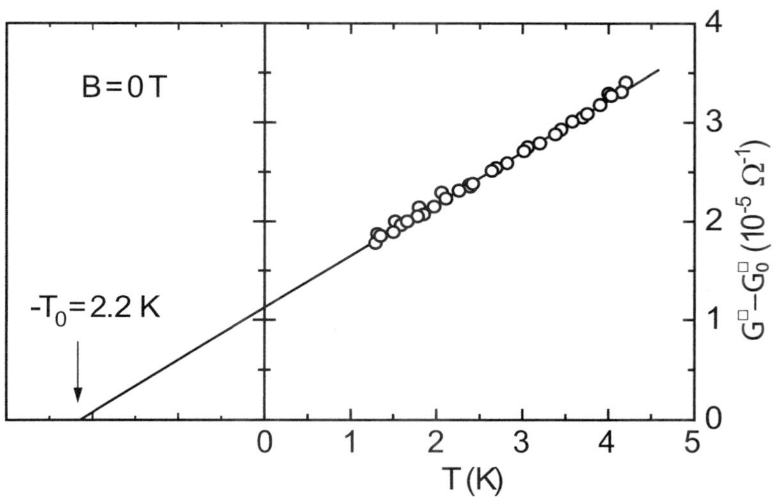

Fig. 3. Temperature dependence of quantum correction to conductivity of InP:S based δ-layer.

conductivity" $G - G_0$ as observed in a two-dimensional InP:S based δ -layer which is due just to the effect of weak localization.[140] The curve, which should theoretically go through the origin of coordinates, is, however, shifted significantly. Assuming that the sample itself acts as a dielectric cavity for the buried δ-layer where the weakly localized orbits controlling transport exist, the value of parameter T_0 characterizing this shift can be estimated from the relative permittivity ϵ and dimensions L of the sample by relation

$$T_0 \approx \frac{4\pi c\hbar}{kL\sqrt{\epsilon - 1}}. \tag{21}$$

As for InP $\epsilon \sim 12$ and the largest dimension of the sample was $L \sim 4 \times 10^{-3}$ m this formula derived within SED theory gives a value $T_0 = 2.1$ K in excellent agreement with experiment.

6.3. *Unconventional superconductivity in Mott's metals*

It is worth to noticing, that the weakly localized states can be, under a rather special conditions and at low temperatures, chained into quite macroscopic (i. e. $\sim$ 1 cm) collective state responsible for unconventional superconductivity observed in highly boron doped polycrystalline diamond.[141]

This is an example of existence of stable macroscopic system represented by collective wave function maintained by the ZPR.

7. Conclusions

Summarizing, there are two alternative theories, namely QED and SED, covering practically the same field of so called quantum phenomena. While SED is based on the reality of zero-point electromagnetic fluctuations in vacuum, this peculiar entity re-appears in QED only as an important virtual construction. It is a historical fact that the quantum mechanics and its continuation, QED, became a governing theory in this field. Moreover, as the ZPR plays in most practical applications of quantum mechanics no or only marginal role, the question of reality of ZPR seems to be of rather academic interest. We are convinced, however, that the decision between reality and non-reality of zero-point fluctuations in vacuum is a significant epistemological problem which is solvable only by experimental means. Therefore, brief accounts of some quantum effects which can be interpreted in the frame of SED using the reality of ZPR as a starting point are given. Our attention was devoted mainly to the points which are almost neglected in the literature, for example, to the close connection between measurement of the black-body radiation spectrum and measurement of noise in an electronic circuit with emphasis put on the conditions under which the ZP component of radiation or noise can be observed in principle. We have further turned our attention to the discussion of experiments with weakly localized orbits in solid state systems the stability of which is controlled by the flux of ZPR. In spite of the fact that these experiments have not the character of experimentum crucis, we are convinced that they may contribute essentially to the solution of the question of the real existence of ZPR in vacuum.

Acknowledgements

The work was partially supported by the Grant Agency of ASCR Contract No IAA1010404 and by the Czech Science Foundation Contract Nos 202/04/585 and 202/06/0040. Furthermore, the research work at the Institute of Physics was supported by Institutional Research Plan No AV0Z10100521.

References

1. M. Planck, *Ann. Phys.* **1**, 69 (1900).
2. M. Planck, *Ann. Phys.* **37**, 642 (1912).
3. A. Einstein and O. Stern, *Ann. Phys.* **40**, 551 (1913).
4. W. Nernst and Ver. Deutsch, *Phys. Ges.* **18**, 83 (1916).
5. H. B. G. Casimir, *Comments on Modern Physics* **1** Part D, 175 (2000).
6. P. W. Milloni, *The Quantum Vacuum - An Introduction to Quantum Electrodynamics*, (Academic Press, Inc, N. Y. 1994).
7. L. de la Pena and A. M. Cetto, *The Quantum Dice - An Introduction to Stochastic Electrodynamics*, (Kluwer Academic Publishers, Dordrecht, 1996).
8. E. A. Power, *Am. J. Phys.* **34**, 516 (1966).
9. B. Davies, *Am. J. Phys.* **34**, 516 (1966).
10. G. Barton, *Phys. Rev. A* **5**, 468 (1972).
11. R. J. Adler, B. Casey, and O. C. Jacob, *Am. J. Phys.* **63**, 620 (1995).
12. S. Weinberg, *Rev. Mod. Phys.* **61**, 1 (1989).
13. P. J. E. Peebles, B. Ratra, *Rev. Mod. Phys.* **75**, 559 (2003).
14. J. A. Stratton, *Electromagnetic Theory*, (Mc Graw-Hill book Co, New York, 1941).
15. G. G. Comisar, *Phys. Rev.* **138**, B 1332 (1965).
16. E. Nelson, *Phys. Rev.* **150**, 1079 (1966).
17. R. Schiller, H. Tesser, *Phys. Rev. A* **3**, 2035 (1971).
18. P. Schramm, and H. Grabert, *Phys. Rev.* **34**, 4515 (1986).
19. G. Y. Hu, R. F. O'Connell, *Phys. Rev. B* **46**, 14219 (1992).
20. M. Corregi, G. Morchio, *Ann. Phys.* **296**, 371 (2002).
21. L. Feligioni, O. Panella, Y. N. Srivastava, and A. Widom, *Eur. Phys. J. B* **48**, 233 (2005).
22. A. Abe, S. Thurner, *Physica A* **356**, 403 (2005).
23. A. Abe, A. K. Rajagopal, *Physica A* **375**, 129 (2007).
24. E. Schrödinger, *Ann.Inst. Henri Poincar*, **11** (1932) 300.
25. R. Fürth, *Zeitschr. f. Physik* **81** (1933) 143.
26. A. Einstein, L. Hopf, *Ann. Phys.* **33**, 1105 (1910).
27. T. A. Welton, *Phys. Rev.* **74**, 1157 (1948).
28. T. W. Marshall, *Proc. Roy. Soc. London A* **276**, 475 (1963).
29. T. W. Marshall, *Proc. Cambridge Phil. Soc.* **61**, 537 (1965).
30. T. W. Marshall, *Nuovo Cimento* **38**, 475 (1965).
31. T. H. Boyer, *Phys. Rev.* **174**, 1764 (1968).
32. T. H. Boyer, *Phys. Rev.* **182**, 1374 (1969).
33. T. H. Boyer, *Phys. Rev. D* **11**, 790 (1975).
34. T. H. Boyer, *Phys. Rev. D* **30**, 1228 (1984).
35. T. H. Boyer, *Sci. Am.* **253**, 70 (1985)

36. T. H. Boyer, *Found. Phys.* **19**, 1371 (1989).
37. T. H. Boyer, *Found. Phys.* **32**, 1 (2002).
38. D. C. Cole, *Phys. Rev. A* **42**, 1847 (1990).
39. D. C. Cole, *Found. Phys. A* **20**, 225 (1990).
40. D. C. Cole, *Phys. Rev. A* **42**, 7006 (1990).
41. D. C. Cole,*Phys. Rev. A* **45**, 8471 (1992).
42. D. C. Cole, *Phys. Rev. A* **45**, 8953 (1992).
43. D. C. Cole, *Phys. Rev. E* **51**, 1663 (1995).
44. D. C. Cole, A. Rueda, and K. Danley, *Phys. Rev. A* **63**, 054101-1 (2001).
45. D. C. Cole, *Found. Phys.* **30**, 1849 (2000).
46. D. C. Cole and Y. Zou, *Phys. Rev. E* **69**, 016601(2003).
47. D. C. Cole and Y. Zou, *Phys. Lett. A* **317**, 14 (2004).
48. D. C. Cole and Y. Zou, *J. Sci. Computing* **21**(2004).
49. L. de la Peña and A. M. Cetto, *Found. Phys* **12**, 1017 (1982).
50. A. M. Cetto, L.de la Peña, *Phys. Rev. A* **37**, 1952 (1988).
51. A. M. Cetto, L. de la Peña, *Phys. Rev. A* **37**, 1960 (1988).
52. A. M. Cetto, L. de la Peña, and R. M. Velasco, *Phys. Rev. A* **39**, 2747 (1989).
53. L. de la Peña and A. M. Cetto, *Found. Phys.* **24**, 753 (1994).
54. L. de la Peña and A. M. Cetto, *Found. Phys* **24**, 917 (1994).
55. H. M. Franca, T. W. Marshall, *Phys. Rev. A* **38**, 3258 (1988).
56. H. M. Franca, T. W. Marshall, E. Santos, *Phys. Rev. A* **45**, 6436 (1992).
57. K. Dechoum, H. M. Franca, *Found. Phys.* **25**, 1599 (1995).
58. H. M. Franca, H. Franco, and C. P. Malta, *Eur. J. Phys.* **18**, 343 (1997).
59. R. Blanco, H. M. Franca, E. Santos, R.C. Sponchiado, *Phys. Lett. A* **282**, 349 (2001).
60. H. E. Puthoff, *Phys. Rev.D* **35**, 3266 (1987).
61. H. E. Puthoff, *Phys. Rev. A* **39**, 2333 (1989).
62. H. E. Puthoff, *Phys. Rev. A* **40**, 4857 (1989).
63. H. E. Puthoff, *Phys. Rev. A* **44**, 3385 (1991).
64. H. E. Puthoff, *Found. Phys. Lett.* **6**, 75 (1993).
65. H. E. Puthoff, *Found. Phys.***32**, 927 (2002).
66. A. Rueda, *Found. Phys. Lett.* **6**, 75 (1993).
67. A. Rueda, *Found. Phys. Lett.* **6**, 139 (1993).
68. B. Haisch, A. Rueda, H. E. Puthoff, *Phys. Rev. A* **49**, 678 (1994).
69. M. Ibson, B. Haisch, *Phys. Rev. A* **54**, 2737 (1996).
70. Y. Dobyns, A. Rueda, B. Haisch, *Found. Phys.* **30**, 59 (2000).
71. B. Haisch, A. Rueda, and Y. Dobyns,*Ann. Phys.*(Leipzig) **10**, 393 (2001).
72. R. Blanco, L. Pesquer, and E. Santos,*J. Math.Phys.* **28**, 1749 (1987).
73. A. Casado, T. W. Marshall, R. Risco-Delgado and E. Santos, *Eur. Phys. J. D* **13**, 109 (2001).
74. F. Guerra, *Phys. Rep.* **77**, 263 (1981).
75. M. Surdin, *Int.J. Theor. Phys.* **14**, 207 (1975).
76. P. C. W. Davies, *J. Phys.A: Math. gen.* **8**, 609, (1975).
77. W. G. Unruh, *Phys. Rev. D* **14**, 870 (1976).
78. R. H. Koch, D. J. VanHarlingen, and J. Clarke, *Phys. Rev. Lett.* **45**, 2132 (1980).
79. R. H. Koch, D. J. VanHarlingen, and J. Clarke, *Phys. Rev. Lett.* **47**, 1216 (1981).
80. R. H. Koch, D. J. VanHarlingen, and J. Clarke, *Phys. Rev. B* **26**, 74 (1982).
81. J. Ambjorn, S. Wolfram,*Ann. Phys.* **147**, 1 (1983).
82. J. Ambjorn, S. Wolfram, *Ann. Phys.* **147**, 33 (1983).
83. L. J. Landau, *Phys. Rev. A* **37**, 4449 (1988).
84. J. Ford, G. Mantica, *Am. J. Phys.* **60**, 1086 (1992).

85. A. Kyprianidis, *Found. Phys.* **22**, 1449 (1992).
86. A. F. Kracklauer, *Physics Essays* **5**, 226 (1992).
87. J. P. Dowling, *Found. Phys.* **23**, 895 (1993).
88. G. Manfredi, S. Mola, and M. R. Feix, *Eur. J. Phys.* **14**, 101 (1993).
89. M. Seidl, P. O. Lipas, *Eur. J. Phys.* **17**, 25 (1996).
90. S. Chaturvedi and P. D. Drummond, *Phys. Rev. A* **55**, 912 (1997).
91. D. T. Pope, P. D. Drummond, and W. J. Munro, *Phys. Rev. A* **55**, 912 (1997).
92. A. F. Kracklauer, *Found. Phys. Lett.* **12**, 441 (1999).
93. L. H. Ford, N. F. Svatier, *Phys. Rev. A* **62**, 062105-1 (2000).
94. H. J. Carmichael, and H. Nha, *J. Opt. B* **6**, S645 (2004)
95. P. W. Milloni, *Am. J. Phys.* **49**, 177 (1981).
96. M. Cray, Me-Li Shih, and P. W. Milloni, *Am. J. Phys.* **50**, 1016 (1982).
97. P. W. Milloni, *Am. J. Phys.* **52**, 340 (1984).
98. P. W. Milloni, *Phys. Scripta T* **21**, 102 (1988).
99. P. W. Milloni, M. L. Shih, *Am. J. Phys.* **59**, 684 (1991).
100. P. W. Milloni and M. L. Shih, *Cont. Phys.* **33**, 313 (1992).
101. P. W. Milloni, M. L. Shih, *Phys. Rev. A* **45**, 4241 (1992).
102. P. W. Milloni, A. Smith, *Phys. Rev. A* **53**, 3484 (1996).
103. P. W. Milloni, M. Schaden, and L. Spruch, *Phys.Rev. A* **59**, 4259 (1999).
104. A. A. Chumak, P. W. Milloni, and G. P. Berman, *Phys. Rev. B* **70**, 085407 (2004).
105. P. W. Milloni, *Phys. Rep.* **25** 1 (1976).
106. D. C. Cole, and A. Rueda, *Found. Phys.* **26**.
107. A. Rueda, B. Haisch, *Ann. Phys.* (Leipzig) **14**, 479 (2005).
108. L. de la Peña and A. M. Cetto, *http://arXiv.org/abs/quantum-ph 0501011*, (2005).
109. D. C. Cole, *Quantum Theory- Reconsideration of Foundations-3*, (Proceedings of the Symposium, Växjö, Sweden, June 2005), (AIP Conference Proceedings **810**, 99, Melville, New York, 2006).
110. Th. M. Nieuwenhuizen, *Quantum Theory- Reconsideration of Foundations-3*, (Proceedings of a Symposium, Växjö, Sweden, June 2005), (AIP Conference Proceedings **810**, 198, Melville, New York, 2006).
111. L. de la Peña and A. M. Cetto, *Quantum Theory- Reconsideration of Foundations-3*, (Proceedings of a Symposium, Växjö, Sweden, June 2005), (AIP Conference Proceedings **810**, 131, Melville, New York, 2006).
112. L. de la Peña and A. M. Cetto, *Found. Phys.* **36**, 350 (2006).
113. E. W. Davis, V. L. Teofilo, B. Haisch, H. E. Puthoff, L. J. Nieksch, A. Rueda, and D. C. Cole, *Review of Experimental Concepts for Studying the Quantum Vacuum Field*, (CP813, Space Technology and Aplications Forum-STAIF 2006), (Am. Inst. Phys, 2006) pp.1390,
114. R. L. Forward, *Phys. Rev. B* **30**, 1700 (1984).
115. S. K. Lamoreaux, *Phys. Rev. Lett.* **78**, 5 (1997).
116. M. E. Bowers and C. R. Hagen, *Phys. Rev. D* **59**, 025007 (1998).
117. F. Pinto, *Phys. Rev. B* **60**, 14740 (1999).
118. F. Chen, G. L. Klimchitskaya, U. Mohideen, and V. M. Mostepanenko, *Phys. Rev. A* **69**, 022117-1 (2004).
119. F. Chen, U. Mohideen, G. L. Klimchitskaya, and V. M. Mostepanenko, *Phys. Rev. A* **74**, 022103 (2006).
120. E. Elizalde and A. Romeo, *Am. J. Phys.* **59**, 711 (1991).
121. S. K. Lamoreaux, *Am. J. Phys.* **67**, 850 (1999).
122. M. Kardar, R. Golestanian, *Rev. Mod. Phys.* **71**, 1223 (1999).
123. M. Bordag, U. Mohideen, and V. M. Mostepanko, *Phys. Rep.* **205**, 353 (2001).

124. P. W. Milloni, *Sol. State. Comm.* **136**, 433 (2005).
125. S. K. Lamoreaux, *Rep. Prog. Phys.* **68**, 201 (2005).
126. K. A. Milton, *The Casimir effect*, (World Scientific, River Edge, N. J., 2001).
127. H. Weber,*Kleines Lehrbuch der Algebra*, (Vieweg und Sohn, Braunschweig, 1921), p.78.
128. J. J. Mareš, V. Špička, J. Krištofik, P. Hubík, and J. Šesták, *Physica E* **29**, 213 (2005).
129. G. Kirchhoff, *Ann. der Phys. und Chemie* **219**, 275 (1860).
130. H. Nyquist, *Phys. Rev.* **32**, 110 (1928).
131. J. Weber, *Phys. Rev.* **90**, 977 (1953).
132. O. Lummer und E. Pringsheim, *Ann. der Phys. und Chemie* **63**, 395 (1897).
133. D. Park and H. T. Epstein, *Am. J. Phys.* **17**, 301 (1949).
134. F. F. Runge, R. E. Liesegang, B. P. Belousov, and A. M. Zhabotinsky in: *Selbsorganisation Chemischer Strukturen*, Eds. by L. Kuhnert and U. Niedersen (Ostwald's Klassiker, Verlag H. Deutsch, Frankfurt am Main, 1987)
135. J. J. Mareš, J. Stávek, J. Šesták, *J. Chem. Phys.* **121**, 1499 (2004).
136. J. J. Mareš, J. Šesták, J. Stávek, H. Ševčíková, J. Krištofik, and P. Hubík, *Physica E* **29**, 145 (2005).
137. L. F. Abbott, M. B. Wise, *Am. J. Phys.* **49**, 37 (1981).
138. G. Bergmann, *Phys. Rev. B* **28**, 2914 (1983).
139. J. Jackson, *Classical Electrodynamics*, (John Wiley and sons, 1975).
140. J. J. Mareš, J. Krištofik, P. Hubík, and V. Špička,*Physica E* **29**, 375 (2005).
141. J. J. Mareš, P. Hubík, M. Nesládek, D. Kindl, J. Krištofik, *Diamond. Relat. Mater* **15**, 1863 (2006).

ON THE ERGODIC BEHAVIOUR OF ATOMIC SYSTEMS UNDER THE ACTION OF THE ZERO–POINT RADIATION FIELD

L. DE LA PEÑA

Instituto de Física, Universidad Nacional Autónoma de México
Apartado postal 20-364, 01000 México, DF.
E-mail: luis@fisica.unam.mx

A. M. CETTO

Instituto de Física, Universidad Nacional Autónoma de México
Apartado postal 20-364, 01000 México, DF.
and
Deputy Director-General, IAEA. Vienna, Austria.

We study anew the behaviour of an otherwise classical bound particle immersed in a radiation field that includes the zero-point field component of average energy $(1/2)\hbar\omega$ per mode. The presence of this field introduces an essential stochasticity into the dynamics of the particle, characterized by Planck's constant $\hbar$; this has been the basis for stochastic electrodynamics. Both the near field and the particle are affected substantially by their continuous interaction. Stationary solutions are in principle possible when a balance is achieved between the mean powers emitted and absorbed by the particle. By demanding that the ensuing approximate stationary solutions satisfy an ergodic principle, we are led to a resonant response that is linear in the Fourier amplitudes of the field; this is the essence of linear stochastic electrodynamics. The connection with the matrix formulation of quantum mechanics can be readily made, with the resonance frequencies of the ergodic solutions corresponding to the quantum mechanical transition frequencies. Some implications of these results for the understanding of quantum phenomena are briefly discussed.

Keywords: Stochastic electrodynamics; Linear stochastic electrodynamics; Quantum mechanics; Semi-classic; Ergodicity.

1. Introduction

In the current quantum description we find both an effective formal apparatus of widespread and far-reaching applications, and the need to break several deep and basic rules of (non-quantum) physics to interpret the formalism. Indeed, learning quantum mechanics is tantamount to learning a new philosophy of physics along with its new and quite sophisticated formalism. In saying this we are not revealing anything new: every student learns this truth the hard way in her/his first encounters with the subject. As he is told that there is not much to understand beyond the formalism even Feynman, the brilliant archetype of intuitive physical thinking, said that nobody understands quantum mechanics he ends up limiting his curiosity to learning how to use and exploit it. No wonder that in some sectors discomfort with this state of affairs sprung almost from the birth of the theory itself. The best known example of the uneasiness produced by the demand to abandon locality, realism, determinism, causality, even the possibility of getting an image of the piece of physical world under scrutiny, is of course the decade long debate between Einstein and Bohr about the principles of quantum mechanics. It has become commonplace to hear that Einstein lost the debate, that he was already too old to understand quantum mechanics. But it would be a mistake to reduce the issue to such a simplistic scheme. This is not a question of who wins or who loses, it is a matter of profound physics: do we have to renounce forever to a real (not merely formal) understanding of what is happening outside there, in the physical world, not just in our minds?

Stochastic electrodynamics (SED) is a systematic attempt to improve our understanding of the meaning of quantum mechanics, and one that has survived several decades of both research and criticism. It has been blamed and this seems to be the single serious criticism to it of being a semi-classical theory. With some reason, since the methods we all used in the old attempts to develop the theory impressed upon it a semi-classical gist. What we briefly present here is a version of it, called linear stochastic electrodynamics LSED for short, free of the basic problems that beset the original theory of SED and characterized by as little speculation as a new fundamental theory could possibly have.

A couple of remarks are in place. Firstly, this is not yet a finished theory; thus whilst we are dealing with its fundamentals, many details are still unknown to us. Although whether the theory can be developed in full is for the moment a matter of speculation, we do not foresee unsurmountable difficulties. We should also add that this form of SED does not represent the only possibility to go ahead; indeed, there are several different attempts underway of a very different nature and differing also from ours. One is the work that D. Cole and co-worker are carrying out by direct and careful computation of the electron's motion in the H atom, with very interesting statistical results.[1–3] A second extensive and long-standing effort is stochastic optics, which represents the optical branch of SED and is being developed with notable success by T. Marshall, E. Santos and co-workers.[4–6] Another attempt is that by T. H. Boyer, who looks for a careful account of the relativistic electron that complies with the full conformal symmetry.[8] Another attempt of a related nature is that of Th. M. Nieuwenhuizen.[9]

2. The Fundamentals of Linear Stochastic Electrodynamics

2.1. *Definition of the system under study*

The key idea behind SED is that the quantum system is recognized as being stochastic. Thus the existence of (apparently irreducible) fluctuations, as manifested e.g. by the Heisenberg uncertainties, is interpreted as a manifestation (or measure, if preferred) of such stochasticity. It is concurrently assumed that there is a physical cause for such fluctuations, so they are neither physically irreducible in principle nor unexplainable, and the question about such cause becomes central. In SED (and in LSED) one identifies this cause with the vacuum radiation field, or zero-point field (zpf), with which the atomic system is in permanent interaction. The electromagnetic zpf is of course only one among several such vacua, but it is assumed that for the kind of systems considered in quantum mechanics, this is the essential one, the effects of other possible vacua being negligibly small for them, at least in the nonrelativistic description.

The field to be here considered may or may not contain an external electromagnetic radiation, but it *always* comprises at least the vacuum field. This latter is a stochastic field, solution of the Maxwell equations with appropriate boundary conditions at infinity, and with all its elementary oscillators in the ground state, which means an energy per mode of value $\hbar\omega/2$. This corresponds to a spectral energy density proportional to ω^3, which is the single relativistic invariant spectrum (a detailed discussion and references to early work can be seen in,[10] hereafter referred to as *The Dice*).

Owing to the mathematical difficulties of this general problem, it has been common practice in SED to limit the studies to the mechanical (atomic) part, and make "reasonable" assumptions about the statistical properties of the field, assumed to be stationary. However, even with this approximation the problem cannot be explicitly solved in the general case, so some further approximations became necessary. In the customary form of SED (say, until 1982) the employment of perturbative methods and the simultaneous use of a Markovian approximation were usual. Even if this procedure eventually turned out to be inappropriate for dealing with nonlinear problems, a series of satisfactory, and in some cases very satisfactory results were obtained (basically for linear problems), which signaled towards the soundness of the principles of the theory. When the procedure was applied to more interesting nonlinear problems, as the H-atom, the results were disappointing. (The interested reader can found detailed accounts and references to earlier work in *The Dice*, so we skip them here.) In order to escape from the intrinsic limitations manifested by such methods, in LSED we investigate the same system but avoiding the use of any perturbative or Markovian approximation.

The theory of LSED has been discussed in several other places[10–14] ; here we develop it in a new form which we hope is more cogent, using a minimum of principles for its formulation. An important characteristic of LSED that distinguishes it from the earlier versions of SED , is that certain statistical properties of the random field are not assumed fixed from the beginning, as if they were just those of the free vacuum field, but they are left to be determined by the theory itself from the demand of self-consistency under stationary conditions, thus allowing the field to be modified by its interaction with matter. The key element that we introduce here is what we call, to use a broad term, the principle of ergodicity. In our previous discussions, in place of this principle we used another one, variously called of minimum stochasticity or of independence from the field realizations. We feel that the use of this principle makes the theory simpler and much more transparent, hence more appealing.

2.2. *Definition of the mechanical component and the quantum regime*

Due to the limitations of space we give here only a very reduced squetch of the fundamentals of the theory, details of which can be found in *The Dice.* From the full Hamiltonian of the SED system one derives the equation of motion for its mechanical part by eliminating the canonical variables that refer to the field, a procedure that leads to the Abraham-Lorentz equation (usually called in SED the Braffort-Marshall equation), namely

$$m\ddot{\mathbf{x}} = \mathbf{f}(\mathbf{x}) + m\tau\dddot{\mathbf{x}} + e\mathbf{E}(\mathbf{x},t). \tag{1}$$

In the nonrelativistic approximation the magnetic contribution to the Lorentz force can be disregarded and by the same token, the electric field can be written in the long-wavelength approximation in terms of a Fourier time transform

$$\mathbf{E}(t) = \int_{-\infty}^{+\infty} \widetilde{\mathbf{E}}(\omega)a(\omega)e^{-i\omega t}d\omega \tag{2}$$

with $\widetilde{\mathbf{E}}^*(\omega) = \widetilde{\mathbf{E}}(-\omega)$, $a^*(\omega) = a(-\omega)$ for $\omega \geq 0$. Each of the coefficients $\widetilde{\mathbf{E}}(\omega)$ contains the infinity of modes of the field that belong to the same frequency ω, with all possible orientations of the wave vector $|\mathbf{k}| = \omega/c$ and both polarizations, so that the (dimensionless) stochastic variables $a(\omega)$ describe the mean behavior of the modes of frequency ω. In the usual form of SED all statistical properties of the stochastic variables $a(\omega)$ are assumed to be those of the *free* vacuum field (when no external field is present). However, as said above we are here interested in the properties of $a(\omega)$ in *presence* of matter, which are to be determined by the theory itself.

We shall focus on the stationary solutions, which are characterized as follows. Multiplying Eq. (1) by $\dot{\mathbf{x}}$ and rearranging we get for the average over the field realizations,

$$\left\langle \frac{dH}{dt} \right\rangle = -m\tau \left\langle \ddot{\mathbf{x}}^2 \right\rangle + e \left\langle \dot{\mathbf{x}} \cdot \mathbf{E} \right\rangle, \tag{3}$$

where H stands for the Hamiltonian of the particle, including the Schott energy,

$$H = \tfrac{1}{2}m\dot{\mathbf{x}}^2 + V(\mathbf{x}) - m\tau\dot{\mathbf{x}} \cdot \ddot{\mathbf{x}}, \tag{4}$$

and $V(\mathbf{x})$ is the potential associated with the external force $\mathbf{f}(\mathbf{x})$. The contribution $-m\tau \left\langle \ddot{\mathbf{x}}^2 \right\rangle$ represents the mean power radiated by the charge and is due basically to the orbital motion, whereas $e \left\langle \dot{\mathbf{x}} \cdot \mathbf{E} \right\rangle$ gives the mean power absorbed from the field, whose principal contribution comes from the highly irregular motion impressed upon the particle by the field. Stationary states are reached when these two contributions cancel out, so that

$$\left\langle \frac{dH}{dt} \right\rangle = 0 \tag{5a}$$

and

$$m\tau \left\langle \ddot{\mathbf{x}}^2 \right\rangle = e \left\langle \dot{\mathbf{x}} \cdot \mathbf{E} \right\rangle . \tag{5b}$$

When condition (5b) is met, the system has reached a stationary regime in the average. However, in LSED a more exacting condition must be satisfied, namely detailed energy (or power) balance, i.e., balance for each separate frequency, as is to be expected when the acting field is stationary, which is assumed to be the case for the zpf (for details see *The Dice*). This is the mechanism proposed by LSED to explain the existence of atomic stationary states despite radiation by the accelerated electron — or rather, thanks to it and to the compensating effect of the background field. Of course in the presence of the zpf alone the single equilibrium solution is the ground state, the excited states decaying with their characteristic lifetimes. Below we see how and when both the higher-order effects of the zpf and radiation reaction can be neglected once their zero-order effects (i.e., steering the system to the stationary state) have been taken into account in writing the appropriate solution to Eq. (1), thus allowing us to treat the excited states as if they were stationary, as is done in quantum mechanics.

3. The Principle of Ergodicity

3.1. *Resonances*

In what follows we use instead of Eq. (2) a discrete Fourier expansion, so we write considering for simplicity the one dimensional case

$$E(t) = \sum_\beta \widetilde{E}_\beta a_\beta e^{-i\omega_\beta t}. \tag{6}$$

We now write the solutions to Eq. (1) in the form

$$x(t) = \sum_\beta \widetilde{x}(\omega_\beta) a(\omega_\beta) e^{-i\omega_\beta t} \equiv \sum_\beta \widetilde{x}_\beta a_\beta e^{-i\omega_\beta t}, \tag{7}$$

and so on. To perform the Fourier expansion of the external force we assume that it can be expressed as a power series in $x(t)$ (without constant term); then using Eq. (7) we write

$$f(x(t)) = k_1 x(t) + k_2 x^2(t) + k_3 x^3(t) + \ldots \tag{8}$$

$$= k_1 \sum_\beta \widetilde{x}_\beta a_\beta e^{-i\omega_\beta t} + k_2 \sum_{\beta',\beta''} \widetilde{x}_{\beta'} \widetilde{x}_{\beta''} a_{\beta'} a_{\beta''} e^{-i(\omega_{\beta'}+\omega_{\beta''})t} + \ldots \tag{9}$$

The component of frequency ω_β is therefore

$$\widetilde{f}_\beta = \widetilde{f}(\omega_\beta) = k_1 \widetilde{x}_\beta a_\beta + k_2 \sum_{\beta',\beta''} \widetilde{x}_{\beta'} \widetilde{x}_{\beta''} a_{\beta'} a_{\beta''} + \ldots, \tag{10}$$

where the summations are to be performed over the set of indices $\beta', \beta'', \beta'', \ldots$ for which

$$\omega_{\beta'} + \omega_{\beta''} + \omega_{\beta'''} + \ldots = \omega_\beta. \tag{11}$$

Introducing Eq. (7) into (1) we may write for the component of frequency ω_β

$$-m\omega_\beta^2 \widetilde{x}_\beta a_\beta e^{-i\omega_\beta t} = \widetilde{f}_\beta e^{-i\omega_\beta t} + im\tau\omega_\beta^3 \widetilde{x}_\beta a_\beta e^{-i\omega_\beta t} + e\widetilde{E}_\beta a_\beta e^{-i\omega_\beta t}, \quad \text{or} \tag{12}$$

$$\widetilde{x}_\beta = -\frac{e}{m} \frac{\widetilde{E}_\beta}{\omega_\beta^2 + i\tau\omega_\beta^3 + \frac{\widetilde{f}_\beta}{m\widetilde{x}_\beta a_\beta}} \equiv -\frac{e}{m} \frac{\widetilde{E}_\beta}{\Delta_\beta}, \tag{13a}$$

$$\Delta_\beta = \omega_\beta^2 + i\tau\omega_\beta^3 + \frac{\widetilde{f}_\beta}{m\widetilde{x}_\beta a_\beta}. \tag{13b}$$

We see that the important contributions to $x(t)$ come from those frequencies that are zeros of Δ_β, i.e., the poles of $\widetilde{x}_\beta$. The resonances at these values are extremely sharp due to the very small value of τ (for all frequencies of interest for atoms we have $\tau\omega \lesssim \alpha^3$) Thus in

$$x(t) = \sum_\beta \widetilde{x}_\beta a_\beta e^{-i\omega_\beta t} = -\frac{e}{m} \sum_\beta \frac{\widetilde{E}_\beta}{\Delta_\beta} a_\beta e^{-i\omega_\beta t} \tag{14}$$

we can separate quite naturally the sum into the contributions that come from the resonances and the remaining noise. It is important to observe that the frequency of a resonance, which is very nearly given by

$$\omega_{r\beta}^2 = -\frac{\widetilde{f}_\beta}{m\widetilde{x}_\beta a_\beta}, \tag{15}$$

depends on the values of the set of $\widetilde{x}_\beta$ for the given response. Hence for different responses the set of those frequencies is different. To single out the selected response and resonant frequency we need an extra index, so we write $\omega_{\alpha\beta}$ instead of ω_β and so on, transforming Eq. (14) into

$$x_\alpha(t) = -\frac{e}{m} \sum_\beta \frac{\widetilde{E}_{\alpha\beta}}{\Delta_{\alpha\beta}} a_{\alpha\beta} e^{-i\omega_{\alpha\beta} t}. \tag{16}$$

This change should be introduced everywhere, particularly in Eq. (11). It simply evinces the fact that to each possible $x_\alpha(t)$ there corresponds a full set of components $\widetilde{x}_{\alpha\beta}$ and frequencies of resonance $\omega_{\alpha\beta}$. Eq. (14) can be seen to represent a set of oscillators of extremely high Q tuned to the resonance frequencies $\omega_{\alpha\beta}$.

3.2. *The ergodic principle*

It is important to observe that the above results depend on the specific realization of the random field. If we identify the current realization with an index (i) then we should write $\omega_{\alpha\beta}$ more explicitly as $\omega_{\alpha\beta}^{(i)}$, and so on, so that

$$x_\alpha^{(i)}(t) = \sum_\beta \widetilde{x}_{\alpha\beta}^{(i)} a_{\alpha\beta}^{(i)} e^{-i\omega_{\alpha\beta}^{(i)} t} = -\frac{e}{m} \sum_\beta \frac{\widetilde{E}_{\alpha\beta}^{(i)}}{\Delta_{\alpha\beta}^{(i)}} a_{\alpha\beta}^{(i)} e^{-i\omega_{\alpha\beta}^{(i)} t}. \tag{17}$$

We have written $\widetilde{E}_{\alpha\beta}^{(i)}$ because in general this quantity is a function of the frequency and depends on (i) through it. From the last equation we may write

$$\left(x_\alpha^{(i)}(t)\right)^2 = \sum_{\beta,\beta'} \widetilde{x}_{\alpha\beta'}^{(i)*} \widetilde{x}_{\alpha\beta}^{(i)} a_{\alpha\beta'}^{(i)*} a_{\alpha\beta}^{(i)} e^{i(\omega_{\alpha\beta'}^{(i)} - \omega_{\alpha\beta}^{(i)})t}, \tag{18}$$

so that separating the diagonal elements from the rest we get

$$\left(x_\alpha^{(i)}(t)\right)^2 = \sum_\beta \left|\widetilde{x}_{\alpha\beta}^{(i)}\right|^2 \left|a_{\alpha\beta}^{(i)}\right|^2 + \sum_{\beta'\neq\beta} \widetilde{x}_{\alpha\beta'}^{(i)*} \widetilde{x}_{\alpha\beta}^{(i)} a_{\alpha\beta'}^{(i)*} a_{\alpha\beta}^{(i)} e^{i(\omega_{\alpha\beta'}^{(i)} - \omega_{\alpha\beta}^{(i)})t}. \tag{19}$$

Since the poles given by Eq. (15) are in general complex numbers, the oscillating terms in the last expression should decay for solutions that satisfy Eq. (5b). Thus for sufficiently long times Eq. (19) reduces to

$$\left(x_\alpha^{(i)}(t)\right)^2 = \sum_\beta \left|\widetilde{x}_{\alpha\beta}^{(i)}\right|^2 \left|a_{\alpha\beta}^{(i)}\right|^2, \quad t \longrightarrow \infty. \tag{20}$$

From Eq. (19) it follows that

$$\left\langle \left(x_\alpha^{(i)}(t)\right)^2 \right\rangle = \sum_\beta \left\langle \left|\widetilde{x}_{\alpha\beta}^{(i)}\right|^2 \left|a_{\alpha\beta}^{(i)}\right|^2 \right\rangle + \sum_{\beta'\neq\beta} \left\langle \widetilde{x}_{\alpha\beta'}^{(i)*} \widetilde{x}_{\alpha\beta}^{(i)} a_{\alpha\beta'}^{(i)*} a_{\alpha\beta}^{(i)} e^{i(\omega_{\alpha\beta'}^{(i)} - \omega_{\alpha\beta}^{(i)})t} \right\rangle \tag{21a}$$

$$= \sum_\beta \left\langle \left|\widetilde{x}_{\alpha\beta}^{(i)}\right|^2 \left|a_{\alpha\beta}^{(i)}\right|^2 \right\rangle \tag{21b}$$

where the average is taken over the realizations of the field and the last equality holds for $t \longrightarrow \infty$. Further, taking the time average of the same equation (19) over long enough times we get

$$\overline{\left(x_\alpha^{(i)}(t)\right)^2} = \sum_\beta \left|\widetilde{x}_{\alpha\beta}^{(i)}\right|^2 \left|a_{\alpha\beta}^{(i)}\right|^2. \tag{22}$$

Comparing these two results we see that in general they are different

$$\left\langle \left(x_\alpha^{(i)}(t)\right)^2 \right\rangle = \langle x_\alpha^2(t)\rangle \neq \overline{\left(x_\alpha^{(i)}(t)\right)^2}, \tag{23}$$

since by construction the left hand side is independent of the realization of the field, whereas the right hand side gives a different result for each realization (i). This means that in general the system is not ergodic, not even in the one dimensional case, where only one integral of the motion exists. Let us examine this behavior more closely.

We first consider the case of the harmonic oscillator, for which Eq. (15) reduces to

$$\left.\frac{\widetilde{f}_\beta}{m\widetilde{x}_\beta a_\beta}\right|_{(i)} = -\omega_0^2, \tag{24}$$

where ω_0 is the frequency of the oscillator, a fixed quantity, the same for all (i). It follows that in this case $\omega_{\alpha\beta}$ and $\widetilde{x}_{\alpha\beta}$ are independent of the realization of the field, and so Eq. (21a) becomes

$$\left\langle (x_\alpha(t))^2 \right\rangle = \sum_\beta |\widetilde{x}_{\alpha\beta}|^2 \left\langle |a_{\alpha\beta}|^2 \right\rangle + \sum_{\beta'\neq\beta} \widetilde{x}^*_{\alpha\beta'}\widetilde{x}_{\alpha\beta} e^{i(\omega_{\alpha\beta'}-\omega_{\alpha\beta})t} \left\langle a^*_{\alpha\beta'} a_{\alpha\beta}\right\rangle, \tag{25}$$

where the now superfluous index (i) has been omitted. To proceed we need a few statistical properties of the stochastic variables $a_{\alpha\beta}$. A conventional rule is to assume that $a_{\alpha\beta}$ and $a_{\alpha\beta'}$ are statistically independent for $\beta \neq \beta'$, so we write

$$\left\langle a^*_{\alpha\beta'} a_{\alpha\beta}\right\rangle = |a_{\alpha\beta}|^2 \delta_{\beta\beta'}, \tag{26}$$

whence Eq. (25) reduces to

$$\left\langle (x_\alpha(t))^2 \right\rangle = \sum_\beta |\widetilde{x}_{\alpha\beta}|^2 \left\langle |a_{\alpha\beta}|^2 \right\rangle. \tag{27}$$

Comparing with Eq. (22), we see that for the condition of ergodicity to be satisfied we must have

$$\left\langle |a_{\alpha\beta}|^2 \right\rangle = \left| a^{(i)}_{\alpha\beta} \right|^2. \tag{28}$$

Since the left hand side is independent of (i), the right hand side must also be independent of the realization, This result tells us that the modulus of $a^{(i)}_{\alpha\beta}$ is independent of (i) and has a sure (non stochastic) value. Therefore, if we write the amplitudes $a_{\alpha\beta}$ in its polar form

$$a_{\alpha\beta} = r_{\alpha\beta} e^{i\varphi_{\alpha\beta}}, \tag{29}$$

the moduli $r_{\alpha\beta}$ are sure numbers. The phases $\varphi_{\alpha\beta}$ are of course random, and it is usual to assume them uniformly distributed over $[0, 2\pi]$ (better over $[-\pi, \pi]$, as will became clear below).

Now we come back to the general case. Since the particle is permanently in contact with the random field, we expect that the former can visit the whole accessible phase space, or, equivalently, that its behaviour is ergodic (in this restricted sense). This image is similar to the one used in the calculation of the transition amplitudes in quantum mechanics using the Feynman path integral method, where for the transition amplitude from point x_i to point x_f all trajectories with all possible momenta p leading from x_i to x_f are considered (not restricting the calculation to the energy surface). As was the case for the harmonic oscillator, it is assumed that also in the general case the behaviour must be ergodic, which means that $\overline{\left(x_\alpha^{(i)}(t)\right)^2}$ should not depend on the specific realization (i) of the field. Under this condition, from Eqs. (21b), (22), and (28) we can write for $t \longrightarrow \infty$,

$$\sum_\beta \left\langle \left|\widetilde{x}_{\alpha\beta}^{(i)}\right|^2 \left|a_{\alpha\beta}^{(i)}\right|^2 \right\rangle = \sum_\beta \left\langle \left|\widetilde{x}_{\alpha\beta}^{(i)}\right|^2 \left\langle \left|a_{\alpha\beta}^{(i)}\right|^2 \right\rangle \right\rangle = \sum_\beta \left\langle \left|\widetilde{x}_{\alpha\beta}^{(i)}\right|^2 \right\rangle \left|a_{\alpha\beta}^{(i)}\right|^2, \tag{30}$$

from which it follows that

$$\left|\widetilde{x}_{\alpha\beta}^{(i)}\right|^2 = \left\langle \left|\widetilde{x}_{\alpha\beta}^{(i)}\right|^2 \right\rangle. \tag{31}$$

Since the right hand side of this equation is independent of the realization (i), also $\widetilde{x}_{\alpha\beta}^{(i)}$ must be independent of it, which leads to

$$\left|\widetilde{x}_{\alpha\beta}^{(i)}\right|^2 = \left|\widetilde{x}_{\alpha\beta}\right|^2, \quad \text{or} \quad \widetilde{x}_{\alpha\beta}^{(i)} = \widetilde{x}_{\alpha\beta}. \tag{32}$$

This implies that also $\omega_{\alpha\beta}$ cannot depend on the specific realization of the field, because the entanglement between the $\widetilde{x}_{\alpha\beta}$ and the resonance frequencies $\omega_{\alpha\beta}$ entailed by Eq. (13) implies that any dependence of the latter on (i) would be echoed in the former; thus $\omega_{\alpha\beta}^{(i)} = \omega_{\alpha\beta}$. Combining this with Eqs. (13) and (15) we conclude that (in full notation)

$$\frac{\widetilde{f}_{\alpha\beta}}{m\widetilde{x}_{\alpha\beta}a_{\alpha\beta}^{(i)}} \text{ is independent of the realization of the field,} \tag{33}$$

since otherwise $\widetilde{x}_{\alpha\beta}$ would depend on such realization in the general nonlinear case. This is our main upshot from the principle of ergodicity.

4. Significance of the Result

The result given in Eq. (33) is central for our arguments. Indeed it has been obtained previously using the demand of minimum stochasticity or, equivalently, independence from the field realization as said above. Since, though through a different path and with a distinct wording, we have reached a situation similar to the one discussed in previous works on LSED, we will be extremely schematic in what follows due to space limitations. A full version will be published elsewhere. Our main conclusion is that Eq. (33) can be satisfied only if the stochastic amplitudes, their phases and the resonance frequencies satisfy the equations

$$a_{\alpha\beta} = a_{\alpha\beta'}a_{\beta'\beta''}a_{\beta''\beta'''}\cdots a_{\beta^{(r)}\beta}, \tag{34a}$$

$$\omega_{\alpha\beta} = \omega_{\alpha\beta'} + \omega_{\beta'\beta''} + \omega_{\beta''\beta'''} + \ldots + \omega_{\beta^{(r)}\beta}, \tag{34b}$$

$$\varphi_{\alpha\beta} = \varphi_{\alpha\beta'} + \varphi_{\beta'\beta''} + \varphi_{\beta''\beta'''} + \ldots + \varphi_{\beta^{(r)}\beta}. \tag{35}$$

These results synthesize what we call *the chain rule.* Eq. (34a) leads to $r_{\alpha\beta} = 1$, whereas Eqs. (34b) and (35) imply relations of the form

$$\omega_{\alpha\beta} = \Omega_\alpha - \Omega_\beta, \tag{36}$$

$$\varphi_{\alpha\beta} = \phi_\alpha - \phi_\beta, \tag{37}$$

which shows that the Bohr transition frequencies correspond to the resonances in the (linear) response of the atomic system. This is an extremely important result because it explains how it is that the electron "knows" in advance the energy of the state where it will land when realizing an atomic transition, since the difference in energies is precisely determined by the resonance, which, as we have seen, is very sharp. Beyond their direct meaning, these relations show that both indices run over the same set of values.

5. The Quantum Regime

It is clear that conditions as those given by Eqs. (34a) and (35) are not realized rigorously, the involved quantities being *stricto senso* random variables. Something similar applies to the sure values of $\widetilde{x}_{\alpha\beta}$ and related variables. However, under the present approximation, which corresponds to the quantum regime, as will be apparent below, they will be realized with sufficient accuracy as to allow the application of the theory. Thus the present account furnishes only an approximate and statistical description of the mechanical subsystem, in which small noisy motions are being neglected. Still further approximations will be introduced below to match the usual quantum mechanical description.

We note from Eq (16), namely

$$x_\alpha(t) = \sum_\beta \widetilde{x}_{\alpha\beta} a_{\alpha\beta} e^{-i\omega_{\alpha\beta} t}, \quad \widetilde{x}_{\alpha\beta} = -\frac{e}{m} \frac{\widetilde{E}_{\alpha\beta}}{\Delta_{\alpha\beta}}, \tag{38}$$

taking into account that $\Delta_{\alpha\beta}$ is now independent of the stochastic amplitudes, that the response of the particle has become proportional to the Fourier components of the field, $\widetilde{E}_{\alpha\beta} a_{\alpha\beta}$. This is the reason for the name of the present theory, *linear stochastic electrodynamics.*

It is now easy to see that the chain rule means matrix mechanics. By applying this rule we see that, for example,

$$\left[x^3(t)\right]_{\alpha\delta} = \sum \widetilde{x}_{\alpha\beta}\widetilde{x}_{\beta\gamma}\widetilde{x}_{\gamma\delta} a_{\alpha\beta} a_{\beta\gamma} a_{\gamma\delta} e^{-i(\omega_{\alpha\beta}+\omega_{\beta\gamma}+\omega_{\gamma\delta})t}$$

$$= \left(\sum_{\beta,\gamma} \widetilde{x}_{\alpha\beta}\widetilde{x}_{\beta\gamma}\widetilde{x}_{\gamma\delta}\right) a_{\alpha\delta} e^{-i\omega_{\alpha\delta} t} = \left(\widetilde{x}^3\right)_{\alpha\delta} a_{\alpha\delta} e^{-i\omega_{\alpha\delta} t}, \tag{39}$$

where we have put

$$\left(\widetilde{x}^3\right)_{\alpha\delta} = \sum_{\beta,\gamma} \widetilde{x}_{\alpha\beta}\widetilde{x}_{\beta\gamma}\widetilde{x}_{\gamma\delta}, \tag{40}$$

which is indeed the rule for matrix multiplication applied to $\widetilde{x}$. It is clear that the rule holds for all integer powers, so that the Fourier amplitudes in equations of the form (6) or (7) become represented by the corresponding matrices. Thus in terms of the elementary oscillators

$$\widetilde{x}_{\alpha\beta}(t) = \widetilde{x}_{\alpha\beta}e^{-i\omega_{\alpha\beta}t}, \quad \widetilde{f}_{\alpha\beta}(t) = \frac{\widetilde{f}_{\alpha\beta}}{a_{\alpha\beta}}e^{-i\omega_{\alpha\beta}t}, \quad \widetilde{E}_{\alpha\beta}(t) = \widetilde{E}_{\alpha\beta}e^{-i\omega_{\alpha\beta}t}, \tag{41}$$

Eq. (12) takes the form

$$m\frac{d^2\widetilde{x}_{\alpha\beta}(t)}{dt^2} = \widetilde{f}_{\alpha\beta}(t) + m\tau\frac{d^3\widetilde{x}_{\alpha\beta}(t)}{dt^3} + e\widetilde{E}_{\alpha\beta}(t), \tag{42a}$$

and in closed matrix notation we get

$$m\frac{d^2\widehat{x}(t)}{dt^2} = \widehat{f}(t) + m\tau\frac{d^3\widehat{x}(t)}{dt^3} + e\widehat{E}(t). \tag{42b}$$

This is the equation of motion that describes the behavior of the mechanical subsystem in the quantum regime; it agrees with the Heisenberg equation of nonrelativistic quantum electrodynamics. In the absence of an external radiation field, $\widehat{E}(t)$ represents the zero point field affected by the presence of matter. Since Eq. (42b) holds anyway only when the quantum regime has been established, and thus both the zpf and radiation reaction have played their fundamental role in carrying the system to that state, the corresponding terms in this equation represent now only small radiative corrections to the motions, so they can be neglected in a first approximation. We thus get the couple of Heisenberg equations

$$\widehat{p} = m\frac{d\widehat{x}}{dt}, \quad \frac{d\widehat{p}}{dt} = \widehat{f}, \tag{43}$$

which are the usual quantum mechanical equations.

From the present analysis we conclude that zero-point field and ergodicity combine to yield quantization. More precisely, the mechanism for quantization appears to be the ergodic behavior of the system under the action of the zpf. We may understand this by considering that the ergodic solutions should be the more stable ones, and thus those that tend to quench other motions that undoubtedly appear.

6. Some Final Words

As there is no space to comment on all the implications of the previous results for the meaning of the usual quantum mechanical formalism and its interpretation, we shall only make a brief general comment. We have shown how the rules of quantum mechanics can be derived from a well defined physical theory that obeys all the desired demands coming from general physics — and, if you prefer, metaphysics — viz. causality, determinism, locality, realism, objectivity, and so on. We are referring of course to the original form of the theory, the one defined by Eq. (1) for a given realization of the field. Once we consider a set of realizations and introduce statistical arguments and approximate calculations, some of these properties are lost or at least weakened in one form or another. From the present perspective, usual quantum mechanics appears thus as an approximate, asymptotic, statistical, partial account of the nonrelativistic spinless system. The final step, that of neglecting the random field (along with radiation reaction) to recover the equations of quantum mechanics, is the definitive blow to causality: the source of stochasticity, and with it the whole mechanism leading to the quantum behavior of the system, are swept away in the emerging statistical theory. From this point on it becomes impossible to understand the physical mechanisms that explain the demeanors of the particle. The account evolves into a noncausal, indeterministic mysterious description, according to which the particle is free to behave in one manner or another in an unpredictable (and indeterministic, nonrealistic, nonlocal) way. According to the present theory, the blame for all these bizarre characteristics of the description is not to be put on nature, but on the kind of approximations leading to quantum mechanics. Our theory is clearly Beyond the Quantum.

References

1. D. C. Cole and Y. Zou (2004), *Phys. Lett. A* **317**, 14 (2003).
2. D. C. Cole and Y. Zou *Phys. Rev. E* **69**, 016601 (2004).
3. D. C. Cole and Y. Zou, *J. Sci. Computing* **21**, 145 (2004) and references therein.
4. T. W. Marshall, E. Santos and A. Vidiella-Barranco, in *Proceedings of the Third International Workshop on Squeezed States and Uncertainty Relations*, (D. Han *et al.*, NASA Conference Publication Series No. 3270, 1994).
5. T. W. Marshall and E. Santos, *Recent Res. Devel. Optics* **2**, 683 (2002) and references therein.
6. A. Casado, T. W. Marshall, R. Risco-Delgado and E. Santos, *Eur. Phys. J. D* **13**, 109 (2001). In these papers and the previous one, the entanglement between a pair of photons is shown to be due to their correlations with the zeropoint field. For a discussion of entanglement in classical brownian motion see Ref. 7.
7. A. E. Allahverdyan, A. Khrennikov and Th. M. Nieuwenhuizen, *Phys. Rev. A* **72**, 032102 (2005).
8. T. H. Boyer, *Found. Phys.* **19**, 349 (1989).
9. Th. M. Nieuwenhuizen, in *Quantum Theory. Reconsideration of Foundationes-3*, (Proceedings of a Symposium, Växjö, Sweden, June 2005, G. Adenier, A. Yu. Khrennikov, Th. M. Nieuwenhuizen (eds.), AIP Conference Proceedings 810, Melville, New York, 2006).
10. L. de la Peña and A. M. Cetto, *The Quantum Dice. An Introduction to Stochastic Electrodynamics*, (Kluwer Academic Publishers, Dordrecht, 1996) We refer in the text to this monograph as *The Dice*, for short.
11. L. de la Peña and A. M. Cetto, in *New Perspectives on Quantum Mechanics*, (Proceedings of the Latin-American School of Physics, Mexico City, 1999 S. Hacyan, R. Jáuregui and R. López-Peña (eds.), AIP Conference Proceedings 464, American Institute of Physics, Woodbury, New York.) For the unabridged version please refer to the authors.
12. L. de la Peña and A. M. Cetto, *Contribution from linear stochastic electrodynamics to the understanding of quantum mechanics*, http://arXiv.org/ abs/quantum-ph 0501011, (2005).
13. L. de la Peña and A. M. Cetto, in *Quantum Theory. Reconsideration of Foundationes-*

3, (Proceedings of a Symposium, Växjö, Sweden, June 2005, G. Adenier, A. Yu. Khrennikov, Th. M. Nieuwenhuizen (eds.), AIP Conference Proceedings 810, Melville, New York, 2006).

14. L. de la Peña and A. M. Cetto, *Festschrift*, (in honour of Prof. E. Santos), Found. Phys. **36**, 350, (2006).

INERTIA AND THE VACUUM–VIEW ON THE EMERGENCE OF THE INERTIA REACTION FORCE

A. RUEDA

Dept. of Electrical Engineering, ECS Bldg.,
Californaia State University, Long Beach,
Long Beach, CA 90840, USA
E-mail: arueda@csulb.edu
www.csulb.edu

H. SUNAHATA

Dept. of Physics and Astronomy,
Californaia State University, Long Beach,
Long Beach, CA 90840, USA
E-mail: hirokis@csulb.edu
www.csulb.edu

The work-in-progress on the conjectured origin of the inertia reaction force (Newton's Second Law) in quantum vacuum fields is discussed and reviewed. It is first pointed out that the inertia reaction force is not a fundamental effect at the particle level, but an emergent macroscopic phenomenon that appears in large condensed aggregates. A brief sketch of the analysis that leads to the derivation of the electromagnetic vacuum contribution to the inertia reaction force is presented, in several complementary ways and also in a fully covariant way. All derivations were initially done within Stochastic Electrodynamics and more recently, we briefly report here for the first time, they have been reformulated within ordinary Quantum Electrodynamics. Analysis leading to an expression for, what we can call, the vacuum electromagnetic field contribution to the inertia reaction force, is briefly reviewed. As an example, the case of an ordinary electromagnetic (microwave) cavity is briefly mentioned with its associated very small but nonnegligible inertial mass of the interior of the microwave cavity case (i.e., the cavity alone not considering its walls). Next, it is briefly mentioned that the results for inertial mass can be passed to passive gravitational mass. Thus some light is thrown on the origin of the Weak Equivalence Principle, which equates inertial mass to passive gravitational mass. Finally we mention the derivation of Newton's gravitational force expression that easily follows from this analysis. Unfortunately, all this has been accomplished just for the electromagnetic vacuum case, as contribution by the other quantum vacuum fields have not been calculated. This specially refers to the gluonic vacuum, which presumably contributes the lion's share of the inertia reaction force in ordinary objects. Furthermore, the origin of what constitutes active gravitational mass has still not been considered within this approach. I.e., why a massive object "bends" space-time still remains unexplained.

Keywords: Quantum vacuum; Inertia reaction force; Zero-point field.

1. Introduction

Newton's Second Law is an expression not found in the physics of elementary particles and fields. Not only that, but the dynamical concept of force — as well as the kinematical concept of acceleration are both absent. It is only when we consider a macroscopic body that $\mathbf{F} = m\mathbf{a}$ becomes applicable. As well-expressed in Ref. [1], Newton's Laws describe emergent phenomena. They only appear for large aggregates of elementary particles. For example in the case of a crystal in condensed matter physics, there is an enormous reduction in the number of degrees of freedom available, that from something like 10^{23} or 10^{24} or so, for a gas of individual atoms, in configuration space, is reduced to just 6, thanks to the relatively large energy barriers that bind the individual atoms to the respective sites of the essentially rigid crystal lattice. Furthermore, many more degrees of freedom are seen to be suppressed when we count also those even larger energy barriers that bind electrons and nucleii within each one of the atoms. And we can continue counting this down by reaching the huge energy barriers that bind protons and neutrons within the nucleus of each one of these atoms and so on. The macroscopic and emergent nature of Newton's Second Law is clear.[1] For further discussion as well as some finer points on the epistemology of the force concept, we refer to three recent enlightening articles by Frank Wilczek.[2]

The analyses that we are going to discuss here were done within this point-of-view: Our objective was to explain the origin of the inertia reaction force which is that force which opposes the acceleration of an essentially rigid macroscopic object that is being accelerated by the action of an external agent. The bulk of the work was first done within Stochastic Electrodynamics (SED)[3–5] and more recently reproduced within Quantum Electrodynamics (QED).[6] As only the electromagnetic field is involved in our calculations (no particles present), low energies nonrelativistic quantum theory is all that is needed.

Prior to the appearance of these papers, an extensive preliminary SED calculation that used well-known semi-classical techniques on the model of Einstein and Hopf, was reported.[7] The analysis of Ref. [7], performed by one of us (AR), derived a sort of inertia reaction force, or better, the contribution to it by the classical electromagnetic zero-point field (ZPF) of SED, for the case of a classical particle, modeled in the manner of a Planck oscillator. Besides referring to a microscopic particle instead of to an extended macroscopic object, as should have been the case, according to the discussion above, the model of Ref. [7] used the Dirac-Abraham-Lorentz equation, which is directly based on Newton's Second Law. This weakened the development, as using results based on Newton's Second Law to derive Newton's Second Law might be seen as tautological. Nevertheless, one interesting point should be observed. At least within the assumptions of SED, the analysis[7] seemed self-consistent: In SED, which is a classical theory based on the classical electrodynamics of Lorentz plus the assumption of a nonzero background electromagnetic field that spectrally looks the same in all inertial frames, the concept of force is viewed as more fundamental. Paraphrasing Wilczek,[2] we say that SED belongs to theories within the "culture of force" (i.e., classical theories where the force concept is used in a regular way in opposition to theories that do not do that like quantum field theory or general relativity). This fact makes it less surprising that a first attempt to examine the physical origin of $\mathbf{F} = m\mathbf{a}$ would come from within the realm of a conceptual framework that belongs to the "culture of force" as clearly is the case of SED.

Now comes a point of clarification. Readers of previous works, e.g., Refs. [3–7] or other papers, have often misconstrued some of the ideas and objectives of this line of research. The main objective of this line of research has always been to find an explanation for the inertia reaction force,[3–7] or at least to establish the contribution to it by the electromagnetic vacuum fields. It is not, and has never been the objective of this research program, to find or evaluate a particle's mass. Expressions for the electromagnetic contribution to inertial mass are important, but they rather come as a byproduct. Questions about finding a spectrum of masses of particles miss the point as clearly, at least in works of Refs. [3,4,6] since 1998, we are dealing with macroscopic objects. Inertia reaction force is an emergent concept that refers to very large aggregates of particles and not to a single one particle. Part of the confusion may arise from our very first development in Ref. [7], which was done before those of Refs. [3–6]: a classical particle interacting with the ZPF was modeled using the model of Einstein and Hopf, where a Planck oscillator coupling the EM field to the classical particle was considered. Furthermore, in the approach of Appendix B in Ref. [3], called the momentum content approach, the resulting procedure may also have the flavor of a renormalization procedure for the mass of a particle, but indeed it is not. The whole point of this research has been to show that Mach's Principle when considered in its strict sense is flawed as the inertia reaction force is generated locally by the action of the quantum vacuum fields and not by the instantaneous action of the distant universe.

2. Comparison of the Quantum with the Stochastic Formalisms

Before proceeding, we digress to make a comparison between the relevant aspects to this research of the quantum formalism of QED with the stochastic one of SED. It is well known that these two theories are different, and only in a certain limited set of cases give the same results.[8] When linear potentials are involved the two theories may give corresponding results.[8]

However, there are two cases when SED and ordinary QED give identical results: Over thirty years ago, T. H. Boyer[9] presented a detailed comparison between SED and QED for the case of *free electromagnetic fields* and for *dipole oscillator systems*. He made a comparison between the averaging methods of SED and those of QED. It was found that in those two cases, if the stochastic averaging is compared with quantum averaging over, in general, unsymmetrized quantum operators, the results are not the same. However, if prior to quantum averaging the QED operators are symmetrized (written in symmetric order), then the stochastic averaging of SED and the quantum averaging of QED yield exactly the same results.

This last point is of much importance for our developments. It so happens that in our SED calculations, e.g., those of Refs. [3], [4], and [5], the SED expressions, when directly re-written in QED form, automatically yield symmetrized expressions over the fields creation and annihilation operators.[6] Consider the following. The SED stochastic averaging over the random phases at each ZPF component in a plane-waves Fourier wave-vector decomposition of the electric field, yields for the electric field autocorrelation function, at two different space-time locations $(\mathbf{r}_1, t_1)$ and $(\mathbf{r}_2, t_2)$, an expression of the form

$$\langle E_i(\mathbf{r}_1, t_1) E_j(\mathbf{r}_2, t_2)\rangle = \int d^3k \left(\delta_{ij} - \hat{k}_i \hat{k}_j\right) \frac{\hbar\omega}{4\pi^2} \cos[\mathbf{k}\cdot(\mathbf{r}_1 - \mathbf{r}_2) - \omega(t_1 - t_2)], \quad (1)$$

where the subindices i and j stand for any two different cartesian space directions, $i,\ j = x,\ y,\ z$ and the $\langle\cdots\rangle$ parentheses mean a stochastic averaging that, after its actual performance, greatly simplifies the algebraic expression finally yielding the right hand side of expression (1). On the other hand, if we do a simple quantum averaging over the vacuum field we get

$$\left\langle 0 \left| \overline{E}_i(\mathbf{r}_1, t_1) \overline{E}_j(\mathbf{r}_2, t_2) \right| 0 \right\rangle = \int d^3k \left(\delta_{ij} - \hat{k}_i \hat{k}_j\right) \frac{\hbar\omega}{4\pi^2} \exp[i\mathbf{k}\cdot(\mathbf{r}_1 - \mathbf{r}_2) - i\omega(t_1 - t_2)] \quad (2)$$

which clearly shows that $\langle\cdots\rangle$ of Eq. (1) and $\langle 0 |\cdots| 0\rangle$ of Eq. (2) are *not the same*. This is of course not surprising. The two procedures are extraordinarily dissimilar, at least on the surface. The stochastic averaging of Eq. (1) involves averaging over the random phases in a manner thoroughly described in Refs. [3, 7, 8, 9]. On the other hand, the averaging described in Eq. (2) is the standard quantum averaging where the $\overline{E}_i$ and $\overline{E}_j$ fields are taken as operators in the Heisenberg picture that are represented by means of creation and annihilation operators. In Eq. (2), the operator nature of the vector components of the field is indicated here by a bar.

Nevertheless, if instead of writing the operator fields as in Eq. (2), we write them in terms of a symmetrized expression, then we have that

$$\begin{aligned} \frac{1}{2}\left[\left\langle 0 \left| \overline{E}_i(\mathbf{r}_1, t_1) \overline{E}_j(\mathbf{r}_2, t_2) \right| 0 \right\rangle + \left\langle 0 \left| \overline{E}_j(\mathbf{r}_2, t_2) \overline{E}_i(\mathbf{r}_1, t_1) \right| 0 \right\rangle\right] \\ = \left\langle 0 \left| \frac{\overline{E}_i(\mathbf{r}_1, t_1) \overline{E}_j(\mathbf{r}_2, t_2) + \overline{E}_j(\mathbf{r}_2, t_2) \overline{E}_i(\mathbf{r}_1, t_1)}{2} \right| 0 \right\rangle \\ = \langle E_i(\mathbf{r}_1, t_1) E_j(\mathbf{r}_2, t_2)\rangle\,. \end{aligned} \quad (3)$$

In the calculations performed in Ref. [3], the expression of Eq. (3) plays a minor role (it appeared in the calculation of the electromagnetic energy-momentum stress tensor that was needed for the covariant calculation[10]). A more prominent role was played by the expressions of the Poynting vector which is proportional to $\langle \overline{\mathbf{E}} \times \overline{\mathbf{B}} \rangle$. For example, we look at the z-component of the Poynting vector $\mathbf{N}$ that is of the form

$$\langle N_z \rangle = \frac{c}{4\pi} \langle E_x B_y - E_y B_x \rangle . \tag{4}$$

So, we will be looking at correlations of the form

$$\left\langle 0 \left| \overline{E}_i(\mathbf{r}_1, t_1)\overline{B}_j(\mathbf{r}_2, t_2) - \overline{E}_j(\mathbf{r}_1, t_1)\overline{B}_i(\mathbf{r}_2, t_2) \right| 0 \right\rangle , \tag{5}$$

where $i,\ j = x,\ y,\ z$. It is easy to see that when transferred to quantum notation in a case like this, though the vector part is indeed the result of an antisymmetrization, the operator part that involves the creation and annihilation operators comes out as a factor that is perfectly symmetrized. In short, though $\mathbf{N}$ remains as expected, vectorially antisymmetric, it comes out as symmetrized in the creation and annihilation operators. So we must have that the stochastic averaging of SED and the quantum averaging of QED in this particular case give identical results, as indeed turns out to be the case when doing a detailed calculation.[6] Also see Appendix at the end. So, we can indeed write

$$\langle \mathbf{N} \rangle = \frac{c}{4\pi} \langle \mathbf{E} \times \mathbf{B} \rangle = \frac{c}{4\pi} \left\langle 0 \left| \overline{\mathbf{E}} \times \overline{\mathbf{B}} \right| 0 \right\rangle = \langle 0 \left| \mathbf{N} \right| 0 \rangle , \tag{6}$$

where as indicated above the first two expressions refer to the stochastic averaging over the phases of SED while the last two refer to the quantum averaging over the EM vacuum of QED.

3. Emergence of Newton's Second Law (Origin of the Inertia Reaction Force)

A laboratory sample of ordinary macroscopic solid has, say, $10^{21} - 10^{24}$ atoms that are practically frozen at their lattice points. Clearly there is a fantastic reduction in number of degrees of freedom. In fact, neglecting the very constrained lattice vibrations, there are only a few degrees of freedom left. (In the rigid body approximation there are 3 degrees left for translation and 3 for rotation). Newton's Second Law and most of Classical Mechanics refer almost exclusively to the kinematics of these few remaining degrees of freedom.

In what follows we consider only these remaining degrees of freedom (in particular the translational ones) of a macroscopic solid body that we call indistinctly the *body* or the *object*.

We consider the object to be uniformly accelerated by a force applied to it by an external agent and such that the object moves rectilinearly along the x-axis with constant proper acceleration $\mathbf{a} = \hat{x}a$. We need only look at the coordinates of the center of mass and for most purposes view the object as punctual. The object performs then so-called hyperbolic motion.[11,12] Assume the body was instantaneously at rest at time $t_* = 0$ in an inertial frame I_* that we call the laboratory frame. Let there be a non-inertial frame S such that its x-axis coincides with that of I_* and let the body be located at coordinates $(c^2/a, 0, 0)$ in S at all times. So this point of S performs hyperbolic motion. The acceleration of the body point in I_* is $\mathbf{a}_* = \gamma_\tau^{-3}\mathbf{a}$

at body proper time τ. We take S as a rigid frame and therefore only neighboring points of S around the body are found to have the same acceleration. The frame S we call the Rindler frame. Let there also be an infinite collection of inertial frames $\{I_\tau\}$ such that at body proper time τ, the body is located at point $(c^2/a, 0, 0)$ of I_τ. The I_τ frames have all axes parallel to those of I_* and their x-axes coincide with that of I_*. We set the proper time τ such that at $\tau = 0$ the corresponding I_τ coincides with I_*. So clearly $I_{\tau=0} = I_*$. If this is so then the hyperbolic motion[11,12] guarantees that

$$x_* = \frac{c^2}{a}\cosh\left(\frac{a\tau}{c}\right), \tag{7}$$

$$t_* = \frac{c}{a}\sinh\left(\frac{a\tau}{c}\right), \tag{8}$$

$$\beta_\tau = \frac{u_x(\tau)}{c} = \tanh(\frac{a\tau}{c}), \tag{9}$$

$$\gamma_\tau = \left(1-\beta^2\right)^{-1/2} = \cosh(\frac{a\tau}{c}). \tag{10}$$

In what follows, we reproduce a brief sketch of the derivation of the electromagnetic contribution to the inertia reaction force.[3,4,6] As the relevant electromagnetic operators (Poynting's vector $\mathbf{N}$, Maxwell stress tensor $\mathbf{T}$ and electromagnetic energy-momentum stress tensor $\mathbf{\Theta}$) have all the proper symmetry (a technical classification here is that in the case of $\mathbf{T}$ and thus in the space-space part of $\mathbf{\Theta}$, it is $\mathbf{T}\cdot\mathbf{v}$ that has the proper symmetry form). In all calculations that follow we use the symbol $\langle\cdots\rangle$ to indicate either the stochastic averaging over the relevant random phases in the SED case or equivalently the quantum averaging $\langle 0\,|\cdots|\,0\rangle$ over the vacuum state in which case $\overline{\mathbf{E}}$ and $\overline{\mathbf{B}}$ represent the corresponding field operators.[6] Both cases yield exactly the same result. So illustrating with the stochastic case is enough: As indicated in Eq. (6), in our case the final averaged results are the same for both developments (SED and QED).

In the classical stochastic SED formalism, the ZPF in the laboratory system I_* is given by

$$\mathbf{E}(\mathbf{R}_*, t_*) = \sum_{\lambda=1}^{2}\int d^3k\hat{\epsilon}(\mathbf{k},\lambda)\sqrt{\frac{\hbar\omega}{2\pi^2}}\cos\left[\mathbf{k}\cdot\mathbf{R}_* - \omega t_* - \theta(\mathbf{k},\lambda)\right], \tag{11a}$$

$$\mathbf{B}(\mathbf{R}_*, t_*) = \sum_{\lambda=1}^{2}\int d^3k\left(\hat{k}\times\hat{\epsilon}\right)\sqrt{\frac{\hbar\omega}{2\pi^2}}\cos\left[\mathbf{k}\cdot\mathbf{R}_* - \omega t_* - \theta(\mathbf{k},\lambda)\right]. \tag{11b}$$

$\mathbf{R}_*$ and t_* denote respectively the space and time coordinates of the point of observation of the field in I_*. The phase term $\theta(\mathbf{k},\lambda)$ is a family of random variables, uniformly distributed between 0 and 2π, or a stochastic process with index set $(\mathbf{k},\lambda)$ whose mutually independent elements are indexed by the wave vector $\mathbf{k}$ and the polarization index λ.

A Lorentz transformation from I_* into I_τ allows us to calculate the EM zero-point field vectors $\mathbf{E}_{zp}$ and $\mathbf{B}_{zp}$ of I_* but as represented in I_τ. We assume that these fields as seen in I_τ, to also correspond to the fields as *instantaneously* seen in S. Though the fields at the object point in S and in the corresponding point of the co-moving frame I_τ that instantaneously coincides with the object point are the same, this does not mean that detectors in S and in I_τ will experience the same radiation-field time evolution. The evolution of the fields in S and I_τ are obviously different. A detector at rest in I_τ and the same detector at rest in S do not experience timewise the same effect. The two fields, namely that of S and that of I_τ, are the same at a given space-time point; however, the time evolution and space distribution of the field in S and those of the field in I_τ are not the same.

All polarization components are understood to be scalars, i.e., directional cosines, but written in the form $\hat{\epsilon}_i(\mathbf{k},\lambda) \equiv \hat{\epsilon}\cdot\hat{x}_i$, where $\hat{x}_i = \hat{x},\hat{y},\hat{z}$; $i = x, y, z$, stands for three unit vectors along the three space directions. The caret in $\hat{\epsilon}_i(\mathbf{k},\lambda)$ means that the directional cosines come from axial projections of the polarization unit vector $\hat{\epsilon}$. We use the same convention for components of the $\hat{k}$ unit vector where, e.g., $\hat{k}_x$ denotes $\hat{k}\cdot\hat{x}$. We can select space and time coordinates and orientation in I_* such that Eqs. (7-10) hold and in particular $x_* = \mathbf{R}_*\cdot\hat{x}$.

After Lorentz-transforming the fields from I_* in Eqs. (11a) and (11b) to those in I_τ and using Eqs. (7-10) we obtain

$$\begin{aligned}\mathbf{E}(0,\tau) &= \sum_{\lambda=1}^{2}\int d^3k\left\{\hat{x}\hat{\epsilon}_x + \hat{y}\cosh\left(\frac{a\tau}{c}\right)\left[\hat{\epsilon}_y - \tanh\left(\frac{a\tau}{c}\right)(\hat{k}\times\hat{\epsilon})_z\right]\right.\\ &\quad\left. + \hat{z}\cosh\left(\frac{a\tau}{c}\right)\left[\hat{\epsilon}_z + \tanh\left(\frac{a\tau}{c}\right)(\hat{k}\times\hat{\epsilon})_y\right]\right\}\\ &\quad\times\sqrt{\frac{\hbar\omega}{2\pi^2}}\cos\left[k_x\frac{c^2}{a}\cosh\left(\frac{a\tau}{c}\right) - \frac{\omega c}{a}\sinh\left(\frac{a\tau}{c}\right) - \theta(\mathbf{k},\lambda)\right],\end{aligned} \tag{12a}$$

$$\begin{aligned}\mathbf{B}(0,\tau) &= \sum_{\lambda=1}^{2}\int d^3k\left\{\hat{x}(\hat{k}\times\hat{\epsilon})_x + \hat{y}\cosh\left(\frac{a\tau}{c}\right)\left[(\hat{k}\times\hat{\epsilon})_y + \tanh\left(\frac{a\tau}{c}\right)\hat{\epsilon}_z\right]\right.\\ &\quad\left. + \hat{z}\cosh\left(\frac{a\tau}{c}\right)\left[(\hat{k}\times\hat{\epsilon})_z - \tanh\left(\frac{a\tau}{c}\right)\hat{\epsilon}_y\right]\right\}\\ &\quad\times\sqrt{\frac{\hbar\omega}{2\pi^2}}\cos\left[k_x\frac{c^2}{a}\cosh\left(\frac{a\tau}{c}\right) - \frac{\omega c}{a}\sinh\left(\frac{a\tau}{c}\right) - \theta(\mathbf{k},\lambda)\right].\end{aligned} \tag{12b}$$

This is the I_* ZPF, at proper object time τ, as instantaneously viewed from the object fixed to the point $(c^2/a, 0, 0)$ of S that is performing the hyperbolic motion.

We consider next the ZPF radiation background of I_* in the act of, to put it graphically, being swept through by the object. Observe that this is not the ZPF of I_τ that in I_τ should be homogeneous and isotropic. For this we fix our attention on a fixed point of I_*, say the point of the observer at $(c^2/a, 0, 0)$ of I_*, that momentarily coincides with the object at the object initial proper time $\tau = 0$, and consider that point as referred to another inertial frame I_τ that instantaneously will coincide with the object at a future generalized object proper time $\tau > 0$. Hence we compute the I_τ-frame Poynting vector, but as instantaneously evaluated at the observer point or the $(c^2/a, 0, 0)$ space point of the I_* inertial frame, namely in I_τ at the I_τ space-time point[a]:

$$ct_\tau = -\frac{c^2}{a}\sinh\left(\frac{a\tau}{c}\right), \tag{13}$$

$$x_\tau = -\frac{c^2}{a}\cosh\left(\frac{a\tau}{c}\right), \quad y_\tau = 0, \quad z_\tau = 0, \tag{14}$$

where the time in I_τ, called t_τ, is set to zero at the instant when S and I_τ (locally) coincide which happens at proper time τ. This Poynting vector we shall denote by $\mathbf{N}_*$. Everything, however, is ultimately referred to the I_* inertial frame as that is the frame of the observer that looks at the object and whose ZPF background the moving object is sweeping through. For further light on this point, see Appendix C of Ref. [3]. In order to accomplish this we first compute

$$\begin{aligned}\langle \mathbf{E}_\tau(0,\tau) \times \mathbf{B}_\tau(0,\tau)\rangle_x &= \langle E_{y\tau}B_{z\tau} - E_{z\tau}B_{y\tau}\rangle \\ &= \gamma^2 \left\langle (E_{y*} - \beta B_{z*})(B_{z*} - \beta E_{y*}) - (E_{z*} + \beta B_{y*})(B_{y*} + \beta E_{z*})\right\rangle \\ &= -\gamma^2\beta \left\langle E_{y*}^2 + B_{z*}^2 + E_{z*}^2 + B_{y*}^2\right\rangle + \gamma^2(1+\beta^2)\left\langle E_{y*}B_{z*} - E_{z*}B_{y*}\right\rangle \\ &= -\gamma^2\beta \left\langle E_{y*}^2 + B_{z*}^2 + E_{z*}^2 + B_{y*}^2\right\rangle, \end{aligned} \tag{15}$$

that we use in the evaluation of the Poynting vector. For briefness, we omitted τ subindex. The averaging symbols denote now the stochastic averaging. Exactly the same can be done for the quantum averaging $\langle 0 |\cdots| 0\rangle$ if we rewrite the fields as operators and in the standard manner of QED.

$$\mathbf{N}_* = \frac{c}{4\pi}\langle \mathbf{E}_\tau \times \mathbf{B}_\tau\rangle_* = \hat{x}\frac{c}{4\pi}\langle \mathbf{E}_\tau(0,\tau) \times \mathbf{B}_\tau(0,\tau)\rangle_x . \tag{16}$$

The integrals are now taken with respect to the I_* ZPF background as this is the background that the I_*-observer considers the object to be sweeping through. This is why we denote this Poynting vector by $\mathbf{N}_*$, with an asterisk subindex instead of a τ subindex, to indicate that it refers to the ZPF of I_*. Observe that in the last equality of Eq. (15) the term proportional to the x-projection of the ordinary ZPF Poynting vector of I_* vanishes as the ZPF of a given frame is seen as homogeneous and isotropic by an observer at rest in that given frame. The net impulse given by the field to the accelerated object, i.e., the total amount of momentum of the ZPF background the object has swept through after a time duration t_*, as judged again from the I_*-frame viewpoint, is

$$\mathbf{p}_* = \mathbf{g}_* V_* = \frac{\mathbf{N}_*}{c^2}V_* = -\hat{x}\frac{1}{c^2}\frac{c}{4\pi}\gamma^2\beta_\tau\frac{2}{3}\left\langle \mathbf{E}_*^2 + \mathbf{B}_*^2\right\rangle V_*. \tag{17}$$

Combining this with Eq. (9), (10), (15) and (16), we obtain

$$\begin{aligned}\mathbf{N}_*(\tau) &= \frac{c}{4\pi}\langle \mathbf{E}\times\mathbf{B}\rangle \\ &= \hat{x}\frac{c}{4\pi}\langle E_yB_z - E_zB_y\rangle = \hat{x}\frac{c}{4\pi}\frac{8\pi}{3}\sinh\left(\frac{2a\tau}{c}\right)\int\frac{\hbar\omega^3}{2\pi^2c^3}d\omega, \end{aligned} \tag{18}$$

[a] Here we correct a sign error in Ref. [3], Eq. (20) where the minus sign in the RHS of Eq. (13) does not appear.

where $\mathbf{E}$ and $\mathbf{B}$ stand for $\mathbf{E}_\tau(0,\tau)$ and $\mathbf{B}_\tau(0,\tau)$ respectively as in the case of Eq. (16) and where as in Eqs. (15), (16) and (17) the integration is understood to proceed over the k-sphere of I_* in the sense described in considerable detail in Appendix C of Ref. [3]. This k-sphere is a subtler point referring to the need to regularize certain *prima facie* improper integrals. $\mathbf{N}_*(\tau)$ represents energy flux, i.e., energy per unit area and per unit time in the x-direction. It also implies a parallel, x-directed momentum density, i.e., field momentum growth per unit time and per unit volume as it is incoming towards the object position, $(c^2/a, 0, 0)$ of S, at object proper time τ and as estimated from the viewpoint of I_*. Explicitly such momentum density is

$$\mathbf{g}_*^{zp}(\tau) = \frac{\mathbf{N}_*^{zp}(\tau)}{c^2} = -\hat{x}\frac{8\pi}{3}\frac{1}{4\pi c}\sinh\left(\frac{2a\tau}{c}\right)\int \eta(\omega)\frac{\hbar\omega^3}{2\pi^2 c^3}d\omega, \tag{19}$$

where we now introduce the henceforth frequency-dependent coupling or interaction coefficient $0 \leq \eta(\omega) \leq 1$, that quantifies the fractional amount of interaction (or fraction of absorption or scattering) at each frequency that the macroscopic body absorbs in, or scatters off, the ZPF radiation. Let V_0 be the proper volume of the object. From the viewpoint of I_*, however, because of Lorentz contraction such volume is then $V_* = V_0/\gamma_\tau$. The amount of momentum due to the radiation inside the volume of the object according to I_*, i.e., the radiation momentum in the volume of the object viewed at the laboratory is

$$\mathbf{p}_*(\tau) = V_*\mathbf{g}_* = \frac{V_0}{\gamma_\tau}\mathbf{g}_*(\tau) = -\hat{x}\frac{4V_0}{3}c\beta_\tau\gamma_\tau\left(\frac{1}{c^2}\int \eta(\omega)\frac{\hbar\omega^3}{2\pi^2 c^3}d\omega\right), \tag{20}$$

which is again Eq. (17).

At proper time $\tau = 0$, the $(c^2/a, 0, 0)$ point of the laboratory inertial system I_* instantaneously coincides and comoves with the object point of the Rindler frame S in which the object is fixed. The observer located at $x_* = c^2/a,\ y_* = 0,\ z_* = 0$ instantaneously, at $t_* = 0$, coincides and comoves with the object but because the latter is accelerated with constant proper acceleration $\mathbf{a}$, the object according to I_* should receive a time rate of change of incoming ZPF momentum of the form:

$$\frac{\mathrm{d}\mathbf{p}_*}{\mathrm{d}t_*} = \frac{1}{\gamma_\tau}\left.\frac{\mathrm{d}\mathbf{p}_*}{\mathrm{d}\tau}\right|_{\tau=0}. \tag{21}$$

We identify this expression with a force from the ZPF on the object. If the object has a proper volume V_0, the force exerted on the object by the radiation from the ZPF as seen in I_* at $t_* = 0$ is then

$$\mathbf{f}_* = \frac{\mathrm{d}\mathbf{p}_*}{\mathrm{d}t_*} = -\left(\frac{4}{3}\frac{V_0}{c^2}\int \eta(\omega)\frac{\hbar\omega^3}{2\pi^2 c^3}d\omega\right)\mathbf{a}. \tag{22}$$

Furthermore

$$m_i = \left(\frac{V_0}{c^2}\int \eta(\omega)\frac{\hbar\omega^3}{2\pi^2 c^3}d\omega\right) \tag{23}$$

is an invariant scalar with the dimension of mass. Observe that in Eq. (23) we have neglected a factor of 4/3. Such factor must be neglected because a fully covariant analysis (See Appendix D of Ref. [3]) shows that it should be replaced by unity. The corresponding form of m_i as written (and without the 4/3 factor) is then the mass of that fraction of the energy of the ZPF radiation enclosed within the object that interacts with the object as parametrized by the $\eta(\omega)$ factor in the integrand. Observe that $\eta(\omega) \to 0$ as $\omega \to \infty$ because all bodies become transparent at sufficiently high frequencies. For further discussions on these developments we refer to the already published literature[3–5,13,14] and in updated form this will appear in Ref. [6].

4. Relativistic Four–Force Expression of Newton's Second Law

This analysis yields not just the nonrelativistic Newtonian case but also a fully relativistic description within special relativity, at least for the case of longitudinal forces, i.e., forces parallel to the direction of motion. Moreover the extension to the more general case, where the accelerating or applied force, **f**, is non-uniform, (i.e., it changes both in magnitude and direction throughout the motion of the object), has been in principle accomplished.[3]

From the definition of the momentum $\mathbf{p}_*$ in Eq. (20), from Eqs. (21) and (22), it easily follows that the momentum of the body is

$$\mathbf{p}_* = m_i \gamma_\tau \boldsymbol{\beta}_\tau c, \tag{24}$$

in agreement with the momentum expression for a moving object in special relativity. The space 3-vector component of the four-force[11] is then

$$\mathbf{F}_* = \gamma_\tau \frac{\mathrm{d}\mathbf{p}_*}{\mathrm{d}t_*} = \frac{\mathrm{d}\mathbf{p}_*}{\mathrm{d}\tau}, \tag{25}$$

and as the force is pure in the sense of Rindler,[11] the correct form for the four-force immediately follows,

$$\mathcal{F} = \frac{\mathrm{d}\mathcal{P}}{\mathrm{d}\tau} = \frac{\mathrm{d}}{\mathrm{d}\tau}(\gamma_\tau m_i c, \mathbf{p}) = \gamma_\tau \left(\frac{1}{c}\frac{\mathrm{d}E}{\mathrm{d}t}, \mathbf{f} \right) = \gamma_\tau \left(\mathbf{f} \cdot \boldsymbol{\beta}_\tau, \mathbf{f} \right) = \left(\mathbf{F} \cdot \boldsymbol{\beta}_\tau, \mathbf{F} \right). \tag{26}$$

Consistency with Special Relativity is established (A more detailed discussion leading to Eqs. (24)-(26) appears in Ref. [3], in particular in its Appendix D).

5. Extension to Gravity

In a recent paper an extension of the inertia analysis described above to some aspects of gravity has been performed.[5] Using the fact, uncovered by Einstein, that a frame fixed in a gravitational field **g** can be identified with a frame S that is uniformly accelerated with acceleration $-\mathbf{g}$ and the concomitant equivalence of local frames that freely fall in that gravitational field with ordinary inertial frames, the analysis for the inertia case can be translated into the case of passive gravity. For example, if in the previous case we encountered an inertial mass m_i, we now have a (passive) gravitational mass m_i. The only provision of course nis that all the new frames have to be taken in a local sense.

The equivalence of the inertial mass m_i, in the previous case, with the passive gravitational mass now constitutes the well-known Weak Equivalence Principle, introduced by Galileo and Newton. If furthermore, in addition to (i) the Weak Equivalence Principle, we invoke two more assumptions that are very natural, namely, (ii) the extended Einstein Relativity Principle: Physics is the same for all local freely-falling inertial frames, plus, formally, (iii) the Uniformity Assumption, which roughly states that the universe is homogenous and isotropic at the cosmological scale, then one obtains the Einstein Strong Equivalence Principle. This is interesting because all theories that satisfy the Strong Equivalence Principle constitute a class called metric theories. Einstein's General Relativity belongs to that class as well as several others, e.g., the Brans-Dicke theory. This suggests that the quantum vacuum inertia hypothesis, that we have discussed so far, is consistent with the metric theories and therefore consistent with General Relativity.

Another interesting feature is that this analysis, that yields passive gravitational mass (as contributed by the electromagnetic ZPF), allows one to derive in a very simple way (See Section 7 of [5]) the classical Newton's inverse square force with distance law for the gravitational force. Its derivation only requires the use of simple considerations based on well-known potential theory. Space limitations prevent us from discussing this at length and we refer the reader to Ref. [5].

6. Conclusion

The fact that there is a contribution to the inertia reaction force by the electromagnetic quantum vacuum, as obtained in Eq. (22), seems well-established. In the particular case of an electromagnetic (microwaves) cavity resonator, this contribution can be estimated in considerable detail.[5] There are two cut-offs in the frequency, one at low frequencies roughly given by the maximal dimension of the cavity, the other at high frequencies given by the plasma frequency of the electrons in the metallic walls. The inertial mass and corresponding inertia reaction force contributed by the ZPF radiation enclosed in the interior of the cavity can be rather easily estimated numerically. It turns out to be, in general, a fairly small but by no means a negligible quantity. In more general matter structures than the interior of a cavity, there is also an electromagnetic contribution to the inertia reaction force but there are other more ponderous contributions presumably due to the quantum vacuum fields of the other interactions.

We should in general think of the quantum vacuum, or just the vacuum, as a medium. There are several kinds of fluctuating fields in that medium. One field in that medium is presumably the Higgs field that according to the Standard Model (SM) permeates all space. And there are vacuum fields of the other SM interactions: weak and strong, that presumably also conspire to oppose the accelerated motion through them of bodies that interact with them. Or alternatively, one can visualize a Dirac vacuum of particle-antiparticle pairs, that strongly couples to the electromagnetic ZPF, and that also opposes the acceleration through it of bodies that interact with it. This concept was independently presented by the late J. P. Vigier.[15] But how such contribution to the inertia reaction force happens has, to our knowledge, never been derived in any detail. The vacua of the strong interaction should also give a contribution to the inertia reaction force. Presumably for the usual objects in our everyday macroscopic world, the gluonic vacuum is the vacuum field that gives most of the inertial mass contribution, and consequently, also most of the contribution to the inertia reaction force for ordinary objects. But how exactly it is that such contribution to the inertia reaction force occurs, has not been calculated in any detail. The contribution to the inertia reaction force by the electromagnetic vacuum, herein discussed,[3–7] is the only one to this day analyzed in some detail.

Nevertheless we have proposed[3–7] that it is the joint contributions of all these vacuum fields, when taken in the sense of a physical vacuum, namely, the medium permeating all of space (see, e.g., Refs. [16,17]), the entity responsible for the totality of the inertia reaction force. This last claim we call the *quantum vacuum inertia hypothesis* and it has been presented primarily[3–7] in order to substitute for the traditional form of Mach's Principle which, by the way, has been shown to violate general relativity[18] and displays several other conceptual difficulties.[19]

Acknowledgements

One of us (AR) thanks the organizers of this Workshop for the chance to present this material and for the opportunity for several illuminating exchanges. AR also thanks stimulating discussions of past years with Doctors D. C. Cole and B. Haisch.

Appendix A. Correspondence Between SED and QED

In this appendix, it is indicated why the Poynting vector, Eq. (6), indeed gives identical results for SED and QED. For this purpose, let us see the case of $\langle 0\,|E_y B_z|\,0\rangle$, one of the two non-vanishing terms. The other seven happen to vanish both in this QED formulation[6] and in SED.[3]

The ZPF in SED after Lorentz-transformation is given by Eqs. (12a) and (12b). To evaluate the Poynting vector component $\langle E_y B_z\rangle$, we multiply the y-component of this ZPF electric field and the z-component of the magnetic field to obtain

$$\begin{aligned}\langle E_y B_z\rangle &= \sum_{\lambda=1}^{2}\sum_{\lambda'=1}^{2}\int d^3k\int d^3k'\sqrt{\frac{\hbar\omega}{2\pi^2}}\sqrt{\frac{\hbar\omega'}{2\pi^2}}\\ &\quad\times\left\{\cosh^2\left(\frac{a\tau}{c}\right)\left[\hat{\epsilon}_y-\tanh\left(\frac{a\tau}{c}\right)(\hat{k}\times\hat{\epsilon})_z\right]\left[(\hat{k}\times\hat{\epsilon})_z-\tanh\left(\frac{a\tau}{c}\right)\hat{\epsilon}_y\right]\right\}\\ &\quad\times\left\langle\cos\left[k_x\frac{c^2}{a}\cosh\left(\frac{a\tau}{c}\right)-\frac{\omega c}{a}\sinh\left(\frac{a\tau}{c}\right)-\theta(\mathbf{k},\lambda)\right]\right.\\ &\qquad\left.\cdot\cos\left[k'_x\frac{c^2}{a}\cosh\left(\frac{a\tau}{c}\right)-\frac{\omega' c}{a}\sinh\left(\frac{a\tau}{c}\right)-\theta'(\mathbf{k}',\lambda')\right]\right\rangle. \end{aligned}\tag{A.1}$$

Since the average over the product of the two cosine functions above gives, (see, e.g., Ref. [3]), $\frac{1}{2}\delta_{\lambda\lambda'}\delta(\mathbf{k}-\mathbf{k}')$, after one integration Eq. (A.1) simplifies to[3]

$$\begin{aligned}\langle E_y B_z\rangle &= \frac{1}{2}\sum_{\lambda=1}^{2}\int d^3k\frac{\hbar\omega}{2\pi^2}\cosh^2\left(\frac{a\tau}{c}\right)\\ &\quad\times\left[\hat{\epsilon}_y-\tanh\left(\frac{a\tau}{c}\right)(\hat{k}\times\hat{\epsilon})_z\right]\left[(\hat{k}\times\hat{\epsilon})_z-\tanh\left(\frac{a\tau}{c}\right)\hat{\epsilon}_y\right].\end{aligned}\tag{A.2}$$

With the help of the polarization relations

$$\sum_{\lambda=1}^{2}\hat{\epsilon}_y\left(\hat{k}\times\hat{\epsilon}\right)_z=\hat{k}_x,\qquad \sum_{\lambda=1}^{2}\left(\hat{k}\times\hat{\epsilon}\right)_z\left(\hat{k}\times\hat{\epsilon}\right)_z=-\hat{k}_z\hat{k}_z,\tag{A.3}$$

and the angular integration $\int d^3k=\int d^2k d\Omega=\int d^2k\int d\sin\theta\int d\phi$, with

$$\int d\Omega=4\pi,\ \text{and}\ \int\hat{k}_x^2 d\Omega=\int\sin^3\theta d\theta\int\cos^2\phi d\phi=\frac{4\pi}{3},\tag{A.4}$$

we find that Eq. (A.2) becomes, as already reported in Appendix A of Ref. [3],

$$\langle E_y B_z\rangle=-\frac{4\pi}{c}\sinh\left(\frac{2a\tau}{c}\right)\int\frac{\hbar\omega^3}{2\pi^2c^3}d\omega.\tag{A.5}$$

The evaluation follows a similar pattern in QED as well. The Lorentz-transformed ZPF in QED is

$$\mathbf{E}(0,\tau) = \sum_{\lambda=1}^{2} \int d^3k \left\{ \hat{x}\hat{\epsilon}_x + \hat{y}\cosh\left(\frac{a\tau}{c}\right)\left[\hat{\epsilon}_y - \tanh\left(\frac{a\tau}{c}\right)(\hat{k}\times\hat{\epsilon})_z\right] \right.$$
$$\left. + \hat{z}\cosh\left(\frac{a\tau}{c}\right)\left[\hat{\epsilon}_z + \tanh\left(\frac{a\tau}{c}\right)(\hat{k}\times\hat{\epsilon})_y\right]\right\}$$
$$\times \sqrt{\frac{\hbar\omega}{2\pi^2}} \left\{\alpha(\mathbf{k},\lambda)\exp[i\Theta] + \alpha^\dagger(\mathbf{k},\lambda)\exp[-i\Theta]\right\}, \quad (A.6)$$

$$\mathbf{B}(0,\tau) = \sum_{\lambda=1}^{2} \int d^3k \left\{ \hat{x}(\hat{k}\times\hat{\epsilon})_x + \hat{y}\cosh\left(\frac{a\tau}{c}\right)\left[(\hat{k}\times\hat{\epsilon})_y + \tanh\left(\frac{a\tau}{c}\right)\hat{\epsilon}_z\right] \right.$$
$$\left. + \hat{z}\cosh\left(\frac{a\tau}{c}\right)\left[(\hat{k}\times\hat{\epsilon})_z - \tanh\left(\frac{a\tau}{c}\right)\hat{\epsilon}_y\right]\right\}$$
$$\times \sqrt{\frac{\hbar\omega}{2\pi^2}} \left\{\alpha(\mathbf{k},\lambda)\exp[i\Theta] + \alpha^\dagger(\mathbf{k},\lambda)\exp[-i\Theta]\right\}, \quad (A.7)$$

with Θ or Θ' given by

$$\Theta^{(\prime)} = k_x^{(\prime)}\frac{c^2}{a}\cosh\left(\frac{a\tau}{c}\right) - \omega^{(\prime)}\frac{c}{a}\sinh\left(\frac{a\tau}{c}\right). \quad (A.8)$$

Comparing the fields (12a) and (12b) with (A.6) and (A.7), we notice that in QED the cosine functions are replaced by the sum of the products of the two exponential functions that are complex conjugates of each other and the annihilation and creation operators, α and $\alpha^\dagger$, respectively. Using the fields above, the expectation value $\langle 0\,|E_yB_z|\,0\rangle$, which is the QED analogue of $\langle E_yB_z\rangle$, can be expressed as

$$\langle 0\,|E_yB_z|\,0\rangle = \sum_{\lambda=1}^{2}\sum_{\lambda'=1}^{2} \int d^3k \int d^3k' \sqrt{\frac{\hbar\omega}{2\pi^2}}\sqrt{\frac{\hbar\omega'}{2\pi^2}}$$
$$\times \cosh^2\left(\frac{a\tau}{c}\right)\left[\hat{\epsilon}_y - \tanh\left(\frac{a\tau}{c}\right)(\hat{k}\times\hat{\epsilon})_z\right]\left[(\hat{k}\times\hat{\epsilon})_z - \tanh\left(\frac{a\tau}{c}\right)\hat{\epsilon}_y\right]$$
$$\times \frac{1}{2}\left\langle 0\left|\left\{\alpha(\mathbf{k},\lambda)\exp[i\Theta] + \alpha^\dagger(\mathbf{k},\lambda)\exp[-i\Theta]\right\}\right.\right.$$
$$\left.\left.\cdot\left\{\alpha(\mathbf{k}',\lambda')\exp[i\Theta'] + \alpha^\dagger(\mathbf{k}',\lambda')\exp[-i\Theta']\right\}\right|0\right\rangle. \quad (A.9)$$

This equation has four terms. However, since only the term proportional to $\left\langle 0\left|\alpha(\mathbf{k},\lambda)\,\alpha^\dagger(\mathbf{k}',\lambda')\right|0\right\rangle$ remains due to the relations

$$\langle 0\,|\alpha(\mathbf{k},\lambda)\,\alpha(\mathbf{k}',\lambda')|\,0\rangle = \left\langle 0\left|\alpha^\dagger(\mathbf{k},\lambda)\,\alpha^\dagger(\mathbf{k}',\lambda')\right|0\right\rangle = 0 \quad (A.10)$$

$$\left\langle 0\left|\alpha(\mathbf{k},\lambda)\,\alpha^\dagger(\mathbf{k}',\lambda')\right|0\right\rangle = \delta_{\lambda,\lambda'}\delta^3(\mathbf{k}-\mathbf{k}') \quad (A.11)$$

$$\left\langle 0\left|\alpha^\dagger(\mathbf{k},\lambda)\,\alpha(\mathbf{k}',\lambda')\right|0\right\rangle = 0, \quad (A.12)$$

the right hand side of Eq. (A.9) becomes identical to that of Eq. (A.2), i.e.,

$$\langle E_yB_z\rangle = \langle 0\,|E_yB_z|\,0\rangle$$
$$= \frac{1}{2}\sum_{\lambda=1}^{2}\int d^3k \frac{\hbar\omega}{2\pi^2}\cosh^2\left(\frac{a\tau}{c}\right)$$
$$\times \left[\hat{\epsilon}_y - \tanh\left(\frac{a\tau}{c}\right)(\hat{k}\times\hat{\epsilon})_z\right]\left[(\hat{k}\times\hat{\epsilon})_z - \tanh\left(\frac{a\tau}{c}\right)\hat{\epsilon}_y\right]. \quad (A.13)$$

It is easy to see, following the same procedures, that this correpondence between SED and QED is achieved in all the other eight Poynting vector components.

References

1. R. B. Laughlin, *A Different Universe—Reinventing Physics from the Bottom Down* (Basic Books, New York, 2005).
2. F. Wilczek, *Physics Today*, October 2004, pg. 11; *Physics Today*, December 2004, pg. 10; *Physics Today*, July 2005, pg. 10.
3. A. Rueda and B. Haisch, *Found. Phys.* **28**, 1057 (1998). arXiv:physics/9802030 v1.
4. A. Rueda and B. Haisch, *Phys. Lett. A* **240**, 115 (1998).
5. A. Rueda and B. Haisch, *Ann. Physik* (Leipzig) **14** (8), 479 (2005). DOI 10.1002/andp.2005.10147. arXiv:gr-qc/0504061 v3.
6. H. Sunahata and A. Rueda, in preparation (2006) and A. Rueda and H. Sunahata, in preparation (2006). See also H. Sunahata, *Interaction of the quantum vacuum with an accelerated object and its contribution to inertia reaction force*, Ph.D. thesis, Claremont Graduate University, (CA, USA, 2006).
7. B. Haisch, A. Rueda and H. E. Puthoff, *Phys. Rev A* **49**, 678 (1994).
8. L. de la Peña and A. M. Cetto, *The Quantum Dice—An Introduction to Stochastic Electrodynamics* (Kluwer Academic Publishers, Dordrecht, 1996). Also for the more recent developments see in these Proceedings the paper by L. de la Peña and A. M. Cetto.
9. T. H. Boyer, *Phys. Rev D* **11**, 809 (1975).
10. See Appendix D of Ref. [3].
11. W. Rindler, *Introduction to Special Relativity* (Clarendon, Oxford, 1991).
12. F. Rohrlich, *Classical Charged Particles* (Addison-Wesley, Reading, 1965).
13. M. Jammer, *Concepts of Mass* (Princeton University Press, Princeton, NJ, 2000). See, in particular, pp. 163-167.
14. Y. Dobyns, A. Rueda and B. Haisch, *Found. Phys.* **30**, 59 (2000).
15. J. P. Vigier, *Found. Phys.* **25**(10), 1461 (1995).
16. T. D. Lee, Is the physical vacuum a medium? in *A Festschrift for Maurice Goldhaber* (Trans. of the NY Academy of Sciences; Series II, Vol 40, 1980) pp. 111-123.
17. B. De Witt, The Casimir Effect in Field Theory, Chapter 9B, *Physics in the Making*, edited by A. Sarlemijn and M. J. Sparnaay, pp. 247-272 (Elsevier Science Publish, 1989).
18. There are many references to this issue, for example, see, e.g., W. Rindler, *Phys Lett. A* **187**, 236 (1994).
19. For an illuminating description on the several difficulties associated with so called Mach's Principle we refer the reader to excellent articles and discussions in the Proceedings of *Mach's Principle—From Newton's Bucket to Quantum Gravity*, edited by J. Barbour and H. Pfister (Birkhausen, Boston, 1995). See, in particular, Section 7, Critical Reflections, articles and discussions there and references therein.

PART F

Models for the Electron

ROTATING HOPF-KINKS: OSCILLATORS IN THE SENSE OF DE BROGLIE

U. ENZ

de Lammert 10, 5662 HW Geldrop, The Netherlands

A specific form of the Hopf map $S3 \rightarrow S2$ is introduced. The map is represented by the fields θ and ϕ describing a (infinitely) extended, topologically stable field structure in three dimensional space performing an internal rotation. That 'bunched' field shows properties that can be identified with properties of a particle: discrete mass and charge, spin, left or right-handed chirality, and a far-reaching field oscillating in phase with the internal rotation of the structure. In this model the particle is not introduced as a basic object but appears as a specific field concentration of the extended field, which is considered to be the fundamental entity.

Keywords: Particle model; Classical field theory; Domain walls.

1. The Model Structure

The present contribution to the workshop 'Beyond the quantum' deals with an extended field structure in three dimensional space. The structure is described by two classical fields $\theta(xyzt)$ and $\phi(xyzt)$. The model is based on a Hopf map where ϕ and ϕ appear as Eulerian angles. In the first instance the structure is introduced as a free construct of purely geometrical character, in the second instance an attempt is made to attribute physical meaning to the structure and its constituting fields.

Our approach is in line with long standing ideas[1,2] that particles should be understood as 'bunched fields', i.e. field concentrations of an extended field, such that high field energies occur near some center, the 'site' of the 'particle', and low field energies far from that center. In such a view there is no room for particles with given properties as basic objects, the only basic entity is the field.

Our particle model has been arrived at by perceiving and exploiting a close factual and formal analogy[3] between moving particles and moving domain walls[4] or magnetic bubbles[5] in magnetic crystals. These well known magnetic structures can be viewed as objects moving in space. The domain walls[3] are governed by the sine-Gordon equation. The breather solution of that equation is of special interest in the context of our model. The following assumptions are made:

1. The model field is assumed to carry energy i.e. the energy density is a function of the gradients of the fields. The total energy (and thus mass) is defined as the total field energy of the structure. The size or the structure is determined by an elementary length l.

2. The field structures are not static, the fields $\theta(xyzt)$ and $\phi(xyzt)$ have a simple oscillatory time dependence. The internal oscillation of the bunched field is an essential property of the structure.
3. Our structure is modeled in analogy to the breather solution of the sine-Gordon equation,[6],[78] which describes a local oscillator in a one-dimensional space x. The postulated structure is a generalization representing a local oscillator in three-dimensional space. Oscillating structures of the type described can be viewed as (moving) oscillators in the sense of de Broglie.[6]

We describe the postulated structure in toroidal coordinates η, ξ, φ by writing $\theta = \theta(\eta, \beta)$ and $\phi = \phi(\eta, \beta)$ with $\beta = \xi + \varphi + \omega t$. (nomenclature according to Margenau and Murphy). The boundary conditions at infinity are $\theta = \pi/2$ and $\varphi = 0$, implying symmetry with respect to $\theta = \pi/2$. (They differ from the boundary conditions proposed before[9] in an earlier attempt to construct a structural particle model). The fields are time-dependent at each point of space except at infinity where they are fixed by the boundary condition given.

Our 'Ansatz' for the functions θ and ϕ is

$$\cos\theta = -f(\eta)\sin\beta \tag{1}$$

and

$$\sin\phi = f(\eta)(1 - f^2 \sin^2\beta)^{-\frac{1}{2}} \cos\beta, \tag{2}$$

where $f(\eta)$ is a function of η alone. A tentative choice of the function f is

$$f = \frac{2\tanh\eta}{\cosh\eta}, \tag{3}$$

which, together with (Eq. 1) and (Eq. 2), describes the entire rotating structure.

The figure provides a visualization of that structure which consists essentially of nested tori: each value of η (and thus of f) defines the surface of a torus. The lines indicated are closed curves known as 'Clifford parallels'. In our description these curves are the lines of constant phases β for any fixed time t. Each of those lines encircles the central axis, defined by $\eta = 0$ and simultaneously the circular axis defined by $\eta \to \infty$. Moreover any two lines are linked to each other. As a function of time the lines of constant phase revolve on the surface of the relevant torus $\eta =$constant.

The function f reaches a maximum of $f = 1$ for a value of η defined by $\sinh\eta_0 = 1$. The coordinate $\eta = \eta_0$ marks the surface of a special torus called 'generating torus'. On that generating torus we find the 'cores' $\theta = \pi$ and $\theta = 0$, which trace that surface while revolving around each other. The cores move in opposition in the sense that the phase difference between them is π at any time. The lateral velocity of the cores on the generating torus is $1/\sqrt{2}$. It should be stressed that the cores are not singularities of the field but are part of an extended regular wave. The above 'Ansatz' specifies the structures as Hopf kinks with Hopf indices ± 1, i.e. structures with opposite chirality.

Conform to our assumption a density u may be defined:

$$u = (\nabla\theta)^2 + (\nabla\phi)^2 \sin^2\theta. \tag{4}$$

We will interprets u as (static) energy density. The integral of u taken over the entire space then yields directly the static contribution W_s to the total energy. The kinetic contribution will be defined below. With the aid of the functions defined the density u (Eq. 4) can be calculated. After straightforward but lengthy summing we find for arbitrary $f(\eta)$:

$$u = b^{-2}(\cosh\eta - \cos\xi)^2 \left(f_\eta^2(1-f^2)^{-1} + f^2(\tanh\eta)^{-2}\right). \tag{5}$$

We remark that (Eq. 5) is the sum of two terms $u = u_c + u_a$. The first one, the central term, reflects the dependence of the energy density on the variable η, the second one, the axial term, its dependence on ξ and φ. The two terms turn out to be equal.
Inserting the function (Eq. 3) delivers:

$$u = 8\rho^{-2}(\tanh\eta)^2 = 32b^2(\rho^2 + z^2 + b^2)^{-2}, \tag{6}$$

where ρ and z are cylindrical coordinates($u_c = u_a = u/2$). For $r \gg b$ the density reduces to the central symmetric expression

$$u = 32b^2 r^{-4}. \tag{7}$$

Departing from (Eq. 6) the static contribution W_s can be calculated:

$$W_s = \int_0^\infty u(r) 4\pi r^2 dr = 128\pi b I_1 = 32\pi^2 b, \tag{8}$$

where $I_1 = \pi/4$.

In order to evaluate the kinetic part of the energy we have to look closer at the rotating structure, which is in fact an extended wave phenomenon. The second part of the density u of (Eq. 5), the axial part u_a, depends on β and contributes to the momentum in both the ξ and φ directions, whereas the first term u_c is independent of time and does not contribute to the kinetic energy. We suppose that the energy flow in each volume element can be treated as a local wave travailing on the surface of any torus with the group velocity $v = (c^2/\omega)k$. The appropriate description of that motion is $v = b/s = \tanh\eta$, with $k = 1/s$ and $\omega = 1/b$. The velocity v then is equal to c on the circular axis of the torus ($s = b$) and decreases to zero for $\eta = 0$. On the generating torus the velocity is $c/\sqrt{2}$.

As the model departs from velocities near c we introduce the momentum in the relativistic form:

$$p^2 = \mu^2 v^2/(1 - v^2), \tag{9}$$

where μ is the field mass contained in a volume element and v the appropriate local velocity of that mass. The local momentum then reeds (with $c = 1$)

$$p^2 = \mu^2 b^2/(s^2 - b^2) = \mu^2 b^2/g^2, \tag{10}$$

which leads to the energy density w_r of kinetic origin:

$$w_r^2 = \mu^2 + p^2 = \mu^2(g^2 + b^2)/g^2 = \mu^2 s^2/g^2 = \mu^2(\cosh\eta)^2 = \mu^2/(1 - v^2). \tag{11}$$

In the space region ($\eta \to 0$) far from the circular torus axis w_r reduces to μ. The total kinetic contribution to the energy follows as the integral of w_r over the entire space. That integral, which has not yet worked out, can be shown to be convergent. As an alternative we may treat the limit for small velocities. The density (Eq. 11) then reduces to

$$w_r \approx \mu(1 + v^2/2) = \mu(1 + b^2/2s^2). \tag{12}$$

Equation (12) includes the classical kinetic energy of the moving mass μ. The geometry of the torus implies $\tanh\eta = b/s$, thus we find from Eqs.(6) and (12) with $\mu = u_a$

$$w_r \approx u_a(1 + (1/2)(\tanh\eta)^2) = u_a + (2/\rho^2)(\tanh\eta)^4). \tag{13}$$

The second part of (13) can be written as classical energy density

$$w_k = \frac{32b^4(r\sin\vartheta)^2}{(r^2 + b^2)^4} \tag{14}$$

with $\rho = r\sin\vartheta$. For $r \gg b$ it has the form of the energy density of a dipole field. The total field energy of that field, the kinetic contribution W_k, follows directly:

$$W_k = 64\pi b^4 \int_0^\pi \int_0^\infty r^4(r^2 + b^2)^{-4}\sin^3\vartheta d\vartheta dr = \frac{256}{3}\pi I_2 b = (\frac{8\pi^2}{3})b, \tag{15}$$

with $I_2 = \pi/32$. The energy (15) is an approximation, valid only in the space region where $v \ll c$. The total energy W in dimensionless form follows from Eqs.(8) and (15)

$$W = W_s + W_k = \pi^2(32 + (8/3))b. \tag{16}$$

For the description of our model we need three constants: a velocity c, a fundamental length l and a constant B having the dimension of an action. The coordinates used are expressed in units l (and l/c), and the energy of the structure takes the form $E = (Bc/l)W$.

2. Interpretation

Up to this point our model is just an exercise in topology and geometry. The properties of the structure, however, are reminiscent of properties of particles. Hence it seems justified to attempt an identification of the structure with a real particle. The electron, generally considered to be the most fundamental stable elementary particle, understood in every detail in the framework of existing theories (with the exception of the discreteness of mass and charge and the internal structure) seems to be a good candidate for such an identification.

Comparing corresponding properties of the electron and the model leads directly to an identification of constants.The first property to be compared is the energy density u of our hypothetical field far from the 'site' of the structure. Equation (7) reflects the remarkable fact that the structure, which as a whole is not central symmetric, has a central symmetric far field indeed. We interprets that far field as the 'electric field' of the model structure, and equate u to the energy density e^2/r^4 of the electric field of the electron, which yields

$$e^2 = 32b^2cB. \tag{17}$$

The identification (17) is in line with our postulate that the field is the basic entity and carries energy.

The next property to by identified is the length l which we relate to the Compton wave length λ_c. We regard that length as a structural element of the electron, an idea that has been argued before and appears even in quantum field theory (Schweber,[10] p.625) where λ_c enters indirectly as the 'size' of the electron. We assume

$$4\pi bl = h/mc. \tag{18}$$

The identification (18) assures that our structure is an oscillator of frequency

$$\omega = c/bl = 2mc^2/\hbar. \tag{19}$$

According to Derrick's theorem[11] a wide class of field structures in three-dimensional space is unstable. However Derrick's reasoning (which moreover applies to static structures only) does not apply here because we do not depart from field equations containing parameters to be optimized, but introduce a specific structure as a whole, including the internal rotation (with $v = c$ where $b = s$). It remains to be demonstrated, however, that our specific structure corresponds to some minimum of energy or action. Nevertheless it seems consequent to view the length l as a fundamental constant, thus we assume $l = \lambda_c$ and $b = \frac{1}{4\pi}$. That assignment completes the identification of our structure with the electron: all of the fundamental constants B, l, c are determined.

Departing from (Eq. 15) and (Eq. 16) the total field energy of the structure, $E = (Bc/l)W$, can now be calculated:

$$E = \frac{We^2}{32b^2\lambda_c} = \frac{e^2mc^2 4\pi^3(32 + 8/3)}{32hc} = \alpha\ 134.36\ mc^2. \tag{20}$$

The energy E turns out to be close to the rest energy of the electron, but that result still depends on the choice of he function f. Yet it is clear that the model yields a large number of the order of magnitude $1/\alpha$.

The structure owns an angular momentum caused by the momentum carried by the waves θ and ϕ. The contributions to the total angular momentum vary from point to point over the entire space. We discern a resulting finite spin component in the z-direction and components in the x, y-plane. The resulting momentum in that plane vanishes for symmetry reasons. The spin properties of the model will be treated in a separate paper.

An important property of our structure is that it comes in two types with opposite chirality. The sign of the electrostatic interaction depends on the chirality:

equal chirality results in repulsion, opposite chirality describes attraction. An isolated structure of one type is topologically stable, for a pair of structures with opposite chirality there is no topological barrier for annihilation. In principle, therefore, pair creation and pair annihilation are inherent in our model. The details of these processes are complex, however.

The present model unifies different aspects of the 'particle': it describes the inner 'bunched' region of the field where the main part of the particle mass is localized, as well as its asymptotic part which represents the force field, and, most important, also the oscillation of the field which is the root of the wave character of the particle. These manifestations result from one unique field, the difference between the above aspects is merely quantitative.

The wave nature of the particle emerges from the internal oscillation as follows: the field of the structure at rest (i.e. without translational motion) oscillates in phase with the fixed frequency ω at all points of space. The field of the structure in uniform translational motion exhibits an additional modulation of the phase with parallel wave fronts corresponding to the wave postulated by De Broglie[12] in his original paper on moving oscillators which led the basis for quantum mechanics. Our structure is infinitely extended, be it with decreasing amplitude of the field far from the central region. Consequently this holds for the electron as well. That insight leads to a simple interpretation of e.g. the two slit experiment: the 'particle' passes simultaneously through both slits! It is interesting to note that Feynman[13] had to exclude explicitly that possibility in his subtle discussion of the two slit experiment.

The question of the entanglement of wave functions may appear in a new light as well. Both of these effects can be interpreted in terms of the 'extension'[14] of 'particles': the 'particle' is presents with the phase of its oscillation, at all points of space simultaneously, even where the amplitude is very small! The existence of a charged 'particle' has a further important consequent. According to the general solution of Liénard and Wiechert[15] the presence of moving charges and the corresponding retarded fields constitutes the validity of Maxwell's equations.

The advocated field model implies that there are no particles at all, but just classical fields: nature behaves 'as if there were particles '.[16] The model unites in a natural way topological stability, infinite range of the fields, and discreteness of mass and charge. We think that it represents a first step towards a new understanding of particles along such lines.

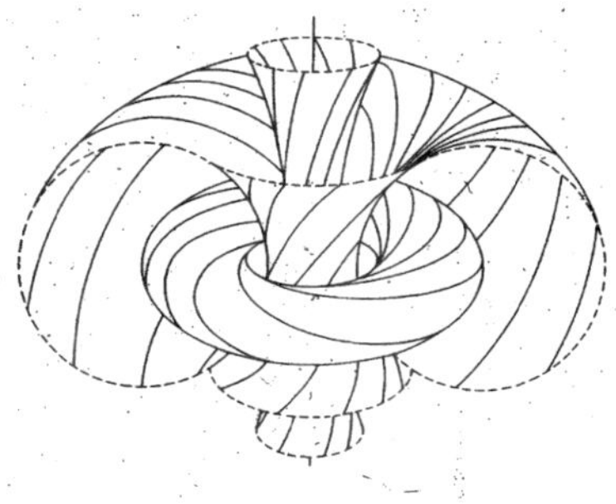

Fig. 1. Structure of the model represented by nested tori. Each torus has a circular axis of radius s and a surface defined by η =constant. The 'thickness' of a torus is $2g$. The drawn lines are known as Clifford parallels: closed lines on the torus surface enclosing once the z-axis, and simultaneously the innermost circular axis of radius $s = b$. The model describes waves θ, ϕ on the torus surfaces, the Clifford parallels are the lines of constant phase β at any fixed time. The z-axis is marked by $\eta = 0$, and $\eta \to \infty$ defines the circular axis of radius b. The following relations are valid: $\tanh \eta = b/s$ and $\sinh \eta = b/g$. The generating torus is defined by $f = 1$ or equivalently by $g = b$.

References

1. R. Mie, *Ann. der Phyzik* **37**, 511 (1912) see Helge Krag in J. Z. Buchwald and A. Warwick, *Histories of the Electron*, (MIT Press, 2001).
   P. A. M. Dirac, *Proc. Roy. Soc. London* **A209**, 291 (1951) and **A268**, 57 (1962).
   T. H. R. Skyrme, *J. Math. Phys.***12**,1735 (1971).
2. A. Einstein and L. Infeld, *Evolution der Physik*, (Zsolnay, Wien, 1950), p. 287.
3. U. Enz, *Phys. Rev.* **131**, 1392 (1963); *Helv. Phys. Act.* **37**, 245 (1964); *Ann. Fond. L. de Broglie* **11**, 87 (1985).
4. F. H. de Leeuw *et al*, *Rep. Prog. Phys* **43**, 689 (1980).
5. A. P. Malozemoff and J. C. Slonczewski, *Magnetic Domain Walls in Bubble Materials* (Acad. Press, New York, 1979).
6. U. Enz, *Physica* **17D**, 116 (1985); *Ann. Fond. L. de Broglie* **15**, (1990).
7. A. C. Scott, *Physica Scripta* **20**, 509 (1979).
8. U. Enz,*Physica* **21D**, 1 (1986).
9. U. Enz, *Math. Phys.* **18**, 347 (1977) and **19**, 1304 (1978); *Physica* **1D**, 329 (1980).
10. S. S. Schweber, *Relativistic quantum field theory*, (Harper & Row, New York, 1961).
11. G. H. Derrick, *J. Math. Phys.* **5**, 1252 (1964).
12. L. de Broglie, *Compt. rend.* **183**, 447 (1926).
13. R. P. Feynman, *Lectures on Physics III* (Addison-Wesley, 1965),Chap. 1-6.
14. U. Enz, *Particle extension:the bridge between wave and particle aspects*, (in Courants, amers, éclueils en microphysique, Foundation L. de Broglie, Paris, 1993).
15. Liénard and Wiechert, see R. P. Feynman, *Lectures on Physics II* (Addison-Wesley, 1965),Chap. 21-9.
16. K. Gottfried and V. F. Weisskopf,*Concepts of particle Physics* (Oxford University Press, 1984), note on p.168.

KERR–NEWMAN PARTICLES: SYMMETRIES AND OTHER PROPERTIES

H. I. ARCOS * and J. G. PEREIRA†

*Physics Department, Universidad Tecnolgica de Pereira
Pereira, Risaralda, Colombia
E-mail: hiarcosv@utp.edu.co

†Intituto de Física Teórica, Universidade Estadual Paulista
Rua Pamplona 145, 01405-900 São Paulo, Brazil

The Kerr-Newman (KN) solution to Einstein's equation shows a gyromagnetic factor $g = 2$, typical of a Dirac spinor. This fact has prompted many attempts to consider this solution as the exterior metric for a fundamental spin 1/2 particle. In the present work, the KN solution is proposed as the exterior and interior solution for a fundamental particle, leading to a redefinition of the particle concept. By considering the extended interpretation of Hawking and Ellis, other properties like the spacetime spinorial structure, mass and charge follow from its non- trivial geometry. A crucial point of the model is the excision of the ring singularity present in the original KN solution. This excision removes non-causal regions of the solution, and is consistent with its metric structure. Although the spacetime dimension of the singularity is of the order of the particles's Compton wavelength, which for the electron is $\lambda = 10^{-11}$cm, the space dimension of the ring is found to vanish. In the three-dimensional space, therefore, it is a point-like object, a property that validates the concept of "fundamental particle" of the model.

Keywords: Kerr-Newman solution; Particle model; Spin.

1. Introduction

With the evolution of particle physics and gravitation, the idea that a fundamental particle could somehow be connected to spacetime began to emerge. This is the case, for example, of the pioneering Wheeler's approach, which was based on the concept of spacetime foam. At the Planck scale, uncertainty in energy allows for large curvature values. At this energy, spacetime can undergo deep transformations, which modify the small scale topology of the continuum. This is where the "foam" notion becomes important. Small regions of spacetime can join and/or separate giving rise to non-trivial topological structures. The simplest of these structures is the so called wormhole, a quite peculiar solution to Einstein's equation. It represents a topological structure that connects spacetime points separated by an arbitrary spatial distance. An interesting property of the wormhole solution is that it can trap an electric field. Since, for an asymptotic observer, a trapped electric field is undistinguishable from a charge distribution, Wheeler introduced the concept of "charge without charge".[1] However, as Wheeler himself stated, these Planckian wormholes could not be related to any particle model for several reasons: charge is not quantized, they are not stable, their mass/charge ratio is very different from that found in known particles, and half-integral spin cannot be defined for a simple wormhole solution. There was the option to interpret a particle as formed by a collective motion of wormholes, in the same way phonons behave as particles in a crystal lattice. Another way out of the drawbacks of the wormhole idea was conceived by Sorkin,[2] who used non-orientability to avoid the linking between the mouths of the wormhole while retaining the "charge without charge" concept.

The discovery of the Kerr-Newman (KN) solution[3–5] in the early sixties opened the door for new attempts to explore spacetime-rooted models for fundamental particles.[6–10] These models tried to give a classical explanation for spin, something that neither the wormhole model nor the point particle model could provide.[a] As is well known, this problem is usually circumvented by saying that spin is a purely quantum property, which cannot be explained by classical physics. However, although shedding some light on the classical spin problem, the first models based on the KN solution failed to explain a fundamental property of spin, namely, that only after a 4π rotation the particle returns to its original state.

By using the Hawking and Ellis[11] extended interpretation of the KN solution, as well as Wheeler's concept of "charge without charge", a KN based model for a fundamental spin 1/2 particle has been proposed, which is able to provide a classical explanation for its "quantum" transformation properties under rotations.[12] Here, we present a glimpse on the main properties of this model, as well as analyze the possible solutions to some questions posed on the validity of the proposal. We will mainly focus on the dimensions of the particle model. We begin by reviewing, in the next section, the main properties and the topological structure of the KN solution.

2. The Extended Kerr-Newman Metric

The stationary axially-symmetric Kerr-Newman (KN) solution of Einstein's equations was found by performing a complex transformation on the tetrad field for the charged Schwarzschild (Reissner-Nordström) solution.[3–5] For $m^2 \geq a^2 + q^2$, it represents a black hole with mass m, angular momentum per unit mass a, and charge q (we use units in which $\hbar = c = 1$). In the so called Boyer-Lindquist coordinates r, θ, ϕ, the KN solution is written as

$$ds^2 = dt^2 - \frac{\rho^2}{\Delta}\, dr^2 - (r^2 + a^2)\sin^2\theta\, d\phi^2 - \rho^2\, d\theta^2 - \frac{Rr}{\rho^2}\,(dt - a\sin^2\theta\, d\phi)^2, \quad (1)$$

where

$$\rho^2 = r^2 + a^2\cos^2\theta, \quad \Delta = r^2 - Rr + a^2, \quad R = 2m - q^2/r.$$

[a]Note that a point particle is, by definition, spherically symmetric, a symmetry violated by the presence of spin.

This metric is invariant under the change $(t, a) \rightarrow (-t, -a)$. It is also invariant under $(m, r) \rightarrow (-m, -r)$ and $q \rightarrow -q$. This black hole is believed to be the final stage of a very general stellar collapse, where the star is rotating and its net charge is different from zero.

The structure of the KN solution changes deeply for $m^2 < a^2 + q^2$. Due to the absence of a horizon, it does not represent a black hole, but a circular naked singularity in spacetime. In fact, it represents a singular disk of radius a, whose border singularity cannot be removed by any coordinate transformation. This means that there is a true singularity at the border. The lack of smoothness of the metric components across the enclosed disk, on the other hand, can be remedied by considering the extended spacetime interpretation of Hawking and Ellis.[11] The basic idea of this extension is to consider that our spacetime is connected to another one through the interior points of the disk. This extended solution does not necessarily implies that the dimensionality of spacetime is greater than four, but rather that the manifold volume is greater than expected. In other words, the disk surface (with the upper points considered different from the lower ones) is interpreted as a shared border between our spacetime, denoted by **M**, and another similar one, denoted by **M'**. According to this construction, the KN metric components are no longer singular

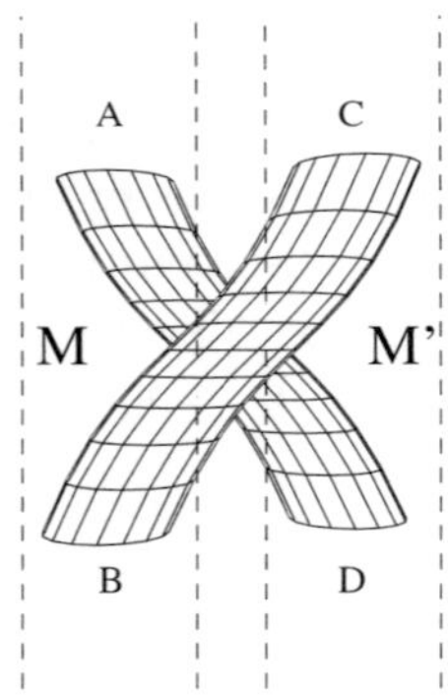

Fig. 1. To better visualize the intrinsic geometry of the KN manifold, the KN disk is drawn as if it presented a finite thickness, and consequently there is a space separation between the upper and lower surfaces of the disk. The left-hand side represents the upper and lower surfaces of the disk in **M**, whereas the right-hand side represents the upper and lower surfaces of the disk in **M'**.

across the disk, making it possible to smoothly join the two spacetimes, giving rise to a single 4-dimensional spacetime, denoted $\mathcal{M}$. This link can be seen in Fig. 1 as solid cylinders going from **M** to **M'**. In this figure, to clearly distinguish the upper from the lower side, the disk was drawn as if it presented a finite thickness. In order to cross the disk, therefore, an electric field line that hits the surface A will forcibly emerge from surface D, in **M'**. Then, it must go through surface C to finally emerge from surface B, in **M**. This picture gives a clear idea of the topological structure underlying the KN solution.

Now, the singular disk is located at $\theta = \pi/2$ and $r = 0$. Therefore, if r is assumed to be positive in **M**, it will be negative in **M'**. Since the KN metric must be the same on both sides of the solution, the mass m will be negative in **M'**. Furthermore, the magnitude of the electric charge q on both sides of the solution is, of course, the same. Taking into account that the source of the KN solution is represented by the electromagnetic potential

$$\boldsymbol{A} = -\frac{qr}{\rho^2}(dt - a\sin^2\theta d\phi), \tag{2}$$

which is clearly singular along the ring, and since r has different signs on different sides of the solution, we see from this expression that, if the charge is positive in one side, it must be negative in the other side.

3. Singularity Excision

As already remarked, the above extended interpretation does not eliminate the singularity at the rim of the disk. However, there are some arguments that can be used to circumvent this problem. First, it is important to observe that there is a torus-like region around the singular ring, in which the coordinate ϕ becomes timelike. Inside this region, defined by

$$r^2 + a^2 + \left(\frac{rR}{\rho^2}\right) a^2 \sin^2\theta < 0, \tag{3}$$

there will exist closed timelike curves.[13] In fact, when crossing the surface of this region, the signature of the metric changes from $(-,-,-,+)$ to $(-,-,+,+)$. This reduction in the number of spatial dimensions is a drawback of the solution.

Now, when the values of a, q and m are chosen to be those of a fundamental particle, as for example the electron, the surface of the torus-like region is separated from the singular ring by a distance of the order of 10^{-34} cm, which coincides roughly with the Planck length. At this scale, as is well known, topology changes are expected to exist, and consequently changes in the connectedness of spacetime topology are likely to occur. One could argue then, that at this level, the excision of a region around the singularity (including it) is justified if both the resulting manifold continues to be a solution of Einstein's equations and the metric structure of the solution is not altered for points outside the excised region. That these two constraints can be fulfilled is not a trivial result. In fact, if one excises the infinitesimal non-causal region defined by Eq. (3), and then glue back the manifold, one can prove that the resulting manifold continues to be a solution to Einstein's equations since it is formally identical to the KN spacetime, and that the gluing procedure does not alter the metric structure.[b] A simple drawing of the region to

[b]This kind of singularity removal is widely used when dealing with naked singularities. For the specific case of the Kerr solution, it has already been explored by Punsly.[14]

be excised can be seen in Fig. 2, where the direction of the gradient of r has been drawn at several points. Notice that the gluing procedure is not trivial; observe, for

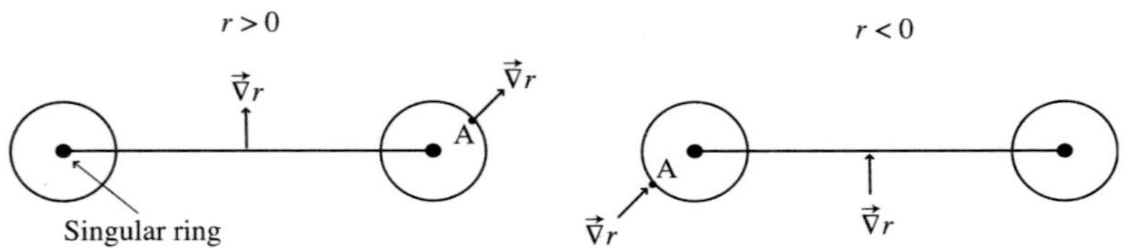

Fig. 2. Tubular–like regions around the singular ring, which is to be excised. Several $\vec{\nabla} r$ directions are also depicted, which show how the borders in the positive and negative r sides can be continuously glued.

example, that the point A on the positive-r side must be glued to the point A on the negative-r side. If we glue all points of the torus border, we obtain a continuous path for the electric field lines that flow through the disk, even for those lines that would hit the disk at the singular ring. Furthermore, since the extrinsic curvature does not change sign when crossing the hypersurface $g_{\phi\phi} = 0$, the above gluing process does not generate stress-energy.[8]

An important point of the above structure is that, after removing the tubular region around the singular ring, the surface delimiting both spacetimes turns out to be defined by a reversed topological-product between two 2-torus. As is well known,[15] this is nothing, but the Klein bottle. This is a crucial property which is in the root of the spinorial behavior of the KN spacetime.

4. The KN Solution as a Dirac Particle

4.1. *Gyromagnetic ratio*

We are going now to explore the possibility of using the extended KN solution as a model for the electron. To begin with, let us observe that the total internal angular momentum L of the KN solution, on either side of $\mathcal{M}$, can be written as

$$L = m\,a. \tag{4}$$

If we take for a, m and q the experimentally known electron values, and considering that, for a spin 1/2 particle $L = 1/2$, it is easy to see that the disk has a diameter equal to the Compton wavelength $\lambda/2\pi = 1/m$ of the electron. Consequently, the angular velocity ω of a point in the singular ring turns out to be

$$\omega = 2\,m, \tag{5}$$

which corresponds to the so called *Zitterbewegung* frequency[16,17] for a point-like electron orbiting a ring of diameter equal to λ. This means that the KN solution has a gyromagnetic ratio $g = 2$.[4,13] This property is one of the reasons for the interest to model the electron by using the KN solution. In prior KN models, however, Wheeler's concept of charge has never been used.

4.2. *Charge, mass and spin*

With our extended KN solution, charge can be interpreted as arising from the multi-connectedness of the spatial section of the solution. In other words, we can associate the electric charge of the KN solution with the net flux of a topologically trapped electric field. In fact, remember that, from the point of view of an asymptotic observer, a trapped electric field is indistinguishable from the presence of a charge distribution. Then, in analogy with the geometry of the wormhole solution, there must exist a continuous path for each electric field line going from one space to the other. Furthermore, the equality of magnetic moment on both sides of $\mathcal{M}$ implies that the magnetic field lines must also be continuous when passing through the disk enclosed by the singularity.

Mass can be associated with the degree of non-flatness of the KN solution. It is given by Komar's integral,[18]

$$m = \int_{\partial\Sigma} \star d\xi, \tag{6}$$

which holds for any stationary, asymptotically flat spacetime. In this expression, $\star$ denotes the Hodge dual operator, ξ is the stationary Killing one-form of the background metric, and $\partial\Sigma$ is a spacelike surface of the background metric. It should be noticed that the mass m is the *total* mass of the system, that is, the mass-energy contributed by the gravitational and the electromagnetic fields.[19]

Finally, spin can be consistently interpreted as an internal rotational motion of the singular ring. Of course, after the excision process, it turns out to be interpreted as an internal rotation of the infinitesimally-sized Klein bottle.

4.3. *Half-integer angular momentum*

The excision process used to eliminate the non-causal region gives rise to highly non-trivial topological structure. Now, it is a well known result that, in order to exhibit gravitational states with half-integral angular momentum, a 3-manifold must fulfill certain topological conditions. These conditions were stated by Friedman and Sorkin,[20] whose results were obtained from a previous work by Hendricks[21] on the obstruction theory in three dimensions. Interesting enough, the KN solution can be shown to satisfy these conditions, which means that it is actually a spacetime spinorial structure.[12]

An alternative way to verify this result is to analyze the behavior of the KN topological structure under rotations. In general, when rotated by 2π, a classical object returns to its initial orientation. However, the topological structure of the KN solution presents a different behavior: it returns to its initial position only after a 4π rotation. This result can be understood from the topology of the 2-dimensional surface that is formed in the excision and gluing procedure. This surface, as we have already seen, is just a Klein bottle. A 2π rotation of the positive r side is equivalent to moving a point on the Klein bottle surface halfway from its initial position. Only after a 4π rotation it returns to its departure point. This is a well known property of Möbius strip, and consequently of the Klein bottle since the latter is obtained by a topological product of two Möbius strips.

4.4. *Particle Dimension*

As already remarked, when the electron values for the mass, charge and spin are used, the diameter of the ring singularity is found to be the electron Compton length $\lambda = 10^{-11}$cm. Now, since we are modeling a fundamental particle, it is reasonable to ask about the consequences of the model for high energy collisions. At first glance one could argue that, since these collisions occur in small regions of spacetime — even smaller than the Compton wavelength of the electron — the model here discussed faces the problem of being too large when compared to the size of the interacting physical particles. We pass then to analyze this question.

To begin with we notice that, although the diameter of the ring singularity is found to be the electron Compton length, this does not mean that the size of the electron is $\lambda = 10^{-11}$cm. The reason behind this statement is that the KN singularity is not a usual singularity in the sense that it is not a static singularity. Actually, it involves an angular momentum, which in turn means that "something" is rotating. It cannot be viewed, therefore, as emerging in a space section of spacetime for a fixed time t. In fact, observe that in Boyer- Lindquist coordinates the position of the singular ring is determined by $r = 0$ and $\theta = \pi/2$. At this region, the KN metric reduces to

$$ds^2 = dt^2 - a^2\, d\phi^2 - \frac{Rr}{\rho^2}\,(dt - a\, d\phi)^2. \tag{7}$$

Since, at the singular ring,

$$\frac{Rr}{\rho^2} \to \infty, \tag{8}$$

we see from the last term of Eq. (7) that the only singular components of the metric tensor are those along the t and the ϕ directions. It is not, therefore, a singular ring in 3-dimensional space, but in spacetime. If the singularity were, let us say, in the xy plane, the angular momentum would be just a component of the *orbital* angular momentum, for which the gyromagnetic factor is well known to be $g = 1$. Since the gyromagnetic factor of the KN solution is $g = 2$, the rotation plane must necessarily involve the time axis. In fact, we know from Noether's theorem that conservation of *spin* angular momentum is related to the invariance of the system under a "rotation" in a plane involving the time axis.

If one tries to compute the size of the KN particle, a remarkable result is obtained. To see it, we write down the spatial metric of the KN solution, which is given by[22]

$$dl^2 = \rho^2 \left[\frac{1}{\Delta} dr^2 + d\theta^2 + \frac{\Delta \sin^2\theta}{\Delta - a^2 \sin^2\theta} d\phi^2\right]. \tag{9}$$

If we use this metric to compute the spatial length $\mathcal{L}$ of the singular ring, we find it to be zero:

$$\mathcal{L} \equiv \int_0^{2\pi} dl = 0. \tag{10}$$

This result is consistent with previous analysis made by some authors,[8,13] who pointed out that an external observer is unable to "see" the KN solution as an extended object, but only as a point-like object. We can then say that the "fundamental particle" concept is validated in the sense that the non-trivial KN structure is seen, by all observers, as a point-like object. Although the spatial dimension of the disk is zero, its spacetime dimension is of the order of the Compton wavelength for the particle, which for the electron is $\lambda = 10^{-11}$cm.

4.5. *Boosted KN solution*

If one tries to compute the interacting size of two colliding KN solutions, one is immediately led to the necessity of computing the form of the boosted KN metric when expressed in laboratory coordinates. Since the KN metric (1) is not written in an appropriate form, we have first to rewrite it in asymptotically Cartesian coordinates. For definiteness, we assume that the direction of the boost coincides with the $+z$ direction in asymptotically Cartesian laboratory coordinates x^μ. We consider also Lorentz-transformed Cartesian coordinates $\bar{x}^\mu = (\bar{x}, \bar{y}, \bar{z}, \bar{t})$, which are related to the laboratory coordinates $x^\mu = (x, y, z, t)$ by

$$\bar{z} = \gamma\,(z - vt), \qquad \bar{y} = y, \qquad \bar{x} = x, \qquad \bar{t} = \gamma\,(t - vz), \tag{11}$$

with $\gamma = (1 - v^2)^{-1/2}$ the relativistic factor. Now, as is well known, in terms of the asymptotically flat coordinates $\bar{x}^\mu$, the KN metric (1) assumes the form[13]

$$ds^2 = ds_0^2 + \frac{2m\bar{r}^3 - e^2\bar{r}^2}{\bar{r}^4 + a^2\bar{z}^2}\left[\frac{\bar{z}}{\bar{r}}\,d\bar{z} + \frac{(\bar{r}\bar{x} + a\bar{y})}{\bar{r}^2 + a^2}\,d\bar{x} + \frac{(\bar{r}\bar{y} - a\bar{x})}{\bar{r}^2 + a^2}\,d\bar{y} - d\bar{t}\right]^2, \tag{12}$$

where

$$ds_0^2 = \eta_{\mu\nu}\,d\bar{x}^\mu d\bar{x}^\nu = d\bar{x}^2 + d\bar{y}^2 + d\bar{z}^2 - d\bar{t}^2, \tag{13}$$

and $\bar{r}$ is written in terms of $\bar{x}, \bar{y}, \bar{z}$ as

$$\frac{\bar{x}^2 + \bar{y}^2}{\bar{r}^2 + a^2} + \frac{\bar{z}^2}{\bar{r}^2} = 1. \tag{14}$$

Thus, in the coordinates $\bar{x}^\mu$, the metric tensor components have the Kerr–Schild form

$$\bar{g}_{\mu\nu} = \bar{\eta}_{\mu\nu} + 2\bar{H}\,\bar{k}_\mu \bar{k}_\nu, \tag{15}$$

where

$$\bar{k}_\mu d\bar{x}^\mu = \frac{\bar{z}}{\bar{r}}\,d\bar{z} + \frac{(\bar{r}\bar{x} + a\bar{y})}{\bar{r}^2 + a^2}\,d\bar{x} + \frac{(\bar{r}\bar{y} - a\bar{x})}{\bar{r}^2 + a^2}\,d\bar{y} - d\bar{t} \tag{16}$$

and

$$\bar{H} = \frac{m\bar{r}^3 - e^2\bar{r}^2}{\bar{r}^4 + a^2\bar{z}^2}. \tag{17}$$

The covariant vector field $\bar{k}_\mu$ is null with respect to $\bar{g}_{\mu\nu}$, and consequently also null with respect to the auxiliary Minkowskian metric tensor $\bar{\eta}_{\mu\nu}$. It should be stressed that m and a are the mass and angular momentum per unit mass as measured by the moving observer.

Using this form of the KN metric, it is possible to study the effect of the boost on the singular ring, now expressed by $\bar{x}^2 + \bar{y}^2 = a^2$. The metric (12) is singular at the points where the function $\bar{H}$ is unbounded, that is, for $\bar{r} = 0$ and $\bar{z} = 0$. In other words, it is singular at the rim of the disk of radius a. The relation between the moving and the laboratory coordinates is given by equations (11), which can be directly replaced into the metric (12) since, due to its Kerr–Schild structure, it is form-invariant. However, an important point must be taken into account: for a laboratory observer, the parameters m and a does not represent anymore the mass and angular momentum per unit mass. The physical meaning of the constant m can be obtained by noticing that, in the moving frame, the integral over a spacelike hypersurface of the normal component of the conserved current associated with the Killing vector field $\partial/\partial\bar{t}$ is conserved. This is just the ADM mass of the solution, which gives exactly m. However, in the laboratory frame, the vector field $\partial/\partial t$ is not a Killing field; in fact, the solution is not even stationary on this frame. Since the solution continues to be asymptotically Minkowskian, one can then consider the weak-field limit, and compare it with the weak-field limit of a boosted Schwarzschild solution. When we do that, we find that the "effective" mass of a point particle static in the laboratory frame transforms like $M \sim \gamma m$,[23] which is the usual relativistic relation.

In the laboratory frame, because the KN metric keeps its Kerr–Schild form

$$g_{\mu\nu} = \eta_{\mu\nu} + 2H(x,y,z,t)\, k_\mu k_\nu, \tag{18}$$

the singular points remain defined by the singular points of the function $H(x,y,z,t)$, which is now given by

$$H = \frac{mr^3 - e^2r^2}{r^4 + a(\gamma)^2[\gamma(z-vt)]^2} \tag{19}$$

with

$$r = \frac{1}{\sqrt{2}}\left[[\gamma(z-vt)]^2 + x^2 + y^2 - a(\gamma)^2 + \sqrt{\left[(\gamma(z-vt))^2 + x^2 + y^2 - a\,\gamma^2\right]^2 + 4\,a\,\gamma^2[\gamma(z-vt)]^2}\right]^{1/2}. \tag{20}$$

Since the boost is along the z-direction, the singularity position turns out to be defined by $z = vt$ and

$$x^2 + y^2 = a(\gamma)^2, \tag{21}$$

where $a(\gamma)$ is the transformed radius of the singularity. This is required because of the total angular momentum conservation: it must be the same in both frames. For a Dirac particle, for example, one must have $L = 1/2$.

Now, considering that the total angular momentum measured in any frame must be the same, the quantity $Ma(\gamma)$ must remain constant under the boost transformation. This means that the *measured* angular momentum per unit mass $a(\gamma)$ must behave like

$$a(\gamma) \sim \gamma^{-1} a, \tag{22}$$

where a is the value measured in the laboratory frame ($\gamma = 1$). This last equation does not mean that the laboratory observer can actually measure the radius of the singular disk, but rather that he will deduce, based on the mass and the angular momentum measurements, that the radius must shrink as the speed becomes higher. This shrinking is not a consequence of the Lorentz boost alone: it is a combined consequence of the boost *and* the conservation of angular momentum. The fact that this phenomena is not expected in a macroscopic body is connected with a non-trivial property of the KN solution: the border of the singular disk rotates with a tangencial velocity equal to c. Lorentz invariance forbids that this tangencial velocity changes in different frames.[c]

The shrinking of the singularity points can be interpreted as a classical realization of the wave-particle duality: the associated wave length decreases as the speed increases. For example, for a typical accelerator energy in the order of GeV, the γ factor is of the order of 10^{-6}, giving rising to $a(\gamma) \approx 10^{-17}$cm. The 3-dimensional size of the solution continues to be zero according to (10), but its associated wave length changes from 10^{-11}cm at rest to a smaller value, depending on the γ factor.

5. Final Remarks

For small values of the mass, or more precisely, for $m^2 < a^2 + q^2$, the KN solution does not represent a black hole, but a circular naked singularity of radius a in spacetime. This solution can be considered to be localized in spacetime, and consequently asymptotically flat. It has a topological structure which behaves as an spinor when analyzed microscopically, that is, in terms of the Boyer-Lindquist coordinates in which it is described at this level. Furthermore, it presents a gyromagnetic factor $g = 2$, which enables it to be considered as a model for spinorial fundamental particles, like for example electrons.

In this paper we have mainly focused on the question of the dimension of the particle. As we have seen, although the spacetime dimension of the singularity is of the order of the Compton wavelength for the particle, which for the electron is

[c]It is important to remark that the transformation property for $a(\gamma)$ and M are not the same as those used in references like [24] and [25]. Those transformations were proposed having in mind the computation of the *ultra*-relativistic limit for the KN solution. This is a singular limit leading to a shock wave solution which is a totally different solution of Einstein's equations.

$\lambda = 10^{-11}$cm, the space dimension of the disk is zero. This property is in consonance with previous results which pointed out that an external observer is unable to "see" the KN solution as an extended object, but only as a point-like object. This result gives some additional support to the idea that the KN structure is consistent with the concept of a "fundamental particle". Furthermore, because it is asymptotically flat, an asymptotic observer can describe its spacetime evolution in terms of the Cartesian coordinates of Minkowski spacetime. Due to the spinorial structure, the spacetime evolution of the KN solution, in Cartesian coordinates, turn out to be governed by the Dirac equation.[12]

Acknowledgments

The authors would like to thank Th. M. Nieuwenhuizen and A. Burinskii for useful discussions. They would like also to thank FAPESP, CNPq, CAPES and COLCIENCIAS for partial financial support.

References

1. J. A. Wheeler, *Geometrodynamics* (Academic Press, New York, 1962).
2. R. Sorkin, *J. Phys. A: Math. Gen.* **10**, 717 (1977).
3. R. P. Kerr, *Phys. Rev. Lett.* **11**, 237 (1963).
4. E. T. Newman and A. I. Janis, *J. Math. Phys.* **6**, 915 (1965).
5. E. T. Newman *et al*, *J. Math. Phys.* **6**, 918 (1965).
6. C. A. Lopez, *Phys. Rev. D* **30**, 313 (1984); C. A. Lopez, *Gen. Rel. Grav.* **24**, 285 (1992).
7. M. Israelit and N. Rosen, *Gen. Rel. Grav.* **27**, 153 (1995).
8. W. Israel, *Phys. Rev. D* **2**, 641 (1970).
9. A. Burinskii, *Sov. Phys. JETP* **39**, 193 (1974).
10. A. Burinskii, *Phys. Rev. D* **68**, 105004 (2003) [hep-th/0308096].
11. S. W. Hawking and G. F. R. Ellis, *The Large Scale Structure of Space-Time* (Cambridge University Press, Cambridge, 1973) p. 161.
12. H. I. Arcos and J. G. Pereira, *Gen. Rel. Grav.* **36**, 2441 (2004) [hep-th/0210103]
13. B. Carter, *Phys. Rev.* **174**, 1559 (1968).
14. B. Punsly, *J. Math. Phys.* **28**, 859 (1987).
15. M. P. do Carmo, *Differential Geometry of Curves and Surfaces* (Prentice-Hall, New Jersey, 1976).
16. A. O. Barut and A. J. Bracken, *Phys. Rev. D* **23**, 2454 (1981).
17. A. O. Barut and W. Thacker, *Phys. Rev. D* **31**, 1386 (1985).
18. A. Komar, *Phys. Rev.* **113**, 934 (1959).
19. H. Ohanian and R. Ruffini, *Gravitation and Spacetime* (Norton & Company, New York, 1994) p. 396.
20. J. L. Friedman and R. Sorkin, *Phys. Rev. Lett.* **44**, 1100 (1980).
21. H. Hendriks, *Bull. Soc. Math. France Memoire* **53**, 81 (1977); see section 4.3.
22. L. D. Landau and E. M. Lifshitz, *The Classical Theory of Fields* (Pergamon, Oxford, 1975).
23. D. W. Olson and R. C. Guarino, *Am. J. Phys.* **53**, 661 (1985).
24. A. Burinskii and G. Magli, *Annals Israel Phys. Soc.* **13**, 296 (1997) [hep-th/9801177].
25. C. Barrabés and P. A. Hogan, *Phys. Rev. D* **67**, 084028 (2003) [gr- qc/0303055].

KERR GEOMETRY BEYOND THE QUANTUM THEORY

A. BURINSKII

Gravity Research Group, NSI Russian Academy of Sciences,
B. Tulskaya 52, Moscow, 115191, Russia
E-mail: bur@ibrae.ac.ru

Dirac electron theory and QED do not take into account gravitational field, while the corresponding Kerr-Newman solution with parameters of electron has very strong stringy, topological and non-local action on the Compton distances, polarizing space-time and deforming the Coulomb field. We discuss the relation of the electron to the Kerr's microgeon model and argue that the Kerr geometry may be hidden beyond the Quantum Theory. In particular, we show that the Foldi-Wouthuysen 'mean-position' operator of the Dirac electron is related to a complex representation of the Kerr geometry, and to a complex stringy source. Therefore, the complex Kerr geometry may be hidden beyond the Dirac equation.

Keywords: Dirac equation; Electron; Kerr geometry; Kerr-Newman solution; Position operator; Geon; String theory; Twistor; Kerr theorem; Complex source; Orientifold.

1. Introduction. Quantum gravity at the Compton level

Superstring theory[1] is based on the extended stringy elementary states: *Points* $\longrightarrow$ *Extended Strings*, and also, on the unification of the Quantum Theory with Gravity on Planckian level of masses M_{pl}, which correspond to the distances of order 10^{-33} cm. Such a penetrating into the deep structure of space-time has been based on the convincing evidences:

a/ The brilliant confirmation of the predictions of QED which ignored gravitational field and has been tested up to the distances of order 10^{-16} cm. It suggests that boundary of Quantum Gravity may be shifted at least beyond the distances 10^{-16} cm.

b/ The dimensional analysis, showing that M_{pl} corresponds to the energies $E_{pl} = \sqrt{\hbar c^5/G}$ which are formed from the fundamental constants relating quantum theory, $\hbar$, special relativity, c, and gravity, G.

c/ Estimation of the masses M_q and distances, where the action of gravity may be comparable with the action of quantum effects, which is done by the comparison of the gravitational radius of the Schwarzschild black hole $r_g = 2M_q$ with the Compton radius of the corresponding quantum particle $r_c = 1/M_q$ (we use here the Planck units $\hbar = c = G = 1$). One sees that the equality $r_g \sim r_c$ is achieved by the Planckian masses $M_q \sim 1$, i.e. by $M_q \sim 10^{-33} cm$. It leads to the conclusion that quantum gravity has to act on the Planckian scale $r_g \approx r_c = \frac{1}{M_q} \sim 1$.

All that is convincing, except for the argument c/ . The Schwarzschild geometry does not take into account spin of quantum particles which is indeed very high with respect to the masses. In particular, for electron $S = 1/2$, while $m \approx 10^{-22}$. So, to estimate gravitational field of spinning particle, one has to use the Kerr, or Kerr-Newman solutions.[2] Of course, there may be objections that quantum processes are strongly non-stationary because of the vacuum fluctuations, and they cannot be described by the stationary Kerr and Kerr-Newman solutions. However, QED tells us that electromagnetic radiative corrections are not too large. On the other hand, we do not know another solution which could better describe the gravitational field of a spinning particle. In any case, estimations on the base Kerr solution have to be much more correct then on the base of Schwarzschild solution.

Performing such estimation, we obtain a striking contradiction with the above scale of Quantum Gravity !

Indeed, for the Kerr and Kerr-Newman solutions we have the basic relation between angular momentum J, mass m and radius of the Kerr singular ring a :

$$J = ma. \tag{1}$$

Therefore, Kerr's gravitational field of a spinning particle is extended together with the Kerr singular ring up to the distances $a = J/m = \hbar/2m \sim 10^{22}$ which are of the order of the Compton length of electron 10^{-11} cm., forming a singular closed string [a] Therefore, in analogy with string theory *the 'point-like' Schwarzschild singularity turns in the Kerr geometry into an extended string of the Compton size.*

Notice, that the Kerr string is not only analogy. It was shown that the Kerr singular ring is indeed the string,[7] and, in the analog of the Kerr solution to low energy string theory,[8] the field around the Kerr string is similar to the field around a heterotic string.[9]

The use of Kerr geometry for estimation of the scale of Quantum Gravity gives the striking discrepancy with respect to the estimation done with the Schwarzschild solution. We arrive at the conclusion that the Kerr geometry has to play an important role in Quantum processes on the Compton distances of electron, of order $\sim 10^{-11}$ cm. The local gravitational field at these distances is extremely small, and the strong field is concentrated near an extremely narrow vicinity of the Kerr singular ring which forms a closed string of the Compton radius.

2. The real structure of the Kerr-Newman solution

The Kerr-Newman metric may be represented in the Kerr-Schild form

$$g_{\mu\nu} = \eta_{\mu\nu} - 2Hk_\mu k_\nu, \tag{2}$$

where $\eta_{\mu\nu}$ is auxiliary Minkowski metric and $H = \frac{mr - q^2/2}{r^2 + a^2\cos^2\theta}$.

[a] See also.[3–5,7]

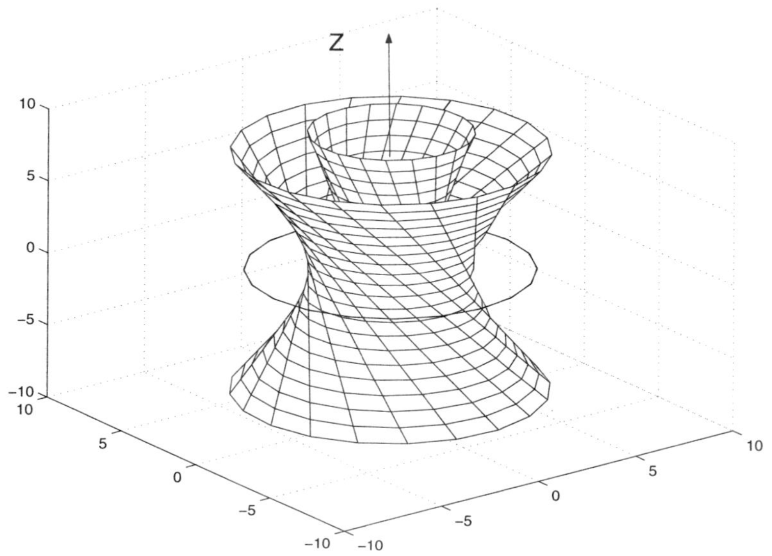

Fig. 1. The Kerr ring as a branch line, and the twistor lines of the Kerr congruence which covers twice the space-time.

One sees that metric is Minkowskian almost everywhere, for exclusion of the negligibly small subset of the space-time. However, this stringy subset has very strong dragging effect which polarizes space-time leading to a very specific polarization of the electromagnetic fields. As a result, the electromagnetic field of the corresponding Kerr-Newman solution $F_{\mu\nu}$, which cannot be consider as a weak one for parameters of charged particles, turns out to be aligned with the Kerr principal null congruence. Electromagnetic and gravitational fields are formed by the twisting vector field $k_\mu(x)$, principal null congruence (PNC), and acquire the Kerr stringy circular singularity as a caustic of PNC.

The explicit form of the field k_μ is determined by the one-form

$$k_\mu dx^\mu = dt + \frac{z}{r}dz + \frac{r}{r^2 + a^2}(xdx + ydy) - \frac{a}{r^2 + a^2}(xdy - ydx). \tag{3}$$

It is a twisting family of null rays, fig.1, forming a vortex which is described by the Kerr theorem in twistor terms.[10–12b]

PNC plays very important role, since the field k_μ determines not only the form of Kerr-Newman metric with mass m and charge q, but also the Kerr-Newman electromagnetic vector potential $A_\mu = \frac{qr}{r^2+a^2\cos^2\theta}\, k_\mu,$, and the flow of radiation in the radiative rotating solutions[c] $T_{\mu\nu} = \Phi(r, \phi, \theta) k_\mu k_\nu$.

[b]Complicate form of the field $k_\mu(x)$ determines the complicate form of the Kerr metric, contrary to the extremely simple Kerr-Schild representation (2).
[c]

The congruence covers spacetime twice, and the Kerr ring is a branch line of the space on two sheets: positive sheet of the 'outgoing' fields ($r > 0$) and negative sheet of the 'ingoing' fields. ($r < 0$). Notice, that for $a^2 >> m^2$ the black hole horizons are absent, and space-time acquires a twofold topology.[3,5]

There appears the Question: "Why Quantum Theory does not feel such drastic changes in the structure of space time on the Compton distances?"

How can such drastic changes in the structure of space-time and electromagnetic field be experimentally unobservable and theoretically ignorable in QED?

The negative sheet of Kerr geometry may be truncated along the disk $r = 0$. In this case, inserting the truncated space-time into the Einstein-Maxwell equation, one obtains on the 'right' side of the equations the source with a disk-like support. This source has a specific matter with superconducting properties.[3,5] The 'negative' sheet of space appears now as a mirror image of the positive one, so the Kerr singular ring is an 'Alice' string related to the mirror world. Such a modification changes interpretation, but does not simplify problem, since it means that Quantum Theory does not feel this 'giant' mirror of the Compton size, while the virtual charges have to be very sensitive to it.[d]

The assumption, that QED has to be corrected taking into account the peculiarities of the space-time caused by the Kerr geometry, may not be considered as reasonable because of the extraordinary exactness of the QED.

There is apparently the unique way to resolve this contradiction: to conjecture that the Kerr geometry is hidden beyond the Quantum Theory, i.e. is already taken into account and play there essential role.

From this point of view there is no need to quantize gravity, since the Kerr geometry may be the source of some quantum properties, i.e. may be primary with respect to the Quantum Theory.

3. Microgeon with spin

Let us consider the Wheeler's model of *Mass Without Mass – 'Geon'*. The photons are moving along the ring-like orbits, being bound by the own gravitational field. Such field configuration may generate the particle-like object having the mass and angular momentum. Could such construction be realized with an unique photon? In general, of course - not, because of the weakness of gravitational field. However, similar idea on 'mass without mass' is realized in the theory of massless relativistic strings and may be realized due to the stringy properties of the Kerr solution with $a >> m$. In the Kerr geometry, one can excite the Kerr circular string by an electromagnetic field propagating along this singular string as along of a waveguide.

$\Phi(r, \phi, \theta) = \frac{1}{r^2+a^2\cos^2\theta}[-6m(\ddot{x}_0^\mu k_\mu) + 2\dot{m}]$ is an angular distribution of the energy density of radiation. Radiation may be related to the loss of mass $\dot{m} < 0$, acceleration $\ddot{x}_0^\mu k_\mu \neq 0$ and to the wave electromagnetic excitations of the Kerr-Newman solution.[7,10,12]

[d]Note, that this disk is relativistically rotating and has a thickness of the order of classical size of electron, $r_e = e^2/2m,$.[5,6]

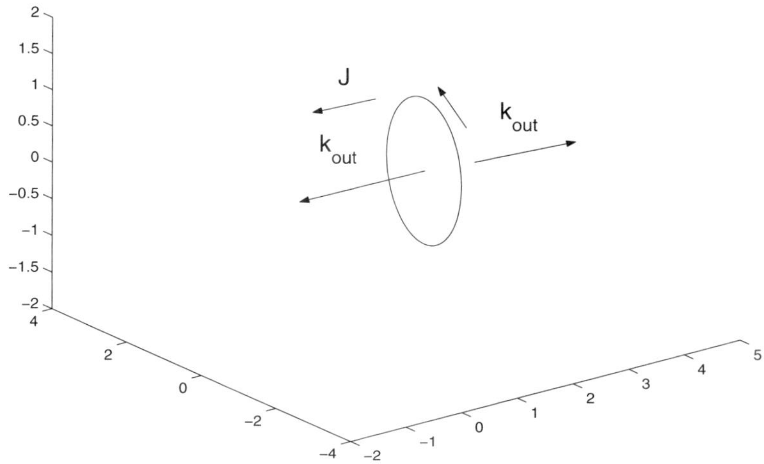

Fig. 2. Skeleton of the Kerr Spinning Particle.

Electromagnetic excitations of the Kerr source with $a >> m$ has the stringy structure, and leads to a contribution to the mass and spin. In particular, the model of microgeon with spin turns out to be self-consistent.[4,7,15]

Analysis of the exact *aligned* electromagnetic excitations on the Kerr background shows an unexpected peculiarity: *the inevitable appearance of two axial singular semi-infinite half-strings of opposite chiralities.* There appears the following stringy skeleton of a spinning particle, fig. 2.

The spin of this microgeon may be interpreted as excitation of the Kerr string by a photon moving along a circular orbit, which is reminiscent of the electron self-energy diagram in QED

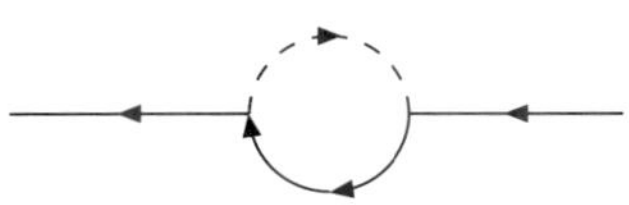

In the Kerr's gravity, the virtual photon line of this diagram does not leave the Compton region of the particle due to the Kerr stringy waveguide. As it was shown in[15] , the axial half-strings are the null-strings (the Schild, or the pp-wave strings) and may be described by the bilinear spinor combinations formed from the solutions of the Dirac equation.[e]

[e]The axial and circular singularities form a specific multisheeted topology of space-time, which admits the spinor two-valuedness.

It should also be mentioned a wonderful fact, that the basic quantum relation $E = \hbar\omega$ is contained in the classical basic relation of the Kerr geometry $J = ma$, (1). Indeed, setting $J = \frac{\hbar}{2}$, one writes (1) as $a = \frac{\hbar}{2m}$. At this step, we can consider the constant $\hbar$ not as a quantum constant, but as an experimentally constant characterizing the spin of electron. Let us consider now the classical fields propagating along the Kerr ring with speed of the light and with the winding phase number $n = 1/2$. The corresponding wave will have the length $\lambda = 2\pi a/n = 2\pi\hbar/m$ and the frequency $\omega = 2\pi c/\lambda = cm/\hbar$. It yields

$$\frac{E}{c} \equiv mc = \hbar\omega. \tag{4}$$

Up to now, we have not used the quantum operators at all. Only the topological properties of space where used, which provided a two-valued representation by rotations and a classical quantization of phase. As a result, we have got the quantum relation (4) from the classical Kerr relation $J = ma$. It suggests that *some Quantum relations may have the origin in topological properties of the 'Kerr geometry'.*

4. Dirac equation and the complex Kerr geometry

4.1. *Dirac equation in the Weyl basis*

In the Weyl basis the Dirac equation demonstrates some interesting peculiarities.[f] The Dirac spinor has the form $\Psi = \begin{pmatrix} \phi_\alpha \\ \chi^{\dot\alpha} \end{pmatrix}$, and the Dirac equation splits into

$$\sigma^\mu_{\alpha\dot\alpha}(i\partial_\mu + eA_\mu)\chi^{\dot\alpha} = m\phi_\alpha, \quad \bar\sigma^{\mu\dot\alpha\alpha}(i\partial_\mu + eA_\mu)\phi_\alpha = m\chi^{\dot\alpha}. \tag{6}$$

Conjugate spinor has the form

$$\bar\Psi = (\chi^+, \phi^+) = (\bar\chi^\alpha, \bar\phi_{\dot\alpha}). \tag{7}$$

The Dirac current

$$J_\mu = e(\bar\Psi\gamma_\mu\Psi) = e(\bar\chi\sigma_\mu\chi + \bar\phi\bar\sigma_\mu\phi), \tag{8}$$

can be represented as a sum of two lightlike components of opposite chirality

$$J^\mu_{\rm L} = e\bar\chi\sigma^\mu\chi\ , \qquad J^\mu_{\rm R} = e\bar\phi\bar\sigma^\mu\phi. \tag{9}$$

The corresponding null vectors $k^\mu_{\rm L} = \bar\chi\sigma^\mu\chi$, and $k^\mu_{\rm R} = \bar\phi\bar\sigma^\mu\phi$, determine the considered above directions of the lightlike half-strings in the microgeon model. The momentum of the Dirac electron is $p^\mu = \frac{m}{2}(k^\mu_{\rm L} + k^\mu_{\rm R})$, and the vector of polarization

---

[f] We use spinor notations of the book,[14] see also.[18] The Dirac matrices in the Weyl basis take the form

$\gamma^\mu = \begin{pmatrix} 0 & \sigma^\mu \\ \bar\sigma^\mu & 0 \end{pmatrix}$, where $\bar\sigma^{\mu\dot\alpha\alpha} = \epsilon^{\dot\alpha\dot\beta}\epsilon^{\alpha\beta}\sigma^\mu_{\beta\dot\beta}$, and $\sigma_0 = \bar\sigma_0, \quad \sigma_k = -\bar\sigma_k,\ k = 1, 2, 3,$

$$\sigma_0 = \begin{pmatrix} 1 & 0 \\ 0 & 1 \end{pmatrix}, \quad \sigma_1 = \begin{pmatrix} 0 & 1 \\ 1 & 0 \end{pmatrix}, \quad \sigma_2 = \begin{pmatrix} 0 & -i \\ i & 0 \end{pmatrix}, \quad \sigma_3 = \begin{pmatrix} 1 & 0 \\ 0 & -1 \end{pmatrix}. \tag{5}$$

of electron[17,18] in the state with a definite projection of spin on the axis of polarization is $n^\mu = \frac{1}{2}(k^\mu_{\rm L} - k^\mu_{\rm R})$. In particular, in the rest frame and the axial z-symmetry, we have $k_{\rm L} = (1, \vec{k}_{\rm L}) = (1, 0, 0, 1)$ and $k_{\rm R} = (1, \vec{k}_{\rm R}) = (1, 0, 0, -1)$, which gives $p^\mu = m(1, 0, 0, 0)$, and $n^\mu = (0, 0, 0, 1)$ and corresponds to the so-called transverse polarization of electron,[17] $\vec{n}\vec{p} = 0$.

The Dirac spinors form a natural null tetrad. The null vectors $k^\mu_{\rm L} = \bar{\chi}\sigma^\mu\chi$ and $k^\mu_{\rm R} = \bar{\phi}\bar{\sigma}^\mu\phi$, may be completed to the null tetrad by two null vectors $m^\mu = \phi\sigma^\mu\chi$, and $\bar{m}^\mu = (\phi\sigma^\mu\chi)^+$ which are controlled by the phase of wave function. Therefore, the de Broglie wave sets a synchronization of the null tetrad in the surrounding space-time, playing the role of an 'order parameter'.

It is well known[19] that the Kerr-Newman solution has the same gyromagnetic ratio ($g = 2$), as that of the Dirac electron. So, there appears a natural question: is it an accidental coincidence, or there is a deep relationship between the Dirac equation and the Kerr-Newman geometry? This problem is related to the problem of description of electron in coordinate representation, and to the problem of localized states in the Dirac theory.[20–22]

4.2. *The problem of position operator in the Dirac theory*

It is known[22] that the position operator $\hat{\vec{x}} = \nabla_{\vec{p}}$ is not Hermitean in any relativistic theory, $(\Psi, \hat{x}\Phi) \neq (\hat{x}\Psi, \Phi)$. In the Dirac theory, the problem of position operator is still more complicate. The plane wave solutions of the Dirac equation correspond to the states with a fixed momentum, and the position of electron $\vec{x}$ is undetermined for plane waves. To get a localization of position, one has to form a wave packet.

The velocity $\dot{x}$ for the operator of coordinate x is $\dot{x} = (xH - Hx) = c\alpha$, and projection of the velocity on any direction yields $\pm c$. Dirac shows[23] that this equation may be integrated, yielding for coordinate x the result

$$x = -\frac{1}{4}c\hbar^2\dot{\alpha}^0_x e^{-2iHt/\hbar}H^{-2} + c^2 p_x H^{-1} t + x_0. \tag{10}$$

Therefore, the velocity of coordinate x consists of the constant term $c^2 p_x H^{-1}$ and the oscillating contribution $\frac{1}{2}ic\hbar\dot{\alpha}^0_x e^{-2iHt/\hbar}H^{-1}$, the so-called 'zitterbewegung'.

In the rest frame, $\vec{p} = 0$, one can get $i\hbar\dot{\alpha}_x = -2H\alpha_x$, and the oscillating part $\tilde{x} = -\frac{1}{4}c\hbar^2\dot{\alpha}^0_x e^{-2iHt/\hbar}H^{-2}$ has the Compton amplitude $\frac{1}{2}ic\hbar\alpha_x H^{-1}$.

Since this expression describes complex oscillations, it is more convenient to consider the complex combination

$$\tilde{x} + i\tilde{y} = \frac{1}{4}ic\hbar^2(\dot{\alpha}^0_x + i\dot{\alpha}^0_y)e^{-2iHt/\hbar}H^{-2} \tag{11}$$

which describes circular motion in the (xy)-plane along a ring of the Compton size $\frac{c\hbar}{2m}$. The plane of oscillations is orthogonal to the direction of polarization of electron. One can see here a correspondence to the structure of the Kerr microgeon model.

The 'zitterbewegung' problem is related to the problem of localized states and caused the treatment of a 'mean-position' operator.[20] One of the best solution on this problem was found by Foldi and Wouthuysen[21,22] which performed a very nontrivial matrix transformation of the wave function Ψ and Hamiltonian H to the so-called Foldi-Wouthuysen (FW) representation[g] $\Psi_{\rm FW} = e^{iS_{\rm FW}}\Psi$ and $H_{\rm FW} = e^{iS_{\rm FW}} H e^{-iS_{\rm FW}}$.

In the new representation the negative frequency modes are suppressed, 'zitterbewegung' is absent, and the new (rather complicate) position operator $\vec{X}_{\rm FW} = e^{iS_{\rm FW}}\vec{x} e^{-iS_{\rm FW}}$ corresponds to the conventional concept of the velocity of particle, $\dot{\vec{X}}_{\rm FW} = i[H_{\rm FW}, \vec{x}] = \beta\vec{p}/E_{\vec{p}}$.

On the other hand, the use of ordinary position operator $\vec{x}$ in the Foldi-Wouthuysen representation may be transformed back to the Dirac or Weyl representation $\hat{\vec{X}} = e^{-iS_{\rm FW}}\vec{x} e^{iS_{\rm FW}}$, leading to a mean-position operator without 'zitterbewegung' which has the form

$$\hat{\vec{X}} = \vec{x} + i\frac{\hbar c}{2E_p}\vec{\gamma} + \frac{i(\vec{\gamma}\cdot\vec{p})\vec{p} + [\Sigma \times \vec{p}]p}{2E_p(E_p + m)p}. \qquad (12)$$

It should be noted that the second term contains a shift in the *complex direction* which is the constant vector $i\frac{\hbar c}{2m}\vec{\gamma}$ for small $\vec{p}$. The corresponding four-vector $i\frac{\hbar c}{2m}\gamma^\mu$ may be represented in terms of the null vectors $k_{\rm L} = (1, \vec{k}_{\rm L}) = (1, 0, 0, 1)$ and $k_{\rm R} = (1, \vec{k}_{\rm R}) = (1, 0, 0, -1)$ in the following form

$$(\bar{\Psi}\hat{X}\Psi) = x + ia(k_{\rm L} + k_{\rm R}), \qquad (13)$$

where x is a center of mass and $a = \frac{\hbar c}{2m}$ is the Compton length. In the Weil representation, the vectors $k_{\rm L}$ and $k_{\rm R}$ transform independently by Lorentz transformations, which indicates that the Dirac particle may be formed by two complex conjugate point-like sources $X_{\rm L}$ and $X_{\rm R}$ propagating along the complex world-lines

$$X^\mu_{\rm L}(t) = x^\mu(t) + ia(1, 0, 0, 1), \quad X^\mu_{\rm R}(t) = x^\mu(t) + ia(1, 0, 0, -1), \qquad (14)$$

where the 3-vector of mean-position is $\vec{X} = \frac{1}{2}(\vec{X}_{\rm L} + \vec{X}_{\rm R}) = \vec{x}(t)$. Such a complex representation turns out to be close related to the known complex representation of the Kerr geometry.[4,10,15,16,24]

4.3. *Complex representation of the Kerr geometry*

In 1887 (!) Appel[26] considered a simple complex transformation of the Coulomb solution $\phi = q/r$, a complex shift $(x, y, z) \to (x, y, z + ia)$ of the origin $(x_0, y_0, z_0) = (0, 0, 0)$ to the point $(0, 0, ia)$. On the real section (real(x, y, z)), the resulting solution

$$\phi(x, y, z) = \Re e\ q/\tilde{r} \qquad (15)$$

[g]Note, that this transformation depends on the momentum $\vec{p}$ of plane wave, and therefore, it deforms spectrum of the wave function.

acquires a complex radial coordinate $\tilde{r} = \sqrt{x^2 + y^2 + (z - ia)^2}$. Representing $\tilde{r}$ in the form

$$\tilde{r} = r - ia\cos\theta \tag{16}$$

one obtains for $\tilde{r}^2$

$$r^2 - a^2\cos^2\theta - 2iar\cos\theta = x^2 + y^2 + z^2 - a^2 - 2iaz. \tag{17}$$

Imaginary part of this equation gives $z = r\cos\theta$, which may be substituted back in the real part of (17). It leads to the equation $x^2 + y^2 = (r^2 + a^2)\sin^2\theta$, which may be split into two conjugate equations $x \pm iy = (r \pm ia)e^{\pm i\phi}\sin\theta$. Therefore, we obtain the transfer from the complex coordinate $\tilde{r}$ to the Kerr-Schild coordinate system

$$\begin{aligned} x + iy &= (r + ia)e^{i\phi}\sin\theta, \\ z &= r\cos\theta, \\ t &= r + \rho. \end{aligned} \tag{18}$$

Here r and θ are the oblate spheroidal coordinates, and the last relation is a definition of the real retarded-time coordinate ρ. The Kerr-Schild coordinates θ, ϕ, ρ fixe a null ray in M^4 (twistor) which is parametrized by coordinate r.

One sees, that after complex shift, the singular point-like source of the Coulomb solution turns into a singular ring corresponding to $\tilde{r} = 0$, or $r = \cos\theta = 0$. This ring has radius a and lies in the plane $z = 0$. The space-time is foliated on the null congruence of twistor lines, shown on fig. 1. It is twofolded, having the ring-like singularity as the branch line. Therefore, for the each real point $(t, x, y, z) \in \mathbf{M^4}$ we have two points, one of them is lying on the positive sheet of space, corresponding to $r > 0$, and another one lies on the negative sheet, where $r < 0$.

It was obtained[4] that the Appel potential corresponds exactly to electromagnetic field of the Kerr-Newman solution written on the auxiliary Minkowski space of the Kerr-Schild metric (2). The vector of complex shift $\vec{a} = (a_x, a_y, a_z)$ corresponds to the direction of the angular momentum J of the Kerr solution[4,25] and $|a| = J/m$.

Newman and Lind[24] suggested a description of the Kerr-Newman geometry in the form of a retarded-time construction, in which it is generated by a complex source, propagating along a *complex world line* $X^\mu(\tau) = x^\mu(0) + u^\mu\tau + ia^\mu$ in a complexified Minkowski space-time $\mathbf{CM}^4$. Here time is complex, $\tau = t + i\sigma$, and u^μ is a unit time-like vector. The rigorous description of this representation may be given in the Kerr-Schild approach[2] based on the Kerr theorem and the Kerr-Schild form of metric (2)[h] In the rest frame one can consider the 'left' and 'right' complex world lines, related to the complex conjugate sources $i\vec{a}$ and $-i\vec{a}$.

$$X^\mu_{\rm L}(\tau_{\rm L}) = x^\mu(\tau_{\rm L}) + i\vec{a}, \quad X^\mu_{\rm R}(\tau_{\rm R}) = x^\mu(\tau_{\rm R}) - i\vec{a}, \tag{19}$$

[h]It is related to the existence of auxiliary Minkowski metric $\eta^{\mu\nu}$, which is necessary for the complex representation, as well as for the conditions of the Kerr theorem.[10,11]

The complex retarded time is determined in analogy with the real one, but is based on the complex null cones, see.[10,15,24]

Let's consider the complex radial distance from a real point x to a complex point $X_{\rm L}$ of the 'left' complex world-line

$$\tilde{r}_{\rm L} = \sqrt{(\vec{x} - \vec{X}_L)^2} = r_{\rm L} - ia\cos\theta_{\rm L}. \tag{20}$$

To determine a retarded-time parameter $\tau_{\rm L}$ one has to write down the light-cone equation $ds^2 = 0$, or

$$\tilde{r}_{\rm L}^2 - (t - \tau_{\rm L})^2 = 0 \tag{21}$$

It may be split into two retarded-advanced-time equations $t - \tau_{\rm L} = \pm\tilde{r}_{\rm L}$. The retarded-time equation corresponds to the sign + and, due to (16), leads to relation

$$\tau_{\rm L} = t - r_{\rm L} + ia\cos\theta_{\rm L}. \tag{22}$$

One sees that $\tau_{\rm L}$ turns out to be complex

$$\tau_{\rm L} = \rho_{\rm L} + i\sigma_{\rm L}, \quad \sigma_{\rm L} = a\cos\theta_{\rm L}. \tag{23}$$

4.4. *The complex worldline as a string*

In the complex retarded-time construction, the left complex world line $X_{\rm L}(\tau_{\rm L})$ has to be parametrized by complex parameter $\tau_{\rm L} = \rho_{\rm L} + i\sigma_{\rm L}$. It has a few important consequences.

i/ Being parametrized by two parameters ρ and σ, a complex world-line $X(\tau)$ is really a world-sheet and corresponds to a *complex string.* This string is very specific, since it is extended in the complex time direction σ.

ii/ A fixed value of $\sigma_{\rm L}$ corresponds to the fixed value of $\cos\theta_{\rm L}$, and, in accordance with Eq. (18), together with the fixed parameter ϕ, it selects a null ray of the Kerr congruence (twistor).

iii/Since $|\cos\theta| \le 1$, parameter σ is restricted by interval $\sigma \in [-a, a]$, i.e. complex string is open and the points $\rho \pm ia$ are positioned at its ends. The world-sheet represents an infinite strip: $(t, \sigma) : -\infty < t < \infty, \sigma \in [-a, a]$.

iv/ From Eq. (19) and Eq. (22) one sees that the left complex point of the Dirac x-coordinate $X_{\rm L} = ia(1,0,0,1)$ has $\Im m\ \tau_{\rm L} = ia\cos\theta_{\rm L}$, which yields $\cos\theta_{\rm L} = 1$. Therefore, this point is the boundary point of the complex world line, and coordinate relations (18) show that the complex light cones positioned at this boundary have the *real section* along the Kerr axial half-string $z = r, \quad x = y = 0$.

Similar treatment for the right complex point of the Dirac x-coordinate $X_{\rm R} = ia(1,0,0,-1)$ show that it is also placed on the same boundary of the stringy strip (the same timelike component ia), however, $\Im m\ \tau_{\rm R} = -ia\cos\theta$, which yields $\cos\theta_{\rm R} = -1$ and corresponds to the axial half-string propagating in opposite direction $z = -r, \quad x = y = 0$. Therefore, in the real space-time, the two complex sources of the Dirac position operator have the real image in the form of the considered above two axial semi-infinite half-strings: left and right.[i]

Note, that there is an asymmetry in the complex left and right coordinates $X_L = ia(1,0,0,1)$ and $X_R = ia(1,0,0,-1)$. The time-like components of the both sources are adjoined to the same right end of the complex string interval $[-ia, ia]$. This asymmetry is removed by a remarkable stringy construction - orientifold.[7,15,16,27]

4.5. *Orientifold*

With respect to the Dirac solutions, the orientifold is analog of the charge conjugation. The models of relativistic strings contain usually two stringy modes: left and right. So, the modes $X_L(\tau_L)$ and $X_R(\tau_R)$ represent only the half-strings on the interval $\sigma \in [-a, a]$. *Orientifold is formed from two open half-strings which are joined forming one closed, but folded string.* The interval $[-a, a]$ is covered by parameter σ twice: the first time from left to right, and (say) the left half-string has the usual parametrization, while the interval $[-a, a]$ is reversed and covers the original one in opposite orientation for the right half-string. Therefore, the string is formed by two half-strings and turns out to be closed, but folded. The right and left string modes are flipping between the the initiate and the reversed intervals. One sees that for the complex interval the revers is equivalent to complex conjugation of the parameter τ. So, one has to put $\tau_R = \bar{\tau}_L$.[j] After orientifolding, the complex timelike coordinates of the points X_L and $\bar{X}_R$ turns out to be sitting on the opposite ends of the interval $[-a, a]$, while their imaginary space-like coordinates will be coinciding, which corresponds to one of the necessary orientifold condition $X_L(\tau_L) = \bar{X}_R(\bar{\tau}_R)$.

5. Conclusion

The above treatment shows that there is a deep internal relationships between the Dirac equation and the complex representation of the Kerr geometry. The Dirac equation works in the complex Minkowski space-time, and electron is not elementary point-like object, but is rather a many-sided particle. It has a nontrivial complex structure which is related to complex structure of the Kerr geometry and its non-trivial real topology. The space-time source of the naked electron represents a very specific complex string with two point-like (quark-like) sources located on the ends of this string. In the same time, after orientifolding this string, the space coordinates of these sources are merging, turning into *one complex point* shifted in the imaginary direction on the Compton distance a. This complex position of the source is, apparently, the origin of the problems with the position operator in the Dirac theory. On the other hand, the real stringy source of the Kerr geometry has the Compton size and is also an evidence of the extended spice-time structure of electron.

---

[i] For more details see.[7,15,16]

[j] Details of this construction may be found in.[7,15,16,27]

The obtained recently multiparticle Kerr-Schild solutions[11] shed some light on the multiparticle structure of the considered in QED dressed electron. This treatment is based on the remarkable properties of the Kerr theorem. There is also remarkable renormalization of the Kerr singularity by gravitational field.[6] However, these questions go out of the frame of this paper.

Acknowledgments

We are thankful to V. Arcos, A. Khrennikov, J. Pereira, I. Volovich, and especially to Prof. G. 't Hooft for very illuminating discussions. We are also thankful to Th. M. Nieuwenhuizen for kind invitation to this conference, attention to this work and stimulating conversations.

References

1. M. Green, J. Schwarz and E. Witten, *Superstring Theory*, v.1, (Cambridge Univ. Press, Cambridge, 1988).
2. G. C. Debney, R. P. Kerr, A. Schild, *J. Math. Phys.* **10** 1842 (1969).
3. W. Israel, Phys. Rev. **D 2** 641 (1970).
4. A. Ya. Burinskii, *Sov. Phys. JETP*, **39** 193 (1974). D. Ivanenko and A. Ya. Burinskii, *Izvestiya Vuzov Fiz.* n.5 135 (1975) (in russian).
5. C.A. López, *Phys. Rev.* **D 30** 313 (1984).
6. A. Burinskii, *J. Phys. A*, **39** 6209 (2006).
7. A. Burinskii, *Phys. Rev.* **D 68** 105004 (2003).
8. A. Sen, *Phys. Rev. Lett.* **69** 1006 (1992).
9. A. Burinskii, *Phys. Rev.* **D 52** 5826 (1995);
10. A. Burinskii, *Phys. Rev.* **D 67**, 124024 (2003).
11. A. Burinskii, *Grav. Cosmol.* **11** 301 (2005), hep-th/0506006; *Grav. Cosmol.* **12** 119 (2006); hep-th/0510246 (to appear in *Int. J. Geom. Methods Mod. Phys. 2007*).
12. D. Kramer, H. Stephani, E. Herlt, M. MacCallum, *Exact Solutions of Einstein's Field Equations*, (Cambridge Univ. Press, 1980).
13. A. Burinskii, E. Elisalde, S. R. Hildebrandt and G. Magli, *Phys. Rev.* **D 74** 021502 (Rapid Com.) (2006).
14. J. Wess and J. Bagger, *Supersymmetry and Supergravity*, (Princeton, New Jersey 1983).
15. A. Burinskii, *Phys. Rev.* **D 70**, 086006 (2004); *Grav. Cosmol.* **10**, 50 (2004); hep-th/0403212; *Czech. J. Phys.* **55**, A261 (2005).
16. A. Ya. Burinskii, *Phys.Lett.* **A 185** 441 (1994); *String-like Structures in Complex Kerr Geometry*, in: *Relativity Today*, (Eds. R. P.Kerr and Z. Perjés, Akadémiai Kiadó, Budapest, 1994, p.149).
17. A. I. Akhiezer and V. B. Berestetsky,*Quantum Electrodynamics*, (Moscow, Nauka, 1981, in russian).
18. V. B. Berestetsky, E. M. Lifshitz, L. P. Pitaevsky, *Quantum Electrodynamics (Course Of Theoretical Physics, 4)*, (Oxford, Uk: Pergamon, 1982).
19. B. Carter, *Phys. Rev.* **174** 1559 (1968).
20. T. D. Newton and E. P. Wigner, *Rev. Mod. Phys.* **21** 400 (1949).
21. L. L. Foldi and S. A. Wouthuysen, *Phys. Rev.* **78** 29 (1950).
22. S. S. Schweber, *An Introduction to Relativistic Quantum Field Theory*, (Row, Peterson and Co.Evanson, Ill., Elmsford, N.Y., 1961).

23. P. A. M. Dirac, *The principles of Quantum Mechanics*, 4-th edn. (Oxford at the Clarendon Press, 1958).
24. E. T. Newman, *J. Math. Phys.* **14** 102 (1973); *Phys. Rev.***D 65** 104005 (2002).
25. A. Ya. Burinskii and G. Magli, *Phys. Rev.* **D 61** 044017 (2000).
26. E.T. Whittacker and G.N. Watson, *A Course of Modern Analysis*, (Cambrige Univ. Press London/New York,p.400, 1969).
27. P. Horava, *Nucl. Phys.* **B327** 461 (1989).

THE ELECTRON AND THE NEUTRINO AS SOLITONS IN CLASSICAL ELECTROMAGNETISM

Th. M. NIEUWENHUIZEN

Institute for Theoretical Physics, University of Amsterdam
Valckenierstraat 65, 1018 XE Amsterdam, The Netherlands
E-mail: nieuwenh@science.uva.nl

In classical electrodynamics, extended with gradients of the electric and magnetic fields, a linear soliton is presented which bears features of the Kerr-Newman electron of electrogravity. This is considered as a model for the electron, having a ring shape, with diameter equal to the Compton length $\hbar/mc$ and thickness smaller by the fine structure constant. The soliton has a finite mass, a spin$-\frac{1}{2}$, a $g = 2$ factor, and an electric quadrupole moment that is also "twice too large". From this setup, all relativistic corrections to the classical version of the Pauli Hamiltonian are derived. There appears an additional, spin-dependent quadrupolar force that may vanish on the average. Particle-antiparticle annihilation may become explained on the basis of electromagnetic attraction.

Keywords: Electron; Classical model; Spin; Charge; Electromagnetism; Electromagnetic mass; Annihilation.

1. Introduction

Since Faraday's insights in elecromagnetic fields were put in mathematical form by Maxwell, nature is described by two antagonistic entities: matter and fields. A triumph for this split was the discovery of the electron by J.J. Thomson in 1897. Still, in those days it was now and then supposed that matter was itself some manifestation of fields, in the very same way that a vortex (a whirl of fluid or vapor), a hurricane and a tsunami are 'solitons' in hydrodynamics.

Models for the electron with extended charge distributions were considered by Abraham[1] in 1903 and Lorentz[2] in 1904. Though this led to a finite energy, they could not satisfactorily explain the criticism that Coulomb repulsion between different parts would prevent stability. Another twist came from the postulate of the electron spin ,[3] which would imply in these models also a surface rotation speed that exceeds the speed of light by a factor of order $1/\alpha$. In those days, classical intuition still held such a firm position that de Kronig refrained from publishing his idea after a discussion with Pauli, while Uhlenbeck begged Ehrenfest to withdraw the manuscript.[4] Only two years later the essential correctness of the idea was demonstrated by Dirac:[5] the spin of the electron is identical to that of a spinning top, but it is an "intrinsically quantum" phenomenon. Nevertheless, here we shall discuss spin on a classical basis.

Our argument starts with considering the Kerr-Newman black hole. It is a solution in electrogravity for a massive, spinning, charged black hole. [14,15] Based on the observation that is has a g-factor equal to 2, Carter proposed to consider it as a model for the electron .[6] For this application, it has a naked singularity with charge located on a ring of diameter equal to the Compton length $\hbar/mc$. Soon several papers elaborated this idea from various perspectives.[7–9] Its spin$-\frac{1}{2}$ structure was elucidated[10] and many particle solutions were discussed.[11]

The aim of the present paper is to regularize this singularity. Because the electron mass is much smaller than the Planck mass, and the radius of the ring is much larger than the Planck length, we may expect that, to a very good approximation, space is not curved. We can thus restrict ourselves to the limit of special relativity and only study its electromagnetic properties. The electromagnetic field is stationary and cylindrically symmetric. In *toroidal* coordinates $x^\mu = (ct, r, \theta, \phi)$, also called Boyer-Lindquist coordinates, [12] one has ${}^0A_1 = {}^0A_2 = 0$ and

$$ {}^0A_0 = -e\frac{r}{\Sigma}, \quad {}^0A_3 = ea\frac{r\sin^2\theta}{\Sigma}, \quad \Sigma \equiv r^2 + a^2\cos^2\theta. \tag{1} $$

Let us recall for completeness the Kerr Newman metric. With

$$ \Delta = r^2 + a^2 + \frac{e^2G}{c^4} - 2\frac{Gmr}{c^2}, \qquad u = a^2\sin^2\theta, \tag{2} $$

it reads:

$$ \begin{aligned} ds^2 = {} & \frac{\Delta - u}{\Sigma}dt^2 - \frac{2a(\Delta - \Sigma - u)}{\Sigma}\sin^2\theta\, dtd\phi - \frac{(\Sigma + u)^2 - \Delta u}{\Sigma}\sin^2\theta d\phi^2 \\ & -\frac{\Sigma}{\Delta}dr^2 - \Sigma d\theta^2. \end{aligned} \tag{3} $$

In our special relativistic approach, $G \to 0$, it becomes the Minkowski metric expressed in toroidal coordinates.

2. Ring Theory

When resolving the physics at distances close to the heart of the ring, the situation looks like a cylindrical problem. We now introduce variables $x^\mu = (ct, \zeta, R, \phi)$ which describe that in a natural manner. The smallest physical distance to the ring is

$$ R \equiv |\mathbf{r} - a(\cos\phi, \sin\phi, 0)| = \sqrt{r^2 + a^2} - a\sin\theta. \tag{4} $$

One may go from the toroidal coordinates (r, θ) to the ring coordinates (ζ, R), where the angle ζ is defined by

$$ \cos\theta = \frac{1}{a}\sqrt{\Sigma}\cos\frac{1}{2}\zeta, \qquad r = \sqrt{\Sigma}\sin\frac{1}{2}\zeta. \tag{5} $$

Inversion brings

$$ \Sigma = -R^2\cos\zeta + \sqrt{4a^2R^2 - R^4\sin^2\zeta}. \tag{6} $$

The disc included by the ring, $x^2 + y^2 \leq a^2$ and $z = 0$, is described by $\zeta = 0$, $0 \leq R \leq a$, with the connection $\sin\theta = 1 - R/a$. The region of large r corresponds to large R and ζ near π. Close to the ring it follows that $\Sigma \approx 2aR - R^2\cos\zeta$ and $(x,y) = [a - R\cos\zeta](\cos\phi, \sin\phi)$, $z = R\sin\zeta$, so the physical angle is indeed ζ.

In these variables one also has ${}^0A_1 = {}^0A_2 = 0$, while

$$ {}^0A_0 = -e\frac{\sin\frac{1}{2}\zeta}{\sqrt{\Sigma}}, \qquad {}^0A_3 = ea\frac{\sin^2\theta\sin\frac{1}{2}\zeta}{\sqrt{\Sigma}}. \tag{7} $$

With its 4π periodicity it displays a spin $\frac{1}{2}$-structure. The range of ζ is just $0 \leq \zeta \leq 2\pi$, while the regime $2\pi \leq \zeta \leq 4\pi$ describes an antiparticle, equivalent to $e \to -e$.

At fixed R and ϕ one can consider changing ζ from 0 to 2π. For $R < a$ this means to start right above the disc, turn around the ring and end up right below the disc. In ring coordinates the electric and magnetic fields read for ζ close to 0 or 2π, ${}^0F_{20} \sim \sin\frac{1}{2}\zeta$, ${}^0F_{23} \sim \sin\frac{1}{2}\zeta$,

$$ {}^0F_{10} \approx \frac{-e\cos\frac{1}{2}\zeta}{2\sqrt{2aR - R^2}}, \qquad {}^0F_{13} \approx \frac{ea\cos\frac{1}{2}\zeta}{2\sqrt{2aR - R^2}}. \tag{8} $$

Through the cosine they exhibit a change in sign when one jumps from $\zeta = 2\pi$ to $\zeta = 0$. When approaching the disc from below ($\zeta \to 2\pi$), the discontinuity is avoided by letting ζ continue in the range $2\pi \leq \zeta \leq 4\pi$. So the ring connects two different three-spaces. This offers a geometrical explanation for the "intrinsically quantum" particle — antiparticle connection of fermions .[10,11]

3. Extended electromagnetism

In order to regularize the above solution, we have to study an extended theory. Let the Maxwell Lagrangian be extended with gradients,

$$ L = -\frac{1}{4}F_{\mu\nu}F^{\mu\nu} - \frac{s_1}{4}F_{\mu\nu;\rho}F^{\mu\nu;\rho} - \frac{s_2}{4}F_{\mu\nu}{}^{\cdot\,\cdot\,;\nu}F^{\mu\rho}{}_{\cdot\,\cdot\,;\rho}, \tag{9} $$

where a semicolon denotes a covariant derivative. As usual, we have defined $F_{\mu\nu} = \partial_\mu A_\nu - \partial_\nu A_\mu$, $F^{\mu}{}_{\cdot\nu} = g^{\mu\rho}F_{\rho\nu}$, etc. In special relativity and for cartesian coodinates, $g_{\mu\nu} = g^{\mu\nu}$ =diag$(-1,1,1,1)$. This leads to the Maxwell equation $F^{\mu\nu}{}_{\cdot\,\cdot\,;\nu} = J^\mu$ with

$$ J^\mu = s_1 F^{\mu\nu;\rho}{}_{\cdot\,\cdot\,\cdot\,;\rho;\nu} + \frac{s_2}{2}(F^{\mu}{}_{\cdot\rho}{}^{\cdot\,;\rho;\nu}{}_{\cdot\,\cdot\,;\nu} - F^{\nu}{}_{\cdot\rho}{}^{\cdot\,;\rho;\mu}{}_{\cdot\,\cdot\,;\nu}). \tag{10} $$

In special relativity, covariant derivatives commute, so the second s_2 term in Eq. (10) vanishes. What remains is a Laplace equation, $\Delta J^\mu \equiv J^{\mu;\nu}{}_{\cdot\,\cdot\,;\nu} = \kappa^2 J^\mu$, where

$$ s_1 = \frac{u_1}{\kappa^2}, \qquad s_2 = \frac{u_2}{\kappa^2}, \qquad u_1 + \frac{1}{2}u_2 = 1. \tag{11} $$

3.1. *The electron as an electromagnetic soliton*

In order that A_μ has the same structure as ${}^0A_\mu$ at larger distances, but be more regular near the ring, we look for a solution

$$ A_\mu = {}^0A_\mu C_\mu[1 - f_\mu(R)], \quad J_\mu = {}^0A_\mu C_\mu \kappa^2 f_\mu(R), \tag{12} $$

with the C_μ constants, $f_\mu(0) = 1$ and $f_\mu(\infty) = 0$. The Laplace equation

$$\frac{(r^2+a^2)\partial_r^2 J_0 + 2r\partial_r J_0 + \frac{1}{\sin\theta}\partial_\theta \sin\theta\partial_\theta J_0}{r^2+a^2\cos^2\theta} = \kappa^2 J_0,$$
$$\frac{(r^2+a^2)\partial_r^2 J_3 + \sin\theta\partial_\theta \frac{1}{\sin\theta}\partial_\theta J_3}{r^2+a^2\cos^2\theta} = \kappa^2 J_3. \tag{13}$$

implies near the ring

$$c_2 f_0''(R) + c_1 f_0'(R) = \kappa^2 f_0(R), \tag{14}$$

with $c_2 = 1$ exactly. For c_1 we have

$$c_1 = \frac{1}{2a}\left(1 - \frac{r^2+3a^2\cos^2\theta}{4a^2} + \cdots\right). \tag{15}$$

It is not a function of R alone, so our solution is approximate. We therefore work in the regime $a \gg 1/\kappa$. Since c_1 is regular at $R = 0$ (i.e., at $r = 0$, $\theta = \pi/2$) and since it is small, we may approximate $c_1 = 1/2a$. This brings

$$f_0(R) = e^{-\kappa_0 R}, \qquad \kappa_0 = \sqrt{\kappa^2 + \frac{1}{16a^2}} + \frac{1}{4a}, \tag{16}$$

and, similarly, for f_3 we find $c_1 = -3/2a$, yielding

$$f_3(R) = e^{-\kappa_3 R}, \qquad \kappa_3 = \sqrt{\kappa^2 + \frac{9}{16a^2}} - \frac{3}{4a}. \tag{17}$$

To keep the charge observed at infinity equal to e, we take $C_0 = 1$. It appears that, in order to have equally strong behavior of A_0 and A_3 at small R, we have to choose $C_3/C_0 = \kappa_0/\kappa_3$. In the combination $F^{\mu\nu;\rho}F_{\mu\nu;\rho}$, that enters the Lagrangian L, see Eq. (9), and the Hamiltonian density $H = T_{00}$, this is needed to cancel the non-integrable $1/R^3$ divergences from the $\nabla\mathbf{E} : \nabla\mathbf{E}$ and $\nabla\mathbf{B} : \nabla\mathbf{B}$ terms. This finally leads to the regularized solution

$$A_0 = -\frac{e\sin\frac{1}{2}\zeta}{\sqrt{\Sigma}}[1 - e^{-\kappa_0 R}], \qquad A_3 = \frac{ea\kappa_0 \sin\frac{1}{2}\zeta\sin^2\theta}{\kappa_3\sqrt{\Sigma}}[1 - e^{-\kappa_3 R}], \tag{18}$$

with Σ taken from Eq. (6) and $\sin\theta$ from (5). These expressions are correct to relative order $1/(\kappa a)^2$. The fields are mildly singular at the heart of the ring $R = 0$.

3.2. *Energy momentum tensor*

In general relativity, the energy momentum tensor is defined in terms of changes of the metric. With $g = \det g_{\mu\nu}$, one has

$$T^{\mu\nu} = \frac{2}{\sqrt{-g}}\frac{\delta\int \mathrm{d}^4x\sqrt{-g}L}{\delta g_{\mu\nu}} = {}^{(0)}T^{\mu\nu} + {}^{(1)}T^{\mu\nu} + {}^{(2)}T^{\mu\nu}. \tag{19}$$

The calculations are somewhat intricate. We give here the results,

$$\begin{aligned}
{}^{(0)}T^{\mu\nu} &= F^{\mu}_{\cdot\,\rho}F^{\nu\rho} - \frac{1}{4}\,g^{\mu\nu}\,F_{\rho\sigma}F^{\rho\sigma}, \\
{}^{(1)}T^{\mu\nu} &= s_1\{-\frac{1}{4}g^{\mu\nu}F_{\rho\sigma;\tau}F^{\rho\sigma;\tau} + \frac{1}{2}F^{\rho\sigma;\mu}F_{\rho\sigma}^{\cdot\,\cdot\,;\nu} - F^{(\mu\sigma;\rho}_{\cdot\,\cdot\,\cdot\,;\rho}F^{\nu)}_{\cdot\,\sigma} - F^{\rho\sigma;(\mu}_{\cdot\,\cdot\,\cdot\,;\rho}F^{\nu)}_{\cdot\,\sigma} \\
&\quad -F^{\sigma(\mu;\nu)}_{\cdot\,\cdot\,\cdot\,;\rho}F^{\rho}_{\cdot\,\sigma} + F^{\rho\sigma;\mu}F^{\nu}_{\cdot\,\rho;\sigma} + F^{\rho(\mu;\nu)}J_{\rho}\}, \qquad (20) \\
{}^{(2)}T^{\mu\nu} &= s_2\{\frac{1}{4}g^{\mu\nu}J^{\rho}J_{\rho} + \frac{1}{2}g^{\mu\nu}F_{\rho\sigma}J^{\rho;\sigma} - \frac{1}{2}J^{\mu}J^{\nu} - J^{(\mu}_{\cdot\,;\rho}F^{\nu)\rho} - F_{\rho}^{\ \cdot\,(\mu}J^{\rho;\nu)}\},
\end{aligned}$$

where $A^{(\mu\cdots\nu)}$ denotes the symmetrization $\frac{1}{2}A^{\mu\cdots\nu} + \frac{1}{2}A^{\nu\cdots\mu}$ and $J^{\mu} = F^{\mu\nu}_{\cdot\,\cdot\,;\nu}$. It can be verified that energy and momentum are conserved, $T^{\mu\nu}_{\cdot\,;\nu} = 0$.

4. Mass and spin

Whereas in the Kerr-Newman solution there is an intrinsic mass and strong gravity, we assume negligible gravity and a mass solely of electromagnetic origin. The inertial mass just arises from the energy density,

$$m_{\rm i}c^2 = \frac{1}{4\pi}\int \mathrm{d}v T_{00}, \qquad (21)$$

with $\mathrm{d}v = \mathrm{d}x\mathrm{d}y\mathrm{d}z \approx \mathrm{d}\zeta\mathrm{d}R\mathrm{d}\phi\, aR(1 - \frac{3R}{2a}\cos\zeta\,)$. There is a spin in the z-direction of magnitude[14,15]

$$S_z = \frac{1}{4\pi c}\int \mathrm{d}v\, T^{0}_{\ \cdot\,3}. \qquad (22)$$

Analyzing the above expressions in ring coordinates, with the help of an algebraic manipulation program, we find for the leading and subleading terms

$$m_{\rm i}c^2 = \frac{\pi e^2\kappa}{16}\,[\,2 - u_2 + \frac{1}{\kappa a}(1 + \frac{1}{2}u_1 - \frac{1}{2}u_2)\,], \qquad (23)$$

$$S_z = \frac{\pi e^2\kappa a}{16c}[\,2 - u_2 + \frac{1}{\kappa a}(\frac{1}{2} + u_1 - \frac{5}{4}u_2)\,]. \qquad (24)$$

In these expressions we have not yet inserted $u_1 + \frac{1}{2}u_2 = 1$. We now equate S_z to $\frac{1}{2}\hbar$. Choosing the charge equal to the electron charge we combine (23) and (24)

$$a = \frac{\hbar}{2m_{\rm i}c}(1 + \frac{\pi\alpha u_2}{8}), \quad \kappa = \frac{8}{\pi u_1\,\ell_*}, \quad \ell_* \equiv \frac{\alpha\hbar}{m_{\rm i}c}. \qquad (25)$$

It is now indeed clear that $\kappa a \sim 1/\alpha \gg 1$. The radius $a \approx \frac{1}{2}\,\hbar/m_{\rm i}c$ equals half the Compton length and it is similar to the one of the Kerr-Newman electron. The scale ℓ_* is known as the Abraham-Lorentz classical radius of the electron. The timescale $\tau_L = (2/3)\ell_*/c = 2\alpha\hbar/(3m_{\rm i}c^2)$ is the Lorentz damping time .[2,16] It now gets a physical interpretation as the time that light needs to traverse the thickness of the ring.

4.1. *Electromagnetic properties*

From the large **r**-behavior of the electromagnetic field tensor, we indeed notice a charge e. Following Israel,[9] we can verify that this charge, enclosed by the surface $r = \infty$, can be written as a volume integral arising from the ring, with a large but finite outcome, $q_{ring} = e\sqrt{\kappa a} \gg e$, and a negative surface integral from the two sides of the disc enclosed by the ring — that add up because of the opposite signs in the fields as in Eq. (8) — yielding indeed $q_{disc} = e - q_{ring}$.

Next, there is a magnetic dipole moment $M_z = ea\kappa_0/\kappa_3$. With the Bohr magneton $\mu_B = \hbar e/(2m_{\rm i}c)$ this defines the magnetic g-factor (Landé factor)

$$g = \frac{2M_z}{\mu_B} = \frac{4amc\kappa_0}{\hbar\kappa_3} = 2(1 + \frac{1}{4}\pi\alpha). \tag{26}$$

Of interest is also the electric quadrupole moment. The quadrupole tensor generally enters as $A^0 \approx q/|\mathbf{r}| + \frac{1}{2}Q^{ij}\partial_i\partial_j(1/|\mathbf{r}|)$. The nonvanishing elements are thus

$$Q^{xx} = Q^{yy} = \frac{1}{2}g_E e a^2, \qquad g_E = 2. \tag{27}$$

The standard answer for a ring would be $g_E = 1$, see, e.g. Jackson.[13] The value $g_E = 2$ is related to the spin$-\frac{1}{2}$ structure, as is long known for the Kerr-Newman solution [7,8] and it just extends here.

4.2. *Behavior in external electromagnetic fields*

The soliton can be characterized by its mass, charge and spin together with its center of mass, its speed and its spin direction. When it moves with a speed $\mathbf{v}$, there will be a Lorentz boost, replacing $m_{\rm i}c^2$ of Eq. (21) by $\gamma m_{\rm i}c^2$, with $\gamma = 1/\sqrt{1-\mathbf{v}^2/c^2}$ the Lorentz factor. In an external electromagnetic field $A^\mu_{\rm ex}$ we have to replace $A^\mu \to A^\mu_{\rm int} + A^\mu_{\rm ex}$ and $F^{\mu\nu} \to F^{\mu\nu}_{\rm int} + F^{\mu\nu}_{\rm ex}$, where $A^\mu_{\rm int}$ is the solution discussed so far. This additionally brings via Eq. (21) the potential energy U and it defines the Hamiltonian H by $H = (\gamma - 1)m_{\rm i}c^2 + U$.

It has been checked that the resulting terms linear in $A^\mu_{\rm ex}$ and $F^{\mu\nu}_{\rm ex}$ are just the ones that can be read off in the far field. The leading term is the charge, yielding the first term [17] of the potential energy, $\delta U = eA^0 \equiv eV$. A spin in the direction $\hat{\mathbf{s}}$ (with $|\hat{\mathbf{s}}| \equiv 1$), has a spin vector $\mathbf{S} = \frac{1}{2}\hbar\hat{\mathbf{s}}$ and magnetic dipole moment $\mathbf{M} = ge\mathbf{S}/(2m_{\rm i}c)$, see (26). The effect in the Hamiltonian is familiar,

$$H_B = -\mathbf{M}\cdot\mathbf{B} = -\frac{ge}{2m_{\rm i}c}\mathbf{S}\cdot\mathbf{B}. \tag{28}$$

When the electron moves with speed $\mathbf{v}$, a Lorentz boost transforms $\mathbf{B} \to \gamma(\mathbf{B} - \mathbf{v}\times\mathbf{E}/c)$. For a centrally symmetric potential V and $|\mathbf{v}| \ll c$, this adds to (28) the well known spin-orbit coupling that involves $\mathbf{L} = m_{\rm i}\mathbf{r}\times\mathbf{v}$.

Including the Thomas factor $\frac{1}{2}(g-1)$,[13] one has

$$\delta H_{so} = \frac{(g-1)e}{2m_{\rm i}{}^2c^2}\frac{V'}{r}\mathbf{S}\cdot\mathbf{L}. \tag{29}$$

Finally, we consider the contribution from the electric quadrupole moment (27). It leads to a potential energy

$$\delta H = -\frac{1}{2}ea^2(\partial_x E_x + \partial_y E_y) \equiv H_D + H_Q. \tag{30}$$

For the first term we add the $\partial_z E_z$ term, which brings

$$H_D = -\frac{e\hbar^2}{8m_{\rm i}{}^2c^2}\nabla \cdot \mathbf{E} = \frac{e\hbar^2}{8m_{\rm i}{}^2c^2}\Delta V. \tag{31}$$

Exactly this expression is known as the Darwin term, a leading relativistic correction from the Dirac equation.[19,20] Notice that $g_E = 2$, see Eq. (27), is crucial.

4.3. *An additional quadrupolar force*

The second term in Eq. (30) arises because we have to undo the addition $\partial_z E_z$. The result is *spin-dependent*. For a centrally symmetric potential and spin direction $\hat{\mathbf{s}}$ this leads to

$$H_Q = -\frac{e\hbar^2}{8m_{\rm i}{}^2c^2}[\frac{V'}{r} + (\hat{\mathbf{r}}\cdot\hat{\mathbf{s}})^2(V'' - \frac{V'}{r})]. \tag{32}$$

The average of (32) vanishes if $\langle(\hat{\mathbf{r}}\cdot\hat{\mathbf{s}})^2\rangle = V'/(V' - rV'')$. For the Coulomb force this is equal to 1/3 (isotropy).

4.4. *Hydrogen-type problems*

For centrally symmetric potentials we collect all terms and arrive after the substitution $m_{\rm i}\mathbf{v} \to \mathbf{p} - e\mathbf{A}/c$ at the Pauli Hamiltonian $H = H_0 + H_1 + H_Q$ with the non-relativistic Hamiltonian $H_0 = (\mathbf{p} - e\mathbf{A}/c)^2/2m_{\rm i} + eV$. The next term,

$$H_1 = -\frac{(\mathbf{p} - e\mathbf{A}/c)^4}{8m_{\rm i}{}^3c^2} - \frac{e\mathbf{S}\cdot\mathbf{B}}{m_{\rm i}c} + \frac{eV'\,\mathbf{S}\cdot\mathbf{L}}{2m_{\rm i}{}^2c^2r} + \frac{e\hbar^2\nabla^2 V}{8m_{\rm i}{}^2c^2},$$

contains all the relativistic corrections known from the Dirac theory .[19,20] But it is still classical, with $\mathbf{v} = (\mathbf{p} - e\mathbf{A}/c)/m_{\rm i}$ the speed and $\mathbf{S} = \frac{1}{2}\hbar\hat{\mathbf{s}}$ the spin vector. On top of this, there is the quadrupole potential H_Q, which may vanish on the average.

4.5. *Towards fixing charge and spin*

In our linear theory, the charge is a free parameter that was adjusted by hand. The same holds for the radius a that was fixed by equating the spin to $\frac{1}{2}\hbar$. Stability questions make sense only in a non-linear theory. To proceed, nonlinear terms have to be imposed, e.g., $\delta L = s_4(F_{\mu\nu}F^{\mu\nu})^2$. It is hoped that this gives a factor $g = 2$ exactly, before fluctuations are taken into account. To 'quantize' charge and spin, different kinds of solitons (electron, muon, etc) are needed.

5. Application to the neutrino

In the literature, the application of Kerr Newman black holes to elementary particles has been limited to the electron. The electromagnetic potentials (1) can be written as a real part,

$$ {}^0A_0 = \Re\frac{-e}{r - ia\cos\theta}, \qquad {}^0A_3 = \Re\frac{ea\sin^2\theta}{r - ia\cos\theta}. \tag{33} $$

Clearly, one can also consider the imaginary parts,

$$ {}^0A_0 = \frac{-ea\cos\theta}{r^2 + a^2\cos^2\theta}, \qquad {}^0A_3 = \frac{ea^2\sin^2\theta\cos\theta}{r^2 + a^2\cos^2\theta}. \tag{34} $$

They also relate to a source-free solution of the Maxwell equations, at least outside the disc discussed above. They describe the far field of a spinning, massive neutral particle, that we identify with the electron neutrino. Now e is not necessarily the elementary charge, but a parameter related to the electric dipole moment d,

$$ d = ea. \tag{35} $$

Since our approach does not need the modified version of the Kerr Newman metric, it can also be applied to this situation. In terms of the ring coordinates the potentials read

$$ {}^0A_0 = \frac{-d\cos\frac{1}{2}\zeta}{a\sqrt{\Sigma}}, \qquad {}^0A_3 = \frac{d\sin^2\theta\cos\frac{1}{2}\zeta}{\sqrt{\Sigma}}, \tag{36} $$

which should be compared with Eq. (7). Because of the cosine, it also displays a spin-$\frac{1}{2}$ structure, so it can be viewed as a model for the neutrino.

The same regularization procedure, as described above for the electron, can be worked out. We shall refrain from presenting details here. To leading order we finally get for the inertial mass and the spin

$$ m_\nu c^2 = \frac{\pi d^2\kappa}{32a^2}(2 + u_2), \qquad S_z^\nu = \frac{\pi d^2\kappa}{32ac}(2 + u_2). \tag{37} $$

This only differs from the electron result (21-22) in an additional factor $\frac{1}{2}$ and the sign of u_2. Equating $S_z^\nu = \frac{1}{2}\hbar$, it appears that the neutrino radius is half of its Compton length,

$$ a_\nu = \frac{1}{2}\,\frac{\hbar}{m_\nu c}. \tag{38} $$

The same connection was observed for the electron, Eq. (25).

The neutrinos electric dipole moment thus reads

$$ d_\nu = \frac{\hbar e}{2m_\nu c}. \tag{39} $$

Estimating the mass as 0.1 eV/c^2 from the cosmological estimate $\sum_\nu m_\nu c^2 \lesssim 0.7$ eV,[24] and keeping e as the electron charge, we arrive at an estimate

$$ d_\nu \lesssim 10^{-21}\ \mathrm{e\!-\!cm}, \tag{40} $$

much less than the upper bound for heavy neutrinos, $d = 10^{-16} - 10^{-15}$ e–cm, discussed in literature.[25]

6. Summary and outlook

In line with the 19th century thinking of, e.g., Abraham, Lorentz, Poincaré and Einstein, we have shown that many aspects of the electron, e.g. its spin$-\frac{1}{2}$, mass, charge and $g = 2$ factor, can be understood within classical electrodynamics. Such "intrinsically quantum" aspects are long known for the Kerr-Newman electron, a singular object with strong gravity.[6] Here they just arise for a ring geometry in classical electrodynamics, and they indeed persist because of the fortunate cancellation of two extra factors two. [21]

To start, we added some gradients of the electric and magnetic fields to the Maxwell action. Then a linear solitonic solution occurs with the shape of a ring having a diameter equal to the Compton wavelength and thickness smaller by a factor $\alpha = 1/137$. The old objections against the Abraham-Lorentz theory do not apply: electrostatic repulsion of different parts of the soliton is prevented by the extra terms we added to Maxwell theory. The spin does not lead to a speed exceeding the speed of light because our structure is a ring rather than a sphere. Loosely speaking, the speed at the ring is $S_z/(m_{\rm i}a) = (\frac{1}{2}\hbar)/[m_{\rm i}\,\hbar/(2m_{\rm i}\,c)]$, so just c.

In an external field our approach reproduces all relativistic corrections to the Pauli Hamiltonian, which determine the fine structure of the H-atom, on a classical basis. In addition, there is a spin-dependent quadrupolar force that may vanish on the average. The question of whether the electric quadrupole moment of the electron has been observed, may now appear to be related to the long established role of the Darwin term in the hydrogen fine structure. The Darwin term is attributed to the Zitterbewegung because the position of the electron cannot be determined with accuracy better than the Compton length. Our picture offers a simpler explanation: the Compton scale is just the physical size of the electron.

Mass and charge arise from electromagnetism alone. Contrary to the Kerr-Newman electron, mass is never negative. Spin is a property related to the currents circling around the ring. In the far field they can be considered as "intrinsic properties" of a "point object". The present experimental upper bound for the electron size is about $10^{-18}m$ at $e^- + e^+$ center-of-mass energy 200 GeV,[22] equivalent to a Lorentz factor $\gamma \sim 2\,10^5$. Our contracted radius $[\hbar/(2m_ec)]/\gamma$ just coincides with that bound. This may indicate that deviations from quantum theory hide themselves at high energies. [23]

Whereas the Kerr-Newman electron can be seen as the real part of a complex potential, we pointed out that the imaginary part of this potential can describe a neutrino. The essential point is then that it has a mass and an electric dipole moment. Our regularization can be carried out as well, and sets the radius of the ring structure again as half of its Compton length.

Another "intrinsically quantum" field theoretic property is particle-antiparticle annihilation. Our picture may explain it as a classical effect. Indeed, when we just superimpose a solution (18), having charge $+e$ and spin parameter $+a$, with another one, having $-e$ and $+a$, we see that the electromagnetic potentials and fields cancel. Such also occurs in the spin-down channel ($+e, -a$ & $-e, -a$), but not in any other channel. There is thus a strong attraction between a particle and an antiparticle with the same spin direction. In a nonlinear approach this may explain on a classical basis why the process $e^- + e^+ \to \gamma$ only occurs in parallel spin channels.

In the s restframe of the Kerr-Newman neutrino, spin reversal amounts to $d \to -d$ and $a \to -a$, replacing ${}^0A_0 \to {}^0A_0$, ${}^0A_3 \to -\ {}^0A_3$. The replacement $d \to -d$ alone, inducing ${}^0A_\mu \to -\ {}^0A_\mu$, defines the antineutrino with spin up in the rest frame. Since neutrinos are mostly left handed and antineutrinos right handed, moving basically with the speed of light, the annihilation process $\nu_\uparrow + \bar{\nu}_\uparrow \to \gamma$ is related to opposite speeds and parallel spins. In our restframe discussion, precisely this situation leads to a cancellation of the electromagnetic fields when the two rings overlap, and is expected produce a photon in an extended version of the approach.

The selected model proves that the spin$-\frac{1}{2}$ structure exists in classical electrodynamics. A more realistic starting point might be that of $U(1) \times SU(2)$ gauge groups in the presence of a Higgs field,[15] which are gauge fields of the electro-weak sector of the standard model. This may open the door for a solitonic description of the other leptons, massive gauge bosons and, after adding $SU(3)$, of gluons and quarks.

The idea that particles are solitons of the fields that surround them is very intriguing. In our case it explains right away why special relativity holds for matter — it already holds for the fields from which they are composed. And because of this composition, identical solitons, that undergo chaotic motion, are indistinguishable, which explains another "intrinsically quantum" property.

Acknowledgments

It is a pleasure to acknowledge discussion with Armen E. Allahverdyan and proofreading by Peter D. Keefe.

References

1. M. Abraham, *Ann. der Phys.* **10**, 105 (1903); Phys. Z. **5**, 576 (1904).
2. H.A. Lorentz, in *Encyclopädie der mathematischen Wissenschaften,* Vol. 5, Physik, Part 2. A. Sommerfeld, Ed., (Teubner, Leipzig, 1904—1922), pp 145.
3. G.E. Uhlenbeck and S. Goudsmit, *Nature* **117**, 264-265, (1926).
4. S. A. Goudsmit, *Ned. Tijdschr. Nat.* **37**, 386 (1971). For an English translation, see http://www.lorentz.leidenuniv.nl/history/spin/goudsmit.html
5. P. A. M. Dirac, *Proc. Roy. Soc. London* **A117**, 610 (1928); Proc. Roy. Soc. London A118, 351 (1928).
6. B. Carter, *Phys. Rev.* 174, 1559 (1968).
7. C. A. López, *Phys. Rev. D* **30**, 313 (1984).
8. C. L. Pekeris and K. Frankowski, *Phys. Rev. A* **36**, 5118 (1987).
9. W. Israel, Phys. Rev. D **2**, 641 (1970).

10. H. I. Arcos and J. G. Pereira, *Gen. Rel. Grav.* **36**, 2441 (2004); in *Beyond the Quantum*, Th. M. Nieuwenhuizen, V. Spicka, B. Mehmani, M. Jafar-Aghdami and A. Yu. Khrennikov, eds, (World Scientific, Singapore, 2007, to appear), pp 000.
11. A. Burinskii, *Class. Quant. Grav.* **16**, 3497 (1999); *Grav. Cosmol.* **10**, 50 (2004); *Phys. Rev. D* **70**, 086006 (2004); gr-qc/0606035, published in *Beyond the Quantum*, Th. M. Nieuwenhuizen, V. Spicka, B. Mehmani, M. Jafar-Aghdami and A. Yu. Khrennikov, eds, (World Scientific, Singapore, 2007, to appear), pp 000.
12. One has $(x, y) = \sqrt{r^2 + a^2} \sin\theta(\cos\phi, \sin\phi), \quad z = r\cos\theta$. The heart of the ring is located at $r = 0$, $\theta = \frac{1}{2}\pi$.
13. J. D. Jackson, *Classical Electrodynamics*, 3rd ed., (Wiley, New York, 1999).
14. R.M. Wald, *General Relativity*, (Univ. of Chicago Press, Chicago, 1984).
15. V. P. Frolov and I. D. Novikov, *Black Hole Physics*, (Kluwer, Dordrecht, 1997).
16. L. de la Pena and A.M. Cetto, *The Quantum Dice: An Introduction to Stochastic Electrodynamics*, (Kluwer, Dordrecht, 1996).
17. We shall from now drop the suffix ex.
18. L. H. Thomas, Nature **117**, 514 (1926).
19. J.J. Sakurai, *Advanced Quantum Mechanics*, (Addison Wesley, Reading, 1967).
20. P.W. Milonni, *Quantum Vacuum: An Introduction to Quantum Electrodynamics.* (Academic, New York, 1994).
21. Th. M. Nieuwenhuizen, unpublished.
22. See e.g. S. Kullander, http://nobelprize.org/physics/articles/kullander/
23. C. O. Loustó and N. Sánchez, *Nucl. Phys. B* **383**, 377 (1992).
24. D.N. Spergel et al., *Astrophys. J. Suppl.* **148**, 175 (2003).
25. M. Sher and S. Nie, *Phys. Rev. D* **65**, 093018 (2002).

PART G

Philosophical Considerations

PROBABILITY IN NON-COLLAPSE INTERPRETATIONS OF QUANTUM MECHANICS

D. DIEKS

Institute for History and Foundations of Science, Utrecht University, Utrecht, The Netherlands
E-mail:dieks@phys.uu.nl
www.phys.uu.nl/ wwwgrnsl/

Non-collapse interpretations of quantum mechanics set themselves the task of developing a self-consistent and empirically adequate version of quantum mechanics that does not make use of the projection postulate (or collapse of the wavefunction). Only unitary evolution is allowed in these interpretations, so that superpositions are always maintained during evolution—even in measurements. In this paper we discuss how this deterministic mathematical scheme can be brought into accordance with the usual statistical predictions of quantum mechanics. In particular, we investigate how the Born probability rule fits in.

Keywords: Non-collapse interpretations; Probability; Born rule; Many worlds; Modal interpretations; Decoherence.

1. The Non-Collapse Point of View

In standard presentations of quantum theory two types of evolution are recognized: unitary evolution, described by the Schrödinger equation or one of its relativistic generalizations—and non-unitary *collapses*. The latter processes are associated with measurements: in a measurement a superposition of states is assumed to be reduced to one of its components, corresponding to the actual outcome. Mathematically, this is achieved by a *projection* of the state onto the component in question.

By contrast, non-collapse interpretations accept only the Schrödinger-type of evolution. This has the consequence that according to these interpretations not only the actual outcome, but also all other outcomes that were possible remain represented in the final state after a measurement. In the notorious Schrödinger cat case, for example, non-collapse interpretations maintain that both the dead cat state and the state corresponding to the living cat occur in the final state (which is a superposition). Clearly some non-standard interpretational step is needed to make this compatible with the empirical fact that only one of the outcomes is actually realized in a measurement—that the cat is either dead or alive, and not both.

The simplest approach is to interpret the total uncollapsed state in a probabilistic way. That is, although only one state of affairs is actual, the total state describes all *possibilities*—it gives rise to a probability distribution that comprises both the actual and the possible. Modal interpretations[1–4] take this point of view, and it also is implicit in approaches based on decoherence.[5,6] An interpretational alternative to this 'probabilism' is to promote the different terms in the superposition all to the level of *actuality*; so that, e.g., both the living and the dead cat are actual. Since these different states of affairs exclude each other, they cannot be realized in one and the same world. Thus the concept of many worlds is born. The many-worlds concept may appear outlandish, but as we can only become conscious of observations in one world, the many-worlds option is *operationally* indistinguishable from the probabilistic approach. Whether the events that were possible but were not realized in our world are *actual* in other worlds *that really exist*, or whether they are *possibilities* that can formally be represented in other *possible* worlds, is a question that cannot be decided by observation. We shall discuss some of the arguments *pro* and *contra* here in Sec. 4. However, the unifying trait, common to all the interpretations we are considering here, is the absence of collapses, and this is the essential element in the next two Sections. We shall focus on the probabilistic viewpoint there, but most results can be carried over to the framework of many worlds without much difficulty.

The *motivation* for investigating non-collapse interpretations comes from the idea that measurement results and measuring devices do not possess a special ontological status: they are physical things with ordinary physical properties, like positions of pointers attached to measuring devices, marks on computer tapes, etc. Moreover, non-collapse interpretations are motivated by the conviction that there is no fundamental difference between microscopic and macroscopic objects: both are subject to the same quantum mechanical principles. In accordance with these ideas, no special status is attributed to measurement interactions. Measurement interactions are treated according to the same principles as all other physical interactions between systems. Indeed, a measurement is just a physical interaction between an object system and a measuring device. Since ordinary interactions are described by means of unitary evolution, it follows that unitary interactions should be the norm across the board, in all processes, in both the macroscopic and the microscopic domains.

Our task in non-collapse interpretations is therefore to endow Hilbert space, equipped with an exclusively unitary time evolution, with physical meaning. We need interpretational rules that tell us how this mathematical formalism relates to physical reality. To this end, we can safely use a number of the usual interpretative principles of quantum mechanics that have proved their mettle; in particular the one saying that physical quantities are represented by hermitian operators (observables). But we cannot accept *everything* from standard textbook interpretations. In particular, since superpositions are always maintained according to the non-collapse scheme, even in measurement interactions, we cannot assume that observables can only have a definite value if the state is an eigenstate of the observable in question. We want to be able to say, for example, that Schrödinger's cat is either dead or alive in spite of the fact that the total state is a superposition of eigenstates corresponding to 'dead' and 'alive', respectively.

An important question thus becomes: which physical quantities—represented by hermitian operators, observables—can be assigned a definite value, when it is given that the physical system is represented by a particular state in Hilbert space. An answer to this question is needed in order to make it possible to define *events*, physical occurrences, to which probabilities can be assigned. In the context of the many worlds interpretation, this becomes the question of how to define the branching process: is there a privileged way of writing down the total state as a superposition, corresponding to the branching into different worlds? In decoherence schemes and modal interpretations the question is how to define a preferred basis that fixes the observables that possess definite values; mathematically, this comes to the same thing. Several proposals have been made to define a preferred basis (e.g., bases that are robust under decoherence or bases with certain entropic properties,[5,6] bases that follow from the Schmidt decomposition[1,2]). In many cases these various proposals lead to effectively the same results. Here, we shall focus on observables singled out by the Schmidt decomposition—these are selected if we impose the requirement that the definite-valued observables should be determined by only the state and the Hilbert space structure.[7–9]

Consider the quantum mechanical treatment of a composite physical system, consisting of two parts. In this case, the total Hilbert space can be decomposed: $\mathcal{H} = \mathcal{H}_1 \otimes \mathcal{H}_2$. According to a famous theorem (Schmidt, Schrödinger) there is a corresponding *biorthogonal decomposition* of every pure state in $\mathcal{H}$:

$$|\psi\rangle = \sum_k c_k |\psi_k\rangle \otimes |R_k\rangle, \tag{1}$$

with $|\psi_k\rangle$ in $\mathcal{H}_1, |R_k\rangle$ in $\mathcal{H}_2$, $\langle\psi_i|\psi_j\rangle = \delta_{ij}$ and $\langle R_i|R_j\rangle = \delta_{ij}$. This decomposition is unique if there is no degeneracy among the values of $|c_k|^2$.

Modal non-collapse interpretations give the following physical interpretation to this state. The system represented by vectors in $\mathcal{H}_1$ possesses exactly *one* of the physical properties associated with the set of projectors $\{|\psi_k\rangle\langle\psi_k|\}$, and definitely does not possess the others. That is, exactly one of the mentioned projectors is assigned the definite value 1, the others get the definite value 0. The interpretation thus selects, on the basis of the form of the state $|\psi\rangle$, the projectors $|\psi_k\rangle\langle\psi_k|$ as definite-valued magnitudes. Given that these projectors are all definite-valued, it is natural to stipulate further that all physical magnitudes represented by maximal hermitian operators with spectral resolution given by $\Sigma a_k|\psi_k\rangle\langle\psi_k|$ are also definite-valued. Since such operators are functions of the definite-valued projectors, their possible values can be taken as given by the functions in question applied to the values assumed by the projections.

In the case of degeneracy, that is $|c_j|^2 = |c_i|^2$, for $i, j \in I_l$ (with I_l a set of indices), the biorthogonal decomposition (1) still determines a unique set of projection operators, but these will now generally be multi-dimensional. The one-dimensional projectors must in this case be replaced by projectors $P_l = \sum_{i \in I_l} |\psi_i\rangle\langle\psi_i|$. The general class of definite physical quantities contains in this case non-maximal hermitian operators in whose spectral resolution such multi-dimensional projectors occur.

We have just stipulated that only *one* of the values that can be assumed by the definite-valued observables is actually realized. This makes it natural to ask for the *probability* that it is the *l-th* possibility that is actual. In accordance with the standard Born rule, we may take this probability as $|c_l|^2$ (in the case of degeneracy this becomes $\sum_{i \in I_l} |c_i|^2$).

The next Section is devoted to the question whether this probability assignment can be further justified within the context of non-collapse interpretations. Should we just *posit* the Born rule, as an axiom, or can it be based on other principles?

2. The Born Measure

Is it possible to *derive* a preferred probability measure on the lattice of definite-valued observables? Any attempt at such a derivation evidently needs premises. If we want to stay as close as possible to the standard mathematics of quantum mechanics, without the introduction of additional elements by hand, it seems natural to require that the measure is to depend only on the state in Hilbert space and the tensor product structure of this Hilbert space. Has this enough bite to single out a definite form of the measure? As we will argue, the answer is 'yes' if we add a whiff of locality or non-contextuality assumptions: the features of system I should be independent of things going on only in $\mathcal{H}_{II}$. The Born measure then becomes the only one that is definable.

Denote the measure to be assigned to the definite-valued projector P, if the state is $|\psi\rangle$, by $\mu(|\psi\rangle, P)$. Write $|\psi\rangle$ in its biorthogonal form:

$$|\psi\rangle = \sum_k c_k |\alpha_k\rangle \otimes |\beta_k\rangle, \tag{2}$$

where we now have taken the non-degenerate case for simplicity. First note that we can take the coefficients c_k to be real numbers: all phase factors can be absorbed into the vectors $|\alpha_k\rangle$ or $|\beta_k\rangle$, without any effect on the observables that are value-definite (the projection operators are invariant under this operation). So if μ is going to depend on the coefficients c_k, only their absolute values or, what amounts to the same thing, only $|c_k|^2$ can enter into the expressions.

An alternative road to this conclusion is to use the transformations $U_I \otimes U_{II}$, with U_I and U_{II} unitaries in $\mathcal{H}_I$ and $\mathcal{H}_{II}$, respectively, under which $|\psi\rangle$ is invariant. As we have seen in the previous Section, both U_I and U_{II} must be pure phase transformations in this non-degenerate case. One could now reason as follows (as Zurek does in his proposed derivation of the Born rule[10]): any physical features pertaining to system I alone should be invariant under the operation of any U_I on $|\psi\rangle$, for the following reason. The effect of U_I can be *undone* by a suitably chosen operator U_{II} (U_I is what Zurek calls an 'envariance' operation); and since this U_{II} operates solely on system II, it cannot affect the physical properties of I. Consequently, any effects U_I may have on the mathematical state of system I

should be immaterial to the physical features of I. In particular, the phases of the coefficients c_k must be irrelevant, so that only their absolute values can count.

As pointed out by Caves,[11] however, the essential element in the reasoning here is already in the starting assumption (also implicitly made by Zurek) that the probabilities and physical properties pertaining to system I only depend on the parts of $|\psi\rangle$ defined in $\mathcal{H}_\mathcal{I}$ and do not depend on the vectors $\{|\beta_k\rangle\}$. Once we make this independence assumption, the additional introduction of the notion of envariance is not really needed. The assumption in question may be regarded as a no-signalling or locality condition: its violation would make it possible to change physical features of I by intervening in the state of II, which would make it possible to signal to I even if system II is far away. It can also be regarded as a non-contextuality condition: it should not make a difference for the physical features of I what unitary operations are taking place in its environment II if this environment is causally disconnected; i.e., what kind of evolutions not involving I are taking place there is irrelevant. This independence assumption already entails invariance of the probabilities under arbitrary U_{II}. In particular, it implies that μ can only depend on system I's definite-valued projectors P and on c_k, and since absorbing all phase factors into $\{|\beta_k\rangle\}$ does not change the projectors, only the absolute values $|c_k|$ can be relevant (for further discussions of Zurek's line of argument see[11–14]). In summary, non-contextuality entails that μ must be invariant under application of arbitrary unitaries $U_I \otimes U_{II}$. So we find that the bases $\{|\alpha_k\rangle\}$ and $\{|\beta_k\rangle\}$ are irrelevant for the probabilities to be defined, and only the values of $|c_k|$ can play a role.

This irrelevance of $\{|\alpha_k\rangle\}$ and $\{|\beta_k\rangle\}$ for the expression of μ can be justified in a yet different (but equivalent) way by an argument in the spirit of the previous Section. We may directly impose the requirement that unitary transformations of the form $U_I \otimes U_{II}$ do not change the values taken by the measure: that these values should remain the same, but should now pertain to transformed projectors (like ${U_I}^{-1} P U_I$). The reason is that these unitary transformations only change the orientation of $|\psi\rangle$ in Hilbert space, but do not change anything in the relation between $|\psi\rangle$ and the definite-valued observables determined by it; all changes are equivalent to those induced by basis transformations in the factor Hilbert spaces. But we want μ to be determined solely by the relation between the state and its associated definite-valued projectors—the choice of a basis in Hilbert space in terms of which the state is expressed should be immaterial. In other words, the same collection of μ values must be associated with the entire class of states that follow from $|\psi\rangle$ by application of arbitrary unitary operations of the form $U_I \otimes U_{II}$. Since the only feature that is common to all these states are the absolute values of all c_k, μ must be a function of these values only. As pointed out before, we can therefore take μ to be a function of $\{|c_k|^2\}$.

Now compare $|\psi\rangle$ with the vector that results from it by erasing the differences between $|\beta_k\rangle$ for $k \geq 2$, and replacing all these vectors by $|\beta_2\rangle$. This erasing process leads to the state

$$|\chi\rangle = c_1|\alpha_1\rangle \otimes |\beta_1\rangle + \sqrt{\sum_{k=2} |c_k|^2}|\alpha\rangle \otimes |\beta_2\rangle, \tag{3}$$

where $|\alpha\rangle$ is a normalized vector. Since distinct and non-overlapping elements of the set of definite-valued projectors have been grouped together in this operation, the measure assigned to $|\beta_2\rangle\langle\beta_2|$ should be the sum of the original measures of the projectors that have coalesced into $|\beta_2\rangle\langle\beta_2|$.

Finally, because we can write $|\chi\rangle$ as

$$c_1|\alpha_1\rangle \otimes |\beta_1\rangle + d|\alpha\rangle \otimes |\beta_2\rangle, \tag{4}$$

with $|d|^2 = 1 - |c_1|^2$, we may write $\mu(|\beta_1\rangle\langle\beta_1|) = g(|c_1|^2, |d|^2) = f(|c_1|^2)$. By parity of reasoning we may write down an analogous formula for the other projectors: $\mu(|\beta_i\rangle\langle\beta_i|) = f(|c_i|^2)$.

Coming back to the just-made observation about the relation between the measures induced by $|\psi\rangle$ and $|\chi\rangle$, respectively, we find that

$$f(\sum |c_k|^2) = \sum f(|c_k|^2). \tag{5}$$

From this it follows that $f(|c_k|^2) = const.|c_k|^2$, and in view of normalization

$$\mu(P_k) = |c_k|^2. \tag{6}$$

This is the Born rule.

3. The Born Measure as a Probability

As explained in the Introduction, modal and decoherence interpretations understand μ as a *probability*: given the state $|\psi\rangle$ in Hilbert space, exactly *one* of the projectors that are singled out as definite-valued by $|\psi\rangle$ possesses the value 1, and the chance that this value is taken by P_k is given by $|c_k|^2$. The relation between the state in Hilbert space on the one hand and the actual physical situation on the other is consequently assumed to be indeterministic. In general, given the state there are more than one possibilities for the actual physical situation (defined by the values of the definite-valued observables), and the state specifies a probability distribution over them. This probability can be taken to quantify our ignorance about the actually obtaining physical situation in cases in which we know the state in Hilbert space and have no additional information. It will also reflect the relative frequencies with which physical properties occur in repetitions of situations corresponding to the same $|\psi\rangle$. In other words, the available range of interpretations of μ is the same as in the case of classical probabilistic theories.

That μ should be given this physical meaning in terms of probability is something that cannot be decided by the mathematical formalism itself (see for a dissenting voice Zurek,[10] and for a critical analysis of this argument Mohrhoff[13]). It is an interpretational postulate that must be judged on the basis of experience and the comparison with alternatives—we shall have more to say about this in the next Section.

According to the probabilistic interpretation the state in Hilbert space is about possibilities, about what *may* be the case; it is about *modalities.* But in addition there is an obvious second aspect to $|\psi\rangle$: it is the theoretical quantity that occurs in the evolution equation, and its time development governs how the set of definite valued quantities changes. This double role of $|\psi\rangle$, on the one hand probabilistic and on the other dynamical and deterministic, is a well-known feature of the Bohm interpretation. The Bohm interpretation can be regarded as a specific version of non-collapse interpretations, namely one in which there is an *a priori* given definite-valued observable (position).[2,9] As we now see, this double deterministic-and-probabilistic aspect of $|\psi\rangle$ is not specific to the Bohm interpretation. It is typical of all interpretations of quantum mechanics in which there are no collapses and in which $|\psi\rangle$ relates in a probabilistic way to the physical world.

4. Non-Collapse and Probability

The non-collapse scheme by itself does not imply anything about probability: it just says that the state vector evolves unitarily. As already emphasized, an *interpretation* that is external to the formalism must be supplied before anything can be stated about what the state represents. It is sometimes suggested, however, in opposition to this, that the non-collapse formalism is capable of providing its own interpretation. What seems to be meant is the claim that there exists a *simplest* interpretation that does most justice to the symmetries inherent in the Hilbert space formalism. In particular, the suggestion is that, granted the usual interpretational links between eigenstates of observables and values of physical quantities, a superposition of such eigenstates should be interpreted as representing the joint existence of the corresponding values. Consistent elaboration of this leads to the many-worlds idea: superpositions represent collections of worlds, in each one of which exactly one value of an observable—corresponding to one term from the superposition—is realized.

In a superposition all terms occur in the same way, i.e. without any markers that single out one, or some, terms as corresponding to what actually is the case. The basic thought of the many-worlds interpretation is that this symmetry signifies that all terms must correspond to reality in the same way: if one term refers to something actually existing, then so must all. The identification of any particular term as representing actuality is regarded as breaking the symmetry present in the state, and therefore as objectionable.

Let us have a closer look at this argument, however. It may be conceded that singling out any particular term from a superposition, and identifying it as the one referring to actuality, breaks the symmetry of the series of superposed terms. But do probabilistic interpretations really work this way; do they single out one term over all others? Consider the analogous situation in classical probability theory:

the same train of thought applied there would also lead to the conclusion that all events to which a probability distribution assigns a value should be simultaneously realized. If one starts from merely the mathematical formalism, one would therefore be led, also in the classical case, to a many worlds ontology as the one that best fits the formalism—e.g. in the case of classical statistical mechanics. But is this interpretation really simpler or more symmetric than the usual probabilistic one?

One way of answering this question is to compare the two different reference relations (between formalism and what the formalism refers to) provided by the many-worlds and probabilistic interpretations, respectively. These are two mappings, both with the mathematical event space as their domain. The probabilistic mapping is from this event space to possibilities; whereas the many-worlds mapping maps all elements of the event space into realities. Apart from this difference in status of the elements of the ranges of the two mappings (possibility and reality), which as far as the mapping itself is concerned is just a difference in labels, everything is the same. It is therefore hard to see how there could be any difference in simplicity, naturalness or symmetry between these two reference relations. One is just as complex or simple as the other.

The notion that there nevertheless is an important difference evidently derives from the impression that the probabilistic interpretation identifies *one* of the possibilities as being actual, and thus violates the symmetry that is present in the many-worlds option. But this impression is incorrect. *Not* singling out such a privileged alternative is precisely what makes an interpretation fundamentally probabilistic. Put in different terms, the probabilistic option treats all elements of the probability space in exactly the same way, by mapping all of them to possibilities that *may* be realized—it does not tell us which possibility *is* realized. Each single element of the interpretation's range may correspond to reality. There is therefore the same symmetry as in the many-worlds option.

Still it has to be admitted, of course, that there is a difference. In the probabilistic interpretation it is stipulated from the outset that exactly one possibility is realized, although there is no indication which one. So even though there is symmetry with respect to which possibility this is, is it not true that the one-world stipulation by itself introduces surplus structure that is not present in the many-worlds interpretation? I do not think this is right. There is perfect equivalence in the following sense: the many-worlds interpretation says that *each* element of the measure space corresponds to an actual states of affairs, whereas the probabilistic alternative tells us that *each* element may correspond to the one actual but unspecified state of affairs. There is consequently no difference in the symmetry properties or simplicity of the interpretations, but rather a difference in the nature of their ranges: in the one case this is a collection of many real worlds, in the other it is a collection of candidates for the one real world. So, in the end the significant difference boils down to the difference between one and many real worlds—and it surely is not a principle of metaphysics that many is simpler than one. General considerations concerning symmetry and simplicity do therefore not favor many worlds over a probabilistic interpretation. If anything, the probabilistic interpretation of the non-collapse scheme is the soberer and more economical one.

5. Conclusion

It is possible to reproduce the usual probabilistic predictions of quantum mechanics without positing any random element in the evolution of the quantum mechanical state. Instead of assuming that the state randomly collapses during a measurement, into a state that uniquely corresponds to the actual outcome, it is consistent to maintain that there is only unitary evolution. It is part and parcel of this non-collapse approach that the state does not only represent what is actually the case in the world we observe, but also contains information about possibilities that have not become actual in our world—there is a 'modal' aspect to the non-collapse approach.

According to this picture we, as observers, are also described by the always unitarily evolving total quantum state. Since we become aware, through experience, only of what is actual, we can in practice only take into account the part of the total state that corresponds to our actual observations. But there are many situations in which measurements, or more generally macroscopic interactions, take place without any human observation being made. In such circumstances the non-collapse point of view generally differs from the collapse scheme, also in a practical respect. According to the non-collapse approach the different terms of a superposition are maintained, even in macroscopic interactions, which makes recombination and interference possible in principle. By contrast, the occurrence of collapses would prevent recombination and the restoration of the original state during an inverse process. Present-day experiments that have verified the possibility of macroscopic interference and recombination thus lend support to the non-collapse position.

The absence of random elements in the evolution of the quantum state is not at odds with a probabilistic interpretation. Quite on the contrary, it turns out (Sec. 2) that the non-collapse scheme is able to justify the Born probability rule on the basis of very general and simple premises.

The issue of whether the non-collapse scheme should be construed strictly in a modal manner, by thinking of one actual state of affairs and many non-realized possibilities; or rather according to the many-worlds idea of many actualities is open to philosophical debate. It seems clear that the modal interpretation is the conceptually simpler one, whereas there are no observational differences between the two options.

References

1. D. Dieks, *Physics Letters A* **142**, 439 (1989).
2. J. Bub, *Interpreting the Quantum World* (Cambridge University Press, Cambridge, 1997).
3. B. van Fraassen, *Quantum Mechanics* (Clarendon Press, Oxford, 1991).
4. D. Dieks and P. Vermaas (eds.), *The Modal Interpretation of Quantum Mechanics* (Kluwer Academic Publishers, Dordrecht, 1988).

5. W.H. Zurek, *Progress of Theoretical Physics* **89**, 281 (1993).
6. M. Schlosshauer, *Rev. Mod. Phys.* **76**, 1267 (2004).
7. J. Bub and R. Clifton, *Studies in the History and Philosophy of Modern Physics* **27**, 181 (1996).
8. J. Bub, R. Clifton and S. Goldstein, *Studies in the History and Philosophy of Modern Physics***31**, 95 (2000).
9. D. Dieks, *Foundations of Physics* **35**, 399 (2005).
10. W.H. Zurek, *Phys. Rev. A* **71**, 052105 (2005).
11. C.M. Caves, *Notes on Zurek's derivation of the quantum probability rule* (2005). Available at info.phys.unm.edu/ caves/reports/ZurekBornderivation.pdf
12. H. Barnum, *No-signalling-based version of Zurek's derivation of quantum probabilities* (2003); arXiv:quant-ph/0312150.
13. U. Mohrhoff, *International Journal of Quantum Information* **2**, 221 (2004).
14. M. Schlosshauer and A. Fine, *Foundations of Physics* **35**, 197 (2005).

THE SCHRÖDINGER-PARK PARADOX ABOUT THE CONCEPT OF "STATE" IN QUANTUM STATISTICAL MECHANICS AND QUANTUM INFORMATION THEORY IS STILL OPEN: ONE MORE REASON TO GO BEYOND?

G. P. BERETTA

Universita di Brescia, via Branze 38, 25123 Brescia, Italy
E-mail: beretta@unibs.it

A seldom recognized fundamental difficulty undermines the concept of individual "state" in the present formulations of quantum statistical mechanics and quantum information theory. The difficulty is an unavoidable consequence of an almost forgotten corollary proved by Schrödinger in 1936 and perused by Park in 1968. To resolve it, we must either reject as unsound the concept of state, or else undertake a serious reformulation of quantum theory and the role of statistics. We restate the difficulty and discuss alternatives towards its resolution.

Keywords: Quantum statistical mechanics; Quantum information theory; Conceptual foundations of quantum mechanics; Nature of quantum states; Quantum thermodynamics.

1. Introduction

In 1936, Schrödinger[1] published an article to denounce a "repugnant" but unavoidable consequence of the present formulation of Quantum Mechanics (QM) and Quantum Statistical Mechanics (QSM). Schrödinger claimed no priority on the mathematical result, and properly acknowledged that it is hardly more than a corollary of a theorem about statistical operators proved by von Neumann[2] five years earlier.

Thirty years later, Park[3] exploited von Neumann's theorem and Schrödinger's corollary to point out quite conclusively an essential tension undermining the logical conceptual framework of QSM (and of Quantum Information Theory, QIT, as well). Twenty more years later, Park returned on the subject in another magistral, but almost forgotten paper[4] in which he addresses the question of "whether an observer making measurements upon systems from a canonical ensemble can determine whether the systems were prepared by mixing, equilibration, or selection", and concludes that "a generalized quantal law of motion designed for compatibility with fundamental thermodynamic principles, would provide also a means for resolving paradoxes associated with the characteristic ambiguity of ensembles in quantum mechanics."

Schrödinger's corollary was "rediscovered" by Jaynes[5] and Gisin,[6] and generalized by Hughston, Jozsa, and Wooters[7] and Kirkpatrick.[8] Also some interpretation has been re-elaborated around it,[9,10] but unfortunately not always the original references have been duly cited.[11] For this reason it is useful once in a while to refresh our memory about the pioneering contributions by Schrödinger and Park. The crystal clear logic of their analyses should not be forgotten, especially if we decide that it is necessary to "go beyond".

The tension that Park vividly brings out in his beautiful essay on the "nature of quantum states" is about the central concept of individual state of a system. The present formulation of QM and QSM implies the paradoxical conclusion that every system is "a quantum monster": a single system concurrently "in" two (and actually even more) different states. We briefly review the issue below (as we have done also in Ref. 12), but we urge everyone interested in the foundations of quantum theory to read the original reference.[3] The problem has been widely overlooked and is certainly not well known, in spite of its periodic rediscoveries. The overwhelming successes of QM and QSM understandably contributed to discourage or dismiss as useless any serious attempt to resolve the fundamental conceptual difficulty.

Here, we emphasize that a resolution of the tension requires a serious re-examination of the conceptual and mathematical foundations of quantum theory. We discuss three logical alternatives. We point out that one of these alternatives achieves a fundamental resolution of the difficulty without contradicting any of the successes of the present mathematical formalism in the equilibrium realm where it is backed by experiments. This alternative originates from a logical implementation of the conjecture — first proposed by Hatsopoulos and Gyftopoulos[13] — that the second law of thermodynamics may be a fundamental physical law valid at the microscopic level. This conjecture is in sharp contrast with the traditional view that the second law is some sort of typical statistical effect that emerges only for macroscopic systems or open subsystems weakly coupled to much larger systems (for references to traditional attempts to resolve the conflict between the second law and the notorious reversibility of the fundamental laws of mechanics, see e.g. Ref. 14, where yet another argument in favor of the traditional lines is discussed).

While entailing all the mathematical successes of equilibrium QSM, the Hatsopoulos-Gyftopoulos Unified Quantum Theory of Mechanics and Thermodynamics, which the present author[12] complemented with the further conjecture of a nonlinear, steepest-entropy-ascent dynamical law (and called it Quantum Thermodynamics), forces a re-interpretation of the fundamental meaning of such successes, but yields the second law as an exact theorem of the new conjectured dynamical law and in the nonequilibrium domain opens to new discoveries, new physics compatible with the second law of thermodynamics,[15–21] including the new theoretical possibility (provided by the nonlinearity of the assumed dynamical law) to distinguish between homogeneous (proper) and heterogeneous (improper) ensembles, by looking at the time-dependent behavior (e.g. by stroboscopic tomography).

As Park says:[3] "problems concerning measurement in quantum physics can be sharpened, and sometimes resolved, by according proper attention to those basic physical characteristics of quantum states." Should the re-interpretation suggested by the careful scrutiny of the Schrödinger-Park paradox and its resolution by conjecturing the validity of our Quantum Thermodynamics, motivate new fundamental experimental tests and prove successful, then once again Thermodynamics would have played a key role in a major step "beyond".

2. Schrödinger-Park quantum monsters

In this section, we review briefly the problem at issue. We start with the seemingly harmless assumption that every (individual) system is always in some definite, though perhaps unknown, state. We will conclude that the assumption is incompatible with the present formulation and interpretation of QSM/QIT. To this end, we concentrate on an important special class of systems that we call "strictly isolated". A system is strictly isolated if and only if (a) it interacts with no other system in the universe, and (b) its state is at all times uncorrelated from the state of any other system in the universe.

The argument that "real" systems can never be strictly isolated and, therefore, that the following discussion should be dismissed as useless is at once counterproductive, misleading and irrelevant, because the concept of strictly isolated system is a keystone of the entire conceptual edifice in physics, particularly indispensable to structure the principle of causality. Hence, the strictly isolated systems must be accepted, at least, as conceivable. It is therefore an essential necessary requirement that, when restricted to such systems, the formulation of a physical theory like QSM be free of internal inconsistencies.

It is useful at this point to emphasize that here, with Schrödinger,[1] von Neumann,[2] and Park,[3] the term "state" is used with reference to the individual system only, and not to indicate generic statistics from (or information about) an "unqualified" (i.e., not necessarily homogeneous[2] or proper[22]) statistical ensemble of systems prepared under identical conditions. In other words, differently from the unfortunate common current use of the term state in Quantum Information (see, e.g., Ref. 23 for a concise account of the statistical interpretation of Quantum Mechanics), here we refer to the traditional concept of state associated with an individual system, another keystone of physical thinking not only in Classical Mechanics but also in Quantum Theory (whenever, for example, we assign a state vector to a single system). From the conceptual point of view, our restrictive use of the term "state" (as thoroughly discussed by Park[3]) is not contradictory with the fact that in Quantum Theory it can be fully reconstructed from measurement results (tomography) only by gathering enough data from a (homogeneous) ensemble of identically prepared systems.

In QM the states of a strictly isolated (noninteracting and uncorrelated) system are in one-to-one correspondence with the one-dimensional orthogonal projection operators on the Hilbert space of the system. We denote such projectors by the symbol P. If $|\psi\rangle$ is an eigenvector of P such that $P|\psi\rangle = |\psi\rangle$ and $\langle\psi|\psi\rangle = 1$ then $P = |\psi\rangle\langle\psi|$. It is well known that differently from classical states, quantum states are characterized by irreducible intrinsic probabilities. We give this for granted here, and do not elaborate further on this point.

The objective of QSM is to deal with situations in which the state of the system is not known with certainty. Such situations are handled, according to von Neumann[2] (but also to Jaynes[5] within the QIT approach) by assigning to each of the possible states of the system an appropriate statistical weight which describes an "extrinsic" (we use this term to contrast it with "intrinsic") uncertainty as to whether that state is the actual state of the system. The selection of a rule for a proper assignment of the statistical weights is not of concern to us here.

To make clear the meaning of the words extrinsic and intrinsic, consider the following non quantal example. We have two types of "biased" coins A and B for which "heads" and "tails" are not equally likely. Say that $p_A = 1/3$ and $1-p_A = 2/3$ are the intrinsic probabilities of all the coins of type A, and that $p_B = 2/3$ and $1 - p_B = 1/3$ those of the coins of type B. Each time we need a coin for a new toss, however, we receive it from a slot machine that first tosses an unbiased coin C with intrinsic probabilities $w = 1/2$ and $1 - w = 1/2$ and, without telling us the outcome, gives us a coin of type A whenever coin C yields "head" and a coin of type B whenever C yields "tail". It is clear that, for such a preparation scheme, the probabilities w and $1 - w$ with which we receive coins of type A or of type B have "nothing to do" with the intrinsic probabilities p_A, $1 - p_A$, and p_B, $1 - p_B$ that characterize the biased coins we will toss. We therefore say that w and $1 - w$ are extrinsic probabilities, that characterize the heterogeneity of the preparation scheme rather than features of the prepared systems (the coins). If on every coin we receive we are allowed only a single toss (projection measurement?), then due to the particular values ($p_A = 1/3$, $p_B = 2/3$ and $w = 1/2$) chosen for this tricky preparation scheme, we get "heads" and "tails" which are equally likely; but if we are allowed repeated tosses (non-destructive measurements, gentle measurements, quantum cloning measurements?) then we expect to be able to discover the trick. Thus, it is only under the one-toss constraint that we would not loose, if we base our bets on a description of the preparation scheme that simply weighs the intrinsic probabilities with the extrinsic ones, i.e., that would require us to expect "head" with probability $p_{\text{head}} = wp_A + (1 - w)p_B = 1/2 * 1/3 + 1/2 * 2/3 = 1/2$.

For a strictly isolated system, the possible states according to QM are, in principle, all the one-dimensional projectors P_i on the Hilbert space of the system. QSM/QIT assigns to each state P_i a statistical weight w_i, and characterizes the extrinsically uncertain situation by a (von Neumann) statistical (or density) operator $W = \sum_i w_i P_i$, a weighted sum of the projectors representing the possible states.

This construction is ambiguous, because the same statistical operator is assigned to represent a variety of different preparations, with the only exception of homogeneous preparations where there is only one possible state P_ψ with statistical weight equal to 1, so that $W = P_\psi$ is "pure". Given a statistical operator W (a nonnegative, unit-trace, hermitean operator on the Hilbert space of the system), its decomposition into a weighted sum of one-dimensional projectors P_i with weights w_i implies that there is a preparation such that the system is in state P_i with probability w_i, to which the QSM/QIT von Neumann construction would assign the statistical operator $W = \sum_i w_i P_i$. The situation described by W has no extrinsic uncertainty if and only if W equals one of the P_i's, i.e., if and only if $W^2 = W = P_i$ (von Neumann's theorem[2]). Then, QSM reduces to QM and no ambiguities arise.

The problem is that whenever W represents a situation with extrinsic uncertainty ($W^2 \neq W$) then the decomposition of W into a weighted sum of one-dimensional projectors is not unique. This is the essence of Schrödinger's corollary[1] relevant to this issue (for a mathematical generalization see Ref. 8 and for interpretation in the framework of non-local effects see e.g. Ref. 9).

For our purposes, notice that every statistical (density) operator W, when restricted to its range $\mathrm{Ran}(W)$, has an inverse that we denote by W^{-1}. If $W \neq W^2$, then $\mathrm{Ran}(W)$ is at least two-dimensional, i.e., the rank of W is greater than 1. Let $P_j = |\psi_j\rangle\langle\psi_j|$ denote the orthogonal projector onto the one-dimensional subspace of $\mathrm{Ran}(W)$ spanned by the j-th eigenvector $|\psi_j\rangle$ of an eigenbasis of the restriction of W to its range $\mathrm{Ran}(W)$ (j runs from 1 to the rank of W). Then, $W = \sum_j w_j P_j$ where w_j is the j-th eigenvalue, repeated in case of degeneracy. It is noteworthy that $w_j = [\mathrm{Tr}_{\mathrm{Ran}(W)}(W^{-1}P_j)]^{-1}$. Schrödinger's corollary states that, chosen an arbitrary vector α_1 in $\mathrm{Ran}(W)$, it is always possible to construct a set of linearly independent vectors $|\alpha_k\rangle$ (k running from 1 to the rank of W, α_1 being the chosen vector) which span $\mathrm{Ran}(W)$ (but are not in general orthogonal to each other), such that the orthogonal projectors $P'_k = |\alpha_k\rangle\langle\alpha_k|$ onto the corresponding one-dimensional subspaces of $\mathrm{Ran}(W)$ give rise to the alternative resolution of the statistical operator $W = \sum_k w'_k P'_k$, with $w'_k = [\mathrm{Tr}_{\mathrm{Ran}(W)}(W^{-1}P'_k)]^{-1}$.

To fix ideas, consider the example of a qubit with the statistical operator given by $W = p|1\rangle\langle 1| + (1-p)|0\rangle\langle 0|$ for some given p, $0 < p < 1$. Consistently with Schrödinger's corollary, it is easy to verify that the same W can also be obtained as a statistical mixture of the two projectors $|+\rangle\langle +|$ and $|a\rangle\langle a|$ where $|+\rangle = (|0\rangle + |1\rangle)/\sqrt{2}$, $|a\rangle = (|+\rangle + a|-\rangle)/\sqrt{1+a^2}$ (note that $|a\rangle$ and $|+\rangle$ are not orthogonal to each other), $|-\rangle = (|0\rangle - |1\rangle)/\sqrt{2}$, $a = 1/(1-2p)$ and $w = 2p(1-p)$ so that $W = w|+\rangle\langle +| + (1-w)|a\rangle\langle a|$. With $p = 1/4$ this is exactly the example given by Park in Ref. 3.

QSM forces on us the following interpretation of Schrödinger's corollary. The first decomposition of W implies that we may have a preparation which yields the system in state P_j with probability w_j, therefore, the system is for sure in one of the states in the set $\{P_j\}$. The second decomposition implies that we may as well have a preparation which yields the system in state P'_k with probability w'_k and, therefore, the system is for sure in one of the states in the set $\{P'_k\}$. Because both decompositions hold true simultaneously, the very rules we adopted to construct statistical operators W allow us to conclude that that the state of the system is certainly one in the set $\{P_j\}$, but concurrently it is also certainly one in the set

$\{P'_k\}$. Because the two sets of states $\{P_j\}$ and $\{P'_k\}$ are different (no elements in common), this would mean that the system "is" simultaneously "in" two different states, thus contradicting our starting assumption that a system is always in one definite state (though perhaps unknown). Little emphasis is gained by noting that, because the possible different decompositions are not just two but an infinity, we are forced to conclude that the system is concurrently in an infinite number of different states! Obviously such conclusion is unbearable and perplexing, but it is unavoidable within the current formulation of QSM/QIT. The reason why we have learnt to live with this issue – by simply ignoring it – is that if we forget about interpretation and simply use the mathematics, so far we always got successful results that are in good agreement with experiments.

Also for the coin preparation example discussed above, there are infinite ways to provide 50% head and 50% tail upon a single toss of a coin chosen randomly out of a mixture of two kinds of biased coins of opposite bias. If we exclude the possibility of performing repeated (gentle) measurements on each single coin, than all such situations are indeed equivalent, and our adopting the weighted sum of probabilities as a faithful representation is in fact a tacit acceptance of the impossibility of making repeated measurements. This limitation amounts to accepting that extrinsic probabilities (w,$1-w$) combine irreducibly with intrinsic ones (p_A,p_B), and once this is done there is no way to separate them again (at least not in a unique way). If these mixed probabilities are indeed all that we can conceive, then we must give up the assumption that each coin has its own possibly unknown, but definite bias, because otherwise we are lead to a contradiction, for we would conclude that there is some definite probability that a single coin has at once two different biases (a monster coin which belongs concurrently to both the box of, say, $2/3 - 1/3$ biased coins and the box of, say, $3/4 - 1/4$ biased coins).

3. Is there a way out?

In this section we discuss three main alternatives towards the resolution of the paradox, that is, if we wish to clear our everyday, already complicated life from quantum monsters. Indeed, even though it has been latent for fifty years and it has not impeded major achievements, the conceptual tension denounced by Schrödinger and Park is untenable, and must be resolved.

Let us therefore restate the three main hinges of QSM which lead to the logical inconsistency:

(1) a system is always in a definite, though perhaps unknown, state;
(2) states (of strictly isolated systems) are in one-to-one correspondence with the one-dimensional projectors P on the Hilbert space $\mathcal{H}$ of the system; and
(3) situations with extrinsic uncertainty as to which is the actual state of the system are unambiguously described by the statistical operators W. The decomposition $W = \sum_i w_i P_i$ implies that the state is P_i with statistical weight w_i.

To remove the inconsistency, we must reject or modify at least one of these statements. But, in doing so, we cannot afford to contradict any of the innumerable successes of the present mathematical formulation of QSM.

A first alternative was discussed by Park[3] in his essay on the nature of quantum states. If we decide to retain statements (2) and (3), then we must reject statement (1), i.e., we must conclude that the concept of state is "fraught with ambiguities and should therefore be avoided." A system should never be regarded as being in any physical state. We should dismiss as unsound all statements of this type: "Suppose an electron is in state ψ ..." Do we need to undertake this alternative and therefore abandon deliberately the concept of state? Are we ready to face all the ramifications of this alternative?

A second alternative is to retain statements (1) and (2), reject statement (3) and reformulate the mathematical description of situations with extrinsic uncertainty in a way not leading to ambiguities. To our knowledge, such a reformulation has never been considered. The key defect of the representation by means of statistical operators is that it mixes irrecoverably two different types of uncertainties: the intrinsic uncertainties inherent in the quantum states and the extrinsic uncertainties introduced by the statistical description.

In Ref. 12, we have suggested a measure-theoretic representation that would achieve the desired goal of keeping the necessary separation between intrinsic quantal uncertainties and extrinsic statistical uncertainties. We will elaborate on such representation elsewhere. Here, we point out that a change in the mathematical formalism involves the serious risk of contradicting some of the successes of the present formalism of QSM. Such successes are to us sufficient indication that changes in the present mathematical formalism should be resisted unless the need becomes incontrovertible.

A third intriguing alternative has been first proposed by Hatsopoulos and Gyftopoulos[13] in 1976. The idea is to retain statement (1) and modify statement (2) by adopting the mathematics of statement (3) to describe the states. The defining features of the projectors P, which represent the states for a strictly isolated system in QM, are: $P^\dagger = P$, $P > 0$, $\mathrm{Tr}P = 1$, $P^2 = P$. The defining features of the statistical (or density) operators W are $W^\dagger = W$, $W > 0$, $\mathrm{Tr}W = 1$. Hatsopoulos and Gyftopoulos propose to modify statement (2) as follows:

(2') States (of every strictly isolated system) are in one-to-one correspondence with the state operators ρ on $\mathcal{H}$, where $\rho^\dagger = \rho$, $\rho > 0$, $\mathrm{Tr}\rho = 1$, without the restriction $\rho^2 = \rho$. We call these the "state operators" to emphasize that they play the same role that in QM is played by the projectors P, according to statement (2) above, i.e., they are associated with the homogeneous (or pure or proper) preparation schemes.

Mathematically, state operators ρ have the same defining features as the statistical (or density) operators W. But their physical meaning according to statement (2') is sharply different. A state operator ρ represents the state of an individual system.

Whatever uncertainties and probabilities it entails, they are intrinsic in the state, in the same sense as uncertainties are intrinsic in a state described (in QM) by a projector $P = |\psi\rangle\langle\psi|$. A statistical operator W, instead, represents (ambiguously) a mixture of intrinsic and extrinsic uncertainties obtained via a heterogeneous preparation. In Ref. 13, all the successful mathematical results of QSM are re-derived for the state operators ρ. There, it is shown that statement (2') is non-contradictory to any of the (mathematical) successes of the present QSM theory, in that region where theory is backed by experiment. However it demands a serious re-interpretation of such successes because they now emerge no longer as statistical results (partly intrinsic and partly extrinsic probabilities), but as non-statistical consequences (only intrinsic probabilities) of the nature of the individual states.

In addition, statement (2') implies the existence of a broader variety of states than conceived of in QM (according to statement (2)). Strikingly, if we adopt statement (2') with all its ramifications, those situations in which the state of the system is not known with certainty stop playing the perplexing central role that in QSM is necessary to justify the successful mathematical results such as canonical and grand canonical equilibrium distributions. The physical entropy that has been central in so many discoveries in physics, would have finally gained its deserved right to enter the edifice from the front door. It would be measured by $-k_{\rm B}{\rm Tr}\rho\ln\rho$ and, by way of statement (2'), be related to intrinsic probabilities, differently from the von Neumann measure $-{\rm Tr}W\ln W$ which measures the state of uncertainty determined by the extrinsic probabilities of a heterogeneous preparation. We would not be anymore embarrassed by the inevitable need to cast our explanations of single-atom, single-photon, single-spin heat engines in terms of entropy, and entropy balances.

The same observations would be true even in the classical limit,[19] where the state operators tend to distributions on phase-space. In that limit, statement (2') implies a broader variety of individual classical states than those conceived of in Classical Mechanics (and described by the Dirac delta distributions over phase-space). The classical phase-space distributions, that are presently interpreted as statistical descriptions of situations with extrinsic uncertainty, can be readily reinterpreted as non-statistical descriptions of individual states with intrinsic uncertainty. Thus, if we accept this third alternative, we must seriously reinterpret, from a new non-statistical perspective, all the successes not only of quantum theory but also of classical theory.

4. Concluding remarks

In conclusion, the Hatsopoulos-Gyftopoulos ansatz, proposed thirty years ago in Ref. 13 and follow up theory,[15,17–20] not only resolves the Schrödinger-Park paradox without rejecting the concept of state (a keystone of scientific thinking), but forces us to re-examine the physical nature of the individual states (quantum and classical), and finally gains for thermodynamics and in particular the second law a truly fundamental role, the prize it deserves not only for having never failed in the past 180 years since its discovery by Carnot, but also for having been and still being a perpetual source of reliable advise as to how things work in Nature.

In this paper, we restate a seldom recognized conceptual inconsistency which is unavoidable within the present formulation of QSM/QIT and discuss briefly logical alternatives towards its resolution. Together with Schrödinger[1] who first surfaced the paradox and Park[3] who first magistrally explained the incontrovertible tension it introduces around the fundamental concept of state of a system, we maintain that this fundamental difficulty is by itself a sufficient reason to go beyond QSM/QIT, for we must resolves the "essential tension" which has sapped the conceptual foundations of the present formulation of quantum theory for almost eighty years.

We argue that rather than adopting the drastic way out provokingly prospected by Park, namely, that we should reject as unsound the very concept of state of an individual system (as we basically do every day by simply ignoring the paradox), we may alternatively remove the paradox by rejecting the present statistical interpretation of QSM/QIT without nevertheless rejecting the successes of its mathematical formalism. The latter resolution is satisfactory both conceptually and mathematically, but requires that the physical meaning of the formalism be reinterpreted with care and detail. Facing the situation sounds perhaps uncomfortable because there seems to be no harmless way out, but if we adopt the Hatsopoulos-Gyftopoulos fundamental ansatz (of existence of a broader kinematics) the change will be at first mainly conceptual, so that practitioners who happily get results everyday out of QSM would basically maintain the *status quo*, because we would maintain the same mathematics both for the time-independent state operators that give us the canonical and grand-canonical description of thermodynamics equilibrium states, and for the time-dependent evolution of the idempotent density operators ($\rho^2 = \rho$), i.e., the states of ordinary QM, which keep evolving unitarily. On the other hand, if the ansatz is right, new physics is likely to emerge, for it would imply that beyond the states of ordinary QM, there are states ("true" states, obtained from preparations that are "homogeneous" in the sense of von Neumann[2]) that even for an isolated and uncorrelated single degree of freedom "have physical entropy" ($-k_B \mathrm{Tr}\rho \ln \rho$) and require a non-idempotent state operator ($\rho^2 \neq \rho$) for their description, and therefore exhibit even at the microscopic level the limitations imposed by the second law,

In addition, if we adopt as a further ansatz that the time evolution of these non-ordinary-QM states (the non-idempotent ones) obeys the nonlinear equation of motion developed by the present author,[13,15,18–20] then in most cases they do not evolve unitarily but follow a path that results from the competition of the Hamiltonian unitary propagator and a new internal-redistribution propagator that "pulls" the state operator ρ in the direction of steepest entropy ascent (maximal entropy generation) until it reaches a (partially) canonical form (or grand canonical, depending on the system). Full details can be found in Ref. 17.

The proposed resolution definitely goes beyond QM, and turns out to be in line with Schrödinger's prescient conclusion of his 1936 article[1] when he writes: "My point is, that in a domain which the present theory does not cover, there is room for new assumptions without necessarily contradicting the theory in that region where it is backed by experiment."

5. Acknowledgements

This paper was written in preparation of a seminar given at MIT in April 1984. At that time, I had enjoyed five years of intense discussions with Elias Gyftopoulos, George Hatsopoulos, James Keck, John Appleton, and Joseph Smith, as well as with Jim Park during his short stay at MIT in 1979. I never published this paper because at that time I thought that Ref. 3 already said the whole story, and Ref. 13 proposed a resolution for both this problem and the long standing question about the second law. However, I have recently realized that on and off the Schrödinger corollary has been forgotten and rediscovered[7–11] and, at least for a small but growing number of physicists concerned with conceptual foundations, the related Schrödinger-Park conceptual paradox remains one of the unresolved knots of quantum statistical mechanics, that might eventually push us beyond it. The author is indebted to Lorenzo Maccone for a recent helpful suggestion.

References

1. E. Schrödinger, *Proc. Cambridge Phil. Soc.* **32**, 446 (1936).
2. J. von Neumann, *Mathematical Foundations of Quantum Mechanics*, Engl. transl. of the 1931 German edition by R. T. Beyer, (Princeton University Press, 1955), pp. 295-346.
3. J. L. Park, *Am. J. Phys.* **36**, 211 (1968).
4. J. L. Park, *Found. Phys.* **18**, 225 (1988).
5. E. T. Jaynes, *Phys. Rev.* **108**, 171 (1957).
6. N. Gisin, *Helvetica Physica Acta* **62**, 363 (1989).
7. L. P. Hughstone, R. Jozsa, and W. K. Wootters, *Phys. Lett. A* **183**, 14 (1993).
8. K. A. Kirkpatrick, *quant-ph/0305068.*
9. N. D. Mermin, *Found. Phys.* **29**, 571 (1999), quant-ph/9807055.
10. O. Cohen, *Phys. Rev. A* **60**, 80 (1999); **63**, 16102 (2001); D. R. Terno, *Phys. Rev. A* **63**, 16101 (2001).
11. The problem at issue in this paper, first raised in Ref. 1, has been acknowledged "in passing" in innumerable other references, but none has to our knowledge gone so deeply and conclusively to the conceptual roots as Ref. 3. See, e.g., W. M. Elsasser, *Phys. Rev.* **52**, 987 (1937); A. E. Allahverdyan and Th. M. Nieuwenhuizen, *Phys. Rev. E* **71**, 066102 (2005). Ref. 1 has been cited by many others, but not about the problem we focus on here, rather for its pioneering contributions to the questions of entanglement, EPR paradox and related nonlocal issues. Both Refs. 1 and 3 have been often cited also in relation to the projection postulate.
12. G. P. Beretta, *On the General Equation of Motion of Quantum Thermodynamics and the Distinction between Quantal and Nonquantal Uncertainties*, Sc. D. thesis, M.I.T., 1981, unpublished, quant-ph/0509116.
13. G.N. Hatsopoulos and E.P. Gyftopoulos, *Found. Phys.* **6**, 15, 127, 439, 561 (1976).
14. J. Gemmer, A. Otte, and G. Mahler, *Phys. Rev. Lett.* **86**, 1927 (2001).
15. G.P. Beretta, in Frontiers of Nonequilibrium Statistical Physics, Proceedings of the NATO Advanced Study Institute, Santa Fe, 1984, edited by G.T. Moore and M. O.

Scully, NATO ASI Series B: Physics (Plenum Press, New York, 1986), Vol. 135, p. 205.

16. J. Maddox, *Nature* **316**, 11 (1985).
17. G. P. Beretta, *quant-ph/0112046.*
18. G. P. Beretta, *Mod. Phys. Lett. A* **20**, 977 (2005); *Phys. Rev. E* **73**, 026113 (2006).
19. G. P. Beretta, *J. Math. Phys.* **25**, 1507 (1984).
20. G. P. Beretta, E. P. Gyftopoulos, J. L. Park, and G. N. *Hatsopoulos, Nuovo Cimento B* **82**, 169 (1984); G. P. Beretta, E. P. Gyftopoulos, and J. L. Park, *Nuovo Cimento B* **87**, 77 (1985); G. P. Beretta, *Found. Phys.* **17**, 365 (1987).
21. S. Gheorghiu-Svirschevski, *Phys. Rev. A* **63**, 022105 (2001); **63**, 054102 (2001).
22. B. d'Espagnat, *Conceptual Foundations of Quantum Mechanics* (Addison-Wesley, New York, 1989), 2nd ed.
23. R. Balian, *Am. J. Phys.* **57**, 1019 (1989).

THE CONJECTURE THAT LOCAL REALISM IS POSSIBLE

E. SANTOS

Departamento de Física, Universidad de Cantabria, Santander, Spain
E-mail: santose@unican.es

It is argued that local realism is a fundamental principle, which might be rejected only if experiments clearly show that it is untenable. Forty years after Bell´s work no experiment has provided a valid, loophole-free, violation of local realism which, in my opinion, is thus reinforced. I study a simple, but wide, family of local realistic models and derive new inequalities almost insensitive to the detection loophole. I argue that quantum mechanics, with some change in the theory of measurement, might be compatible with local realism.

Keywords: Local realism; Bell inequality; Hidden variables.

1. Local Realism vs. Quantum Mechanics

In the present article I shall be concerned with *epistemological realism*, that is the opinion that *physics makes assertions about the world, not merely about the results of the observations or experiments.* This kind of realism was supported by Einstein, who wrote: "Physics is an attempt conceptually to grasp reality as it is thought independently of its being observed. In this sense one speaks of physical relity".[1]

The idea that physics, or more generally natural science, is knowledge about the world was fully accepted before the advent of quantum mechanics in 1925. But soon after this date the founding fathers became aware that a trivial realistic interpretation of quantum mechanics was not possible, and a non-realistic interpretation (the "Copenhagen interpretation") was developed resting upon the idea that the aim of physics is not to provide a picture of the natural world but just predicting the results of experiments.

The difficulty for a realistic interpretation derives from the fact that in quantum mechanics not all observables possessing a meaning for the said system have definite values in a given state, as shown for instance by the Heisenberg uncertainty relations. Indeed, usually quantum mechanics predicts a probability distribution for every observable in a given state rather than a specific value. It would appear that a realistic interpretation is still possible attaching a joint probability distribution of all observables to every state, a distribution whose marginals should correspond to the quantum predictions. This solution was envisaged by several people, amongst them Einstein, who wrote: "The statistical character of the present theory would then have to be a necessary consequence of the incompleteness of the description of the systems in quantum mechanics, and there would no longer exist any ground for the supposition that a future basis of physics must be based upon statistics".[1] Thus the possibility of a realistic interpretation is linked to the incompleteness of quantum mechanics, and incompleteness is popularly expressed as "existence of hidden variables". However the said solution is not possible because the existence of *joint* probability distributions for any state contradicts quantum mechanics, as is shown by the Kochen-Specker theorem (for a clear exposition of the different theorems related to hidden variables, see Mermin[2]). This and other proofs of impossibility of hidden variables have been formulated along the years, the most celebrated in early times being the 1932 von Neumann´s theorem. The problem was studied in 1966 by John Bell,[3] who stated that *contextual hidden variables, and therefore contextual realism, is always possible but local realism is not (*the latter statement is called Bell´s theorem). This has been the common wisdom during the last 40 years.

I think that few people are aware of the acute conflict put by Bell´s theorem. For most physicists any violation of quantum mechanics would be dramatic, but not too many appreciate how dramatic would be a violation of local realism. In fact such a violation would make impossible a realistic interpretation of nature which preserves locality, something making the world rather bizarre. I fully agree with Einstein´s opinion that: "On one supposition we should, in my opinion, absolutely hold fast: the real factual situation of the system S_2 is independent of what is done with the system S_1 which is spatially separated from the former". Indeed I consider that the conflict between local realism and quantum mechanics is the most important open problem in theoretical physics. Of course I do not agree with the common wisdom that local realism has been empirically disproved so that the problem is already solved. Consequently most of my scientific work during the last 30 years has been devoted to the attempt at finding a solution for the said conflict and the present article contains my current opinion after these years of effort. This opinion may be summarized as follows.

I think that John Bell captured the essence of local realism when he proposed the correlation formula (here written in terms of probabilities)

$$p_{12}(a,b) = \int \rho(\lambda) P_1(\lambda, a) P_2(\lambda, b) d\lambda, \tag{1}$$

$$p_1(a) = \int \rho(\lambda) P_1(\lambda, a) d\lambda, \; p_2(b) = \int \rho(\lambda) P_2(\lambda, b) d\lambda, \tag{2}$$

where $p_{12}(a,b)$ is the probability of getting the value a for the observable A of system 1 and the value b for the observable B of system 2 in a simultaneous measurement of both observables, *simultaneous* meaning here that the measurements are performed at space-like separation in the sense of relativity theory. Similarly $p_1(p_2)$ is the probability when only the observable A of system 1 (B of system 2) is performed. λ represents collectively all hidden variables relevant for the problem. The functions involved should possess the properties of probabilities, that is

$$\rho(\lambda) \geq 0, \int \rho(\lambda) d\lambda = 1, \; 0 \leq P_1(\lambda, a), P_2(\lambda, b) \leq 1. \tag{3}$$

From Eq. (1) to Eq. (3) one may derive inequalities which should be fulfilled if local realism is possible. In idealized experiments these *Bell inequalities* are violated by the quantum-mechanical predictions, but no real experiment has shown a true (loophole-free) violation of the inequalities up to now. Therefore I may safely conclude that local realism has not been refuted at the empirical level, in spite of repeated claims of the opposite which minimize the relevance of the existing loopholes by introducing some hypotheses allegedly plausible. In my view the current wisdom is misleading and harmful for the progress of science. Misleading because it attempts answering a fundamental scientific question by means of a subjective assessment of plausibility. Harmful because it discourages people from making the necessary effort to perform a real, loophole-free, test.

On the basis of the failure to perform a loophole-free test of a Bell inequality in the 40 years elapsed from Bell´s work, I conjecture that there are restrictions, not appearing in the quantum formalism (or not obviously), which would prevent the empirical violation of local realism, that is restrictions making impossible any contradiction with Eqs. (1) to (3). Percival[4] has proposed an analogy which, for me, is illuminating. The second law of thermodynamics does not contradict the laws of Newtonian mechanics or Maxwellian electromagnetism but nevertheless it restricts the possible evolutions of actual physical systems. Similarly a new law, not yet known, should exist which, without contradicting quantum mechanics, restricts the available states and/or evolutions in such a way that local realism is always maintained at the empirical level.

2. Local model for correlation experiments

In contrast with the initial hopes, it has been shown that most classes of correlation experiments, which Bell´s correlation formulas Eqs.(1) and (2) refer to, cannot discriminate between LR and QM even with ideal set-ups. In particular :

1) Polarization correlation of photon pairs produced in atomic cascades, which includes Aspect´s.[5]

2) Experiments using high-energy photons, not reliable due to the low efficiency of the polarization measurements.

3) Experiments using K mesons or B mesons (although here the situation is not yet clear[6]).

Other experiments are extremely complex if a true discrimination is desired. This is the case of spin correlation of non-relativistic particles, where "space-like" measurements are very difficult to achieve,[78]

The fact is that *only correlation experiments using photon pairs produced by parametric down conversion appear now as promising.* But the loophole of the low detection efficiency remains a big difficulty. In order to make clear this point I shall study to which extent quantum mechanics is compatible with a simple, but wide, family of local hidden variables models for polarization correlation experiments. I define the family by the following properties of the functions ρ and P_j of Eqs. (1) and (2). I shall assume that a and b are angles, labelled ϕ_1 and ϕ_2 from now on, and λ is the set of two angles, χ_1 and χ_2, associated to the polarization of the two photons of the pair. Also I assume identity of the polarizer-detector systems in the sense that $P_1 = P_2 = P$. Thus any local hidden variables (LHV) model of the said family predicts the following coincidence and single detection probabilities

$$p_{12}(\phi) = \int \rho(\chi_1 - \chi_2) P(\chi_1 - \phi_1) P(\chi_2 - \phi_2) d\chi_1 d\chi_2, \tag{4}$$

$$p_j = \int \rho(\chi_1 - \chi_2) P(\chi_j - \phi_j) d\chi_1 d\chi_2,\ j = 1, 2. \tag{5}$$

where here and below both functions ρ and P are periodic with period π and the integrals go from $-\pi/2$ to $\pi/2$. In addition the functions possess the following properties of positivity, symmetry and normalization (ρ is normalized so that $p_{12}(\phi) = 1$ if $P = 1$ in Eq. (4)

$$\rho(x) = \rho(-x) \geq 0, \int \rho(x) dx = 1/\pi,\ 0 \leq P(x) = P(-x) \leq 1. \tag{6}$$

In typical polarization correlation experiments quantum mechanics predicts

$$p_j^Q = \frac{1}{2}\eta,\ p_{12}^Q(\phi) = \frac{1}{4}\eta^2 (1 + V \cos 2\phi), \tag{7}$$

η being the detection efficiency of photons. (I am assuming symmetry amongst the polarizer-detector set-ups of the two photons and rotational invariance, that is independence of p_{12} on $\phi_1 + \phi_2$. In experimental practice these conditions are not exactly fulfilled, but the lack of symmetry and rotational invariance makes *easier* to construct LHV models reproducing the results of the experiments). Then I will search for the best LHV model of the form Eqs. (4) and (5), defining "best" by the conditions that, for fixed η and V, the following two conditions are fulfilled: 1) the model prediction for the single probability, p_j, agrees with the quantum prediction, p_j^Q, 2) the model coincidence probability, $p_{12}(\phi)$, is as close as possible to $p_{12}^Q(\phi)$. By the latter condition I mean that the quantity S must be a minimum, where

$$S \equiv \int d\phi \left[p_{12}(\phi) - \frac{1}{4}\eta^2 (1 + V \cos 2\phi) \right]^2, \tag{8}$$

with $p_{12}(\phi)$ given by Eq. (4) and the functions $\rho(x), P(x)$ fulfilling the condition (6).

In the solution of the problem there are two cases which should be studied separately (the details of the derivation will be given elsewhere). The first case corresponds to experiments where the values of the detection efficiency, η, and the visibility of the correlation curve, V, fulfil de inequality

$$V \leq \frac{\sin^2(\pi\eta/2)}{(\pi\eta/2)^2} \simeq 1 - \frac{\pi^2\eta^2}{12}, \tag{9}$$

the latter approximation being valid for $\eta << 1$. In this case there are many choices of ρ and P making $S = 0$ (see Eq. (8), for instance)

$$P(x) = 1 \text{ if } |x| \leq \frac{\pi\eta}{4}, \text{ zero otherwise,} \tag{10}$$

$$\rho(x) = \frac{1}{\pi^2}\left[1 + \frac{(\pi\eta/2)^2}{\sin^2(\pi\eta/2)} V \cos 2x\right]. \tag{11}$$

Consequently for any experiment where the inequality (9) is fulfilled there are LHV models of the type (4) giving complete agreement with quantum mechanics. Actually, as far as I know, no experiment has achieved values violating the inequality (9), which means that *our Eqs.* (10) *and* (12) *provide an explicit LHV model for all performed experiments.*

The second case happens when the inequality (9) is violated. In such a situation no model of the family may fully agree with quantum mechanics and the best model corresponds to the choice (10) for P combined with

$$\rho(x) = N\left[\cos 2\varepsilon + \cos 2x\right]_+, \tag{12}$$

where ε is a function of η and V to be specified below, N is a normalization constant (see (6) and $[]_+$ means putting 0 if the quantity inside the parenthesis is negative. It may be shown that the parameter ε possess a value given by the following equality (valid for $\varepsilon << \pi/4$)

$$\frac{(\pi\eta/2)^2}{\sin^2(\pi\eta/2)} V - 1 = 2\varepsilon^2 - \frac{8}{\pi}\varepsilon^3 + O(\varepsilon^4).$$

The disagreement with quantum mechanics may be exhibited writing our *best LHV* prediction in the form

$$p_{12}(\phi) = \frac{1}{4}\eta^2[1 + V\cos(2\phi) + \delta(\phi)], \tag{13}$$

where the term $\delta(\phi)$, which might be expanded in Fourier series of $\cos(2n\phi)$ with $n \geq 2$, shows the departure from a pure cosinus curve. Then it is possible to prove that

$$\left\langle \delta(\phi)^2 \right\rangle^{1/2} \equiv \sqrt{\frac{1}{\pi}\int \delta(\phi)^2 \, d\phi} \tag{14}$$

$$\geq \frac{64\sin^3(2\varepsilon)}{9(\pi + \sin 2\varepsilon - 2\varepsilon\cos 2\varepsilon)} \frac{\sin^2(\pi\eta)}{\pi^2\eta^2} \simeq 2.3\left(V - 1 + \frac{\pi^2\eta^2}{12}\right)^{3/2},$$

the latter equality being valid for not too high efficiency η (the proof will be given elsewhere). This is a new inequality which I propose to be tested in place of a Bell´s inequality. Although the empirical violation of Eq. (14) would refute only a restricted family of LHV theories, namely those defined by Eqs (4) and (5) , it has the advantage that *may be tested easily in a loophole-free experiment.* In fact, it requires only a modest detection efficiency (about 30%). The convenience of the test is reinforced by the fact that a similar but less stringent inequality, derived elsewhere,[9] has been tested with the result that it was fulfilled and the quantum prediction contradicted[10] (although the conditions of the experiment apparently were not good enough to claim a true violation of quantum mechanics).

In recent optical tests of Bell´s inequalities people use two-channel polarizers and four coincidence detection rates are measured, $R_{++}(\phi)$, $R_{+-}(\phi)$, $R_{-+}(\phi)$, $R_{--}(\phi)$. Typically there is symmetry between the two channels, at least approximate. In this case our model may be extended to these experiments by using two functions $P_{+}(\phi)$ and $P_{-}(\phi) = P_{+}(\phi + \pi/2)$ instead of only one, $P(\phi)$, as in Eq. (4). Assuming the form Eq. (10) for $P_{+}(\phi)$ we may get the model prediction for the joint probabilities

$$p_{+-}(\phi + \pi/2) = p_{-+}(\phi + \pi/2) = p_{++}(\phi) = p_{--}(\phi),$$

using trivial generalizations of Eq. (4). These probabilities are proportional to the coincidence detection rates. The quantity usually reported in the experiments is the correlation, defined by

$$E(\phi) = \frac{R_{++}(\phi) + R_{--}(\phi) - R_{+-}(\phi) - R_{-+}(\phi)}{R_{++}(\phi) + R_{--}(\phi) + R_{+-}(\phi) + R_{-+}(\phi)}, \tag{15}$$

for which our model predicts

$$E(\phi) = V\cos(2\phi) - \delta(\pi/2 + \phi),$$

where $\delta(\phi)$ fulfils the inequality (14). Thus for these experiments I propose to test the inequality

$$\sqrt{\pi^{-1}\int [E(\phi) - V\cos(2\phi)]^2 d\phi} \geq 2.3\left(V - 1 + \frac{\pi^2\eta^2}{12}\right)^{3/2},$$

valid for not too high detection efficiency η. This inequality should hold true for any value of V.

3. Is Local Realism Truly Incompatible with Quantum Mechanics?

During the four decades elapsed since John Bell[3] discovery a lot of papers have been written pointing out quantum-theoretical violations of the inequalities in very many different phenomena, but in sharp contrast only a few dozen empirical tests have been actually performed. The results of all performed experiments are compatible with local realism and, with few exceptions, agree with the quantum predictions.[11] This period of 40 years may be compared with the few months elapsed from the conjecture, by Lee and Yang, that parity is not conserved in the weak interactions to the uncontroversial (loophole-free) empirical proof by Wu et al. The logical interpretation of these facts, unbiased by theoretical prejudices, should be that there is no empirical evidence against local realism in spite of quantum mechanics having been confirmed.

It is a historical fact that repeated experimental failures have led to fundamental principles of physics. The best known example is the failure to measure the absolute velocity of Earth, which led (or supported) Einstein´s relativity. Another example is the impossibility of making a steam engine producing work at zero cost, which led Sadi-Carnot, in 1824, to the conclusion that two reservoirs at a different temperature were needed and the efficiency is always (much) less than 100%. This led Clausius, 25 years later, to the second law of thermodynamics.

In my opinion the failure to perform a loophole-free test of a Bell inequality reinforces local realism. Thus I conjecture that a local realist description of nature is possible. However I do not think that quantum mechanics is wrong. But how may be compatible these two assertions with Bell´s theorem, which states that quantum mechanics contradicts local realism? My answer is that the proof of Bell´s theorem involves:

1) Derivation of Bell´s inequalities,

2) Example of violation of a Bell inequality by the quantum predictions.

I fully agree with the former but I question the latter. The reason is that quantum mechanics consists of: 1) the formalism (including the equations) and 2) the theory of measurement. The standard proofs of Bell´s theorem require both and, in my opinion, although *the formalism and the equations are correct* to a high degree of accuracy, *the postulates of measurement are absurd and unnecessarily strong.* Too strong because most measurements reduce to position measurement (e.g. of a pointer) and thus only a postulate about position observable (i.e. Born´s rule) would be required. Absurd because the so called measurement "postulates" cannot be postulates but just practical rules. In fact if I state that "the measurement of the observable O will certainly give one of the eigenvalues of the associated operator", we are implicitly assuming that the measurement is made with good apparti by an expert scientist (otherwise the measurement would not be reliable). But, which is the precise meaning of "good" or "expert" in order to promote the rule to the quality of a postulate?. In conclusion, I think that local realism is likely compatible with the quantum formalism (and the equations) plus some (weak) assumptions about the relation between elements of the theory and the experiments. Maybe only Born´s rule is required and this may be stated in a realistic form as *the modulus squared of the wave function provides the probability distribution of the relevant position coordinates.*

In summary, there is not yet an empirical disproof of local realism. This leads me to assume that the apparently unsurmontable difficulty to make a loophole-free test of Bell´s inequalities reinforces the conjecture that a local realistic interpretation of nature is possible.

References

1. A. Einstein, *Autobiographical notes.* In P. A. Schilpp, Ed., *Albert Einstein: Philosopher-Scientist.* (Open Court, La Salle (Illinois), 1949).

2. D. N. Mermin, *Rev. Mod. Phys.* **65,** 803 (1993). This article contains a clear exposition about hidden variables theories.
3. J. S. Bell, *Speakable and unspeakable in quantum mechanics.* (Cambridge University Press, Cambridge, 1987). This book contains reprints of most of Bell´s papers about foundations of quantum mechanics.
4. I. Percival, *Why do Bell experiments?*, *quant-ph/0008097.*
5. E. Santos, *Phys. Rev. Lett.* **66**, 1388(1991); **68**, 2702 (1992);*Phys. Rev. A* **46**, 3646(1992).
6. A. Bramón and G. Garbarino. Novel Bell´s inequality for entangled $K^o\overline{K}^o$ pairs. *Phys. Rev. Let.* **88**, 040403 (2002).
7. E. Santos,*Phys. Lett. A* **200**, 1 (1995).
8. E. S. Fry, T. Walther and S. Li. Proposal ,*Phys. Rev. A* **52**, 4381 (1995).
9. E. Santos, *Phys. Lett. A* **327**, 33 (2004).
10. M. Genovese, *et al. Optics and Spectroscopy* **99**, 185 (2005).
11. E. Santos,*Found. Phys.* **34**, 1643 (2004).

PART H

Round Table

ROUND TABLE DISCUSSION AT THE LORENTZ WORKSHOP "BEYOND THE QUANTUM"

A. M. CETTO*

Instituto de Física, Universidad Nacional Autónoma de México Apartado Postal 20-364, 01000 México, DF, México;
Department of Technical Cooperation, IAEA.
Wagramer Strasse 5, A-1400 Vienna, Austria.
**E-mail: ana@fisica.unam.mx*

R. BALIAN

Service de Physique Théorique, CEA-Saclay,
91191 Gif-sur-Yvette cedex, France

G. 't HOOFT

Spinoza Instituut, University of Utrecht,
Leuvenlaan 4, PO Box 80.195, 3508 TD Utrecht, The Netherlands

A. Yu. KHRENNIKOV

International Centre for Mathematical Modeling in Physics and Cognitive Sciences,
MSI, University of Växjö, S-35195, Sweden

Th. M. NIEUWENHUIZEN

Institute for Theoretical Physics, University of Amsterdam,
Valckenierstraat 65, 1018 XE Amsterdam, The Netherlands

This is a transcript of the round table discussion, moderated by G. 't Hooft, that took place at the end of the workshop "Beyond The Quantum" in the Lorentz Center of the University of Leiden, the Netherlands, 29 May - 2 June 2006.
It displays current views on foundations of quantum mechanics.

Keywords: Quantum Stochasticity; Nonlocality; Stochastic Electrodynamics; Quantum Field Theory; Higgs Particle; Bell Inequalities; String Theory

Transcript of the discussion

Gerard 't Hooft: From our discussions it is clear that we have different ideas about how to complete quantum theory, or at least about how to understand the nature of the apparently stochastic phenomena that it describes. To my mind, the answer should come from general relativity. Some think the answer is given by string theory, but the formulation of this theory is still far from exact, still too vague. In physics, we are used to descriptions of phenomena that are infinitely precise, and in this regard quantum mechanics works superbly for atoms and molecules.

We have seen here different approaches to the quantum problem. Where do they have common starting points, where do they disagree? For example, Stochastic Electrodynamics (SED) is very close to what I am thinking of. Why do we have to believe that energy has to be an absolutely conserved quantity?

One concept that is important in this context is information loss. Consider the motion of a particle in (x_1, x_2) space; in the conservative case the particle moves on a torus. But if there is information loss, the particle can move toward an attractor in this space, which represents a stable orbit. One could actually model the stable orbits in an H atom this way. Moreover, the particle may be jittering along its trajectory. This is an interesting possible approach. There must be other such approaches.

Christian Maes: In your view, nature is nonlocal?

Gerard 't Hooft: In quantum field theory, nature is local. Bell's theorem strongly suggests it is not. I think the Bell inequalities are based on a misrepresentation of the quantities involved.

Willem de Muynck: The Bell inequalities have nothing to do with nonlocality. When a quadrivariate joint probability distribution $p_{i,j,k,\ell}$ exists, the Bell inequalities are satisfied. Incompatibility of the observables is necessary for the violation of the Bell inequalities. But incompatibility is a local affair, hence also violation of the Bell inequalities is a local affair.

With four variables you can reproduce the Aspect experiments, but the $p_{i,j,k,\ell}$ depends on the experimental setup. There is no single $p_{i,j,k,\ell}$ for all Aspect experiments, but this is not because of mutual disturbance between two parts of the measurement (the arms of the interferometer), which have nothing to do with each other.

Igor Volovich: Indeed, there is no such thing as quantum nonlocality. Quantum field theory is local. Why so much talk about nonlocality in quantum mechanics? Because of an erroneous calculation of Bell. The quantity $\langle\psi_{spin}|\sigma \cdot a \otimes \sigma \cdot b|\psi_{spin}\rangle$ does not correspond to a physical measurement, the starting point is wrong. We must start from a space dependent form of the wavefunction, $\psi_{ab}(x_1, x_2)$ and write $\langle\psi|\sigma \cdot aP_{1a} \otimes \sigma \cdot bP_{2b}|\psi\rangle$ where P_1 is the projection on the measuring part 1, *etc.*

Franck Laloë: Yes, of course you are right: if you start from any quantum state, for instance one with spherical orbital wave functions for the particles, no violation of the Bell inequalities is obtained when the detectors have a limited size. But you can also assume a more appropriate quantum state, for instance one where the particles are localized within the two measurement apparatuses. Then the orbital variables disappear, the only relevant variables are spin variables, and the usual spin calculation applies perfectly well. This is implicitly assumed in most discussions of the Bell inequalities, and explains why Bell's calculation focuses on spins only; nothing is erroneous in this. Of course it would be wrong to assume that the Bell inequalities are violated with all quantum states having any orbital wave functions.

Theo Nieuwenhuizen: When you change the setup, you also change the forces acting on the particles, so the non-existence of the quadrivariate probability mentioned by de Muynck is obvious: Why should experiments that cannot be carried out simultaneously, necessarily have a joint distribution?

Giacomo D'Ariano: There is confusion between acausality and nonlocality. Quantum mechanics satisfies Einstein locality, the structure of the theory guarantees no-signaling. But we can still have violation of the Bell due to correlations.

Gerard 't Hooft: Indeed, Bell inequalities refer to correlations. Let me recall an argument given by Conway and Kochen (quant-ph/0604079) for the case of angular momentum $l = 1$. We have in this case $[L_i^2, L_j^2] = 0$ and $L^2 = \langle L_x^2 + L_y^2 + L_z^2 \rangle = 2$. If the observer could choose the direction of the measurement at the last instant, he could get a different result, but this would violate quantum mechanics. This is a beautiful twist to Bell's theorem.

Walter Philipp: The scheme presented by Maes this morning always leads to a contradiction. Consider the case of two possible results ± 1 for each one of the measured X_i and Y_i, $(i = 1, 2, 3)$. If we start from the two assumptions: $X_i = Y_i$, $P(X_i = Y_j) = \alpha$ for $i \neq j$, then the expectation value of $X_i Y_j$ is $E(X_i Y_j) = \alpha + (-1)(1-\alpha) = 2\alpha - 1$. Now take $S = \sum_{i,j} X_i Y_j = \sum_i X_i \sum_j Y_j$: the expectation value of this quantity is $E(S) = (\sum_i X_i)^2 \geq 1$. On the other hand, $E(S) = (1/9)[3 + 6(2\alpha - 1)] = (1/3)(-1 + 4\alpha)$, hence $(4\alpha/3) > E(S) > 1$ or $\alpha > 3/4$, which is incompatible with the assumed value $\alpha = 1/4$.

Karl Hess added afterwards: The contribution made by Walter goes along the lines of our paper in these proceedings: It is impossible to accomplish what is to be accomplished on one probability space. The point is that when incompatible experiments are involved (in the sense of quantum mechanics, relativity, and Vorob'ev's mathematical criteria), they can not necessarily be described on one probability space. This, as Vorob'ev has shown, is a general feature that also occurs in entirely classical problems and has nothing to do with action at a distance.

Giacomo D'Ariano: This is precisely the point: Bell's theorem means that there cannot be a joint probability distribution.

Igor Volovich: I agree with Prof. Laloë that if we have a quantum state where the particles are localized within the two measurement apparatuses, then one can focus on spins only. But I think that such a state it is almost impossible to prepare by using a remote source. In particular for this reason there are no experiments on Bell inequalities without loopholes. But let me ask a different question for our colleagues: do you consider Stochastic Electrodynamics as an approximation to QED? Is it not quantum mechanics that can be derived from QED?

Luis de la Peña: Of course quantum mechanics can be derived from QED, but this is a formal procedure that does not throw much light on the fundamentals of QM. Long ago A. A. Sokolov made a calculation starting from a classical oscillator interacting with a quantized EM field, and after some approximations he obtained a quantized oscillator. The point is that in our attempts to retrieve quantum mechanics from beyond quantum theory, we are driven by the quest for an understanding of the physical meaning of quantum mechanics, not by a purely formal or mathematical question.

Theo Nieuwenhuizen: Quantum field theory looks at extended objects in the point limit, so your question does not completely cover the issue.

Igor Volovich: There is an old Einstein's programme to describe elementary particles as point like or extended objects in classical field theory. Unfortunately, it is very hard to deal with nonlinear equations. The program is partially realized only for two dimensional solitons and even in this case we have to quantize them. So, even in this case we have to use standard quantum theory.

Theo Nieuwenhuizen: Solitons alone cannot lead to Quantum Mechanics, you also need stochastic forces. This brings us to Stochastic Electrodynamics and the combined program could be called Stochastic Soliton Mechanics. But what are these fluctuations?

Luis de la Peña: I do not see a fundamental contradiction between our view and that of others, like 't Hooft's. According to Stochastic Electrodynamics, the electromagnetic field is enough to explain quantum mechanical phenomena, in particular the inescapable fluctuations, and it is a necessary ingredient. On the other hand, Nelson's approach (mentioned by 't Hooft earlier in the discussion) and other similar ones are purely phenomenological, they do not identify the nature of the force causing the fluctuations.

Gerard 't Hooft: If special relativity is true and quantum mechanics works, the amount of freedom is extremely limited. The ultimate question would then be, in quantum theory, what is beyond the quantum?

Theo Nieuwenhuizen: Could the Higgs particle be an effective way to solve the problem?

Gerard 't Hooft: Maybe nature has solved for us the problem of an unlimited number of particles. Let me briefly explain the present situation in particle physics. The Higgs particle could be still much heavier than usually expected, and there may be also a lot of super-particles. The Higgs particles are needed for the scattering matrices to be unitary, but otherwise there may be more new fields and new particles. Or it may also happen that several Higgs are detected.

Theo Nieuwenhuizen: If there is no Higgs at all, how do particles get their masses?

Gerard 't Hooft: Something must play the role of the Higgs. If its mass were infinite, many couplings would be infinite as well, and our theory would be inconsistent. When going to still higher energies, we might be entering into a new domain, with strong interactions. Dark matter shows that there are still particles missing in our present theories. In brief, quantum field theory - quantum theory in general - shares all the problems of quantum mechanics.

Alexander Burinskii: During our conversations you said that the C-symmetry of the electron is not compatible with the topology of the Kerr spinning particle, which is based on the Kerr singular ring of Compton radius. Could you please clarify this point?

Gerard 't Hooft: The C-symmetry is there: $e \to e^+$, $q \to \bar{q}$, proton $\to$ antiproton, $\nu_L \to \bar{\nu}_R$ (antineutrino). Consider a Feynman diagram, with an electron-positron pair at one vertex and a cascade at the other, and an appropriate boson connecting them. One can consider the electron as a point; the scale is much smaller than Compton's wavelength. Hence the electron obeys locality. But the real picture must be more complicated than that.

Alexander Burinskii: In the Kerr spinning particle, there are other, axial singularities: forming a world-sheet in space-time. The problem is that the Kerr gravitational field of an electron creates a singular ring of the Compton size, and therefore, there have to be drastic changes in topology on the Compton distances. The puzzle is why QED can ignore this, and yet be in beautiful agreement with experiment.

Theo Nieuwenhuizen: Coming back to our original question, the idea that the world is deterministic at the "beyond the quantum" level seems to be shared.

Luis de la Peña: There is some convergence. 't Hooft is elaborating a nice general and deep model; Khrennikov presented an abstract mathematical approach, Cole a more physical (and more classical) one - we still have not seriously explored the common ground between us, but surely it must exist. In Stochastic Electrodynamics we are modest: we are just trying to understand what is beyond the quantum.

Gerard 't Hooft: Any model you want to work out, should be based on solid mathematics. The approach should not be called modest; small improvements are important.

Igor Volovich: String theory also uses standard quantization rules. Here we are trying to understand quantum mechanics.

Gerard 't Hooft: We must ask questions we can answer. String theories came with answers to some important questions.

Giacomo D'Ariano: Can we say something about quantum field theory without being specific? For instance, we do not have a quantum theory of measurement.

Gerard 't Hooft: Quantum field theory is quantum theory, there is no need to be specific.

Roger Balian: In fact we do not need a new general theory of quantum measurements, since a measurement can be treated in the mere framework of standard quantum mechanics, as a dynamical process involving interaction between a system and an apparatus. My talk, included in the proceedings, gives an example of such a process, with an exact solution exhibiting all features of a quantum measurement. Quantum statistical mechanics was needed because a measurement is irreversible.

Giacomo D'Ariano: Think of the quantum theory of a field. Are there, for instance, no-go theorems? Do we have a set of open problems?

Gerard 't Hooft: One burning problem is quantum gravity; another one is the possibility of an unlimited number of fields. Further, there are mathematical difficulties, not all proofs are mathematically solid. For instance, is the small-distance limit taken correctly? Or in the direction of complexity: at extremely high energies you have showers of particles, which is far from the condition where you can make a linear approximation.

Igor Volovich: And then there is the problem of quark confinement, which does not yet have a technical solution. Superstring theory can be interpreted as a quantum field theory with an infinite number of fields.

Gerard 't Hooft: Field theory works fine for four dimensions, not more. In short, quantum field theory is not a closed book. In any case, there is gravity. And superstring theory does not solve the problem.

Igor Volovich: Let us assume that particles have a finite size?

Gerard 't Hooft: Then you make things worse. If it is not a field, what is it? It is much harder to do away with point particles in quantum field theory than you might guess.

Alexander Burinskii: The Kerr ring-like string is an extended object, but it is pointlike from a complex point of view.

Gerard 't Hooft: When you take into account all the constraints, you end up with quantum field theory. It is hard to be crazier than that on these matters, because we do desire mathematical accuracy.

The general conclusion was: It is hard to be crazy enough.
But if Nature decided to go Beyond the Quantum, we must keep on trying.

AUTHOR INDEX

KEYWORD INDEX